Natural Gas Hydrate and Deep-Water Hydrocarbon Exploration

Natural Gas Hydrate and Deep-Water Hydrocarbon Exploration

Editors

Tao Jiang
Ming Su
Zenggui Kuang

MDPI

Basel • Beijing • Wuhan • Barcelona • Belgrade • Novi Sad • Cluj • Manchester

Editors
Tao Jiang
College of Marine Science and Technology
China University of Geosciences
Wuhan, China

Ming Su
School of Marine Sciences
Sun Yat-Sen University
Guangzhou, China

Zenggui Kuang
Guangzhou Marine Geological Survey
Guangzhou, China

Editorial Office
MDPI
St. Alban-Anlage 66
4052 Basel, Switzerland

This is a reprint of articles from the Special Issue published online in the open access journal *Energies* (ISSN 1996-1073) (available at: https://www.mdpi.com/journal/energies/special_issues/natural_gas_hydrate_and_deep_water_hydrocarbon_exploration).

For citation purposes, cite each article independently as indicated on the article page online and as indicated below:

Lastname, A.A.; Lastname, B.B. Article Title. *Journal Name* **Year**, *Volume Number*, Page Range.

ISBN 978-3-0365-8884-1 (Hbk)
ISBN 978-3-0365-8885-8 (PDF)
doi.org/10.3390/books978-3-0365-8885-8

Contents

About the Editors

Tao Jiang

Tao Jiang, Ph.D., is a Professor at the College of Marine Science and Technology, China University of Geosciences. His research interests lie in marine geology and resources and luminescence dating on marine sediments. He received his BSc and PhD degrees in Mineral Exploration and Exploration from China University of Geosciences in Wuhan in 2000 and 2005, respectively. From 2006 to 2008, he was a Postdoctoral Fellow in the Ocean Research Institute, Tokyo University, and began his teaching career in 2005 at China University of Geosciences in Wuhan.

Ming Su

Ming Su, Ph.D., is a Professor at the School of Marine Sciences, Sun Yat-sen University. His research interests lie in deep-water sedimentology and resources, sediment routing systems and source-to-sink systems. He received his BSc in Mineral Resource Prospecting and Exploration, MSc in Resource Exploration Engineering and PhD degrees in Marine Geology from China University of Geosciences in Wuhan, in 2005, 2008 and 2011, respectively. From 2011 to 2016, he worked as an Assistant Researcher and Associate Researcher at Guangzhou Institute of Energy Conversion, Chinese Academy of Sciences. He began his teaching career in 2016 at the School of Marine Sciences, Sun Yat-sen University in Zhuhai.

Zenggui Kuang

Zenggui Kuang is a Senior Engineer at the National Engineering Research Center of Gas Hydrate Exploration and Development, Guangzhou Marine Geological Survey. His research interests lie in gas hydrate accumulation and deep water sedimentology. He received his BSc degrees in Mineral Exploration and Exploration from China University of Geosciences in Wuhan in 2008 and 2010. He began his research work in 2010 at Guangzhou Marine Geological Survey in Guangzhou.

Preface

Natural gas hydrate and deep-water hydrocarbon have become hotspots of global oil and gas exploration in recent years. Natural gas hydrate is an efficient and clean energy source with huge resource potential, generally occurring within submarine sediments at a water depth of more than 300 meters, and also in permafrost areas. It is estimated that the global resource of natural gas hydrate could reach up to the equivalent of 20 trillion tons of oil. The natural gas hydrate resources in submarine sediments warrant more attention for their effective exploitation. The key issue is that conventional geological theories and understanding cannot be applied to natural gas hydrate and deep-water hydrocarbon exploration due to their complex exploration and development processes. Therefore, innovative theories and technologies are needed to achieve the commercial exploitation of natural gas hydrate and deep-water oil and gas under complex conditions. Various discoveries and new methods, comprehensive laboratory experiments and field investigations, theoretical and numerical simulations, and a more in-depth understanding of natural gas hydrate and deep-water hydrocarbon accumulation are partly covered by the research results in this book.

Tao Jiang, Ming Su, and Zenggui Kuang
Editors

energies

MDPI

Article

Reconstruction of Lake-Level Changes by Sedimentary Noise Modeling (Dongying Depression, Late Eocene, East China)

Zhongheng Sun [1,2], Tao Jiang [1,2,*], Hongtao Zhu [3,*], Xinluo Feng [4] and Pengli Wei [5]

1 Hubei Key Laboratory of Marine Geological Resources, China University of Geosciences, Wuhan 430074, China
2 College of Marine Science and Technology, China University of Geosciences, Wuhan 430074, China
3 Key Laboratory of Tectonics and Petroleum Resource of Ministry of Education, School of Earth Resources, China University of Geosciences, Wuhan 430074, China
4 Research Institute of Exploration and Development, Tarim Oilfield Company, PetroChina, Korla 841000, China
5 Hubei Geological and Mining Exploration Company Limited, Wuhan 430074, China
* Correspondence: taojiang@cug.edu.cn (T.J.); htzhu@cug.edu.cn (H.Z.)

Abstract: The late Eocene succession of the Dongying Depression forms a highly productive hydrocarbon source. However, due to lack of an unambiguous fine chronostratigraphic framework for the late Eocene stratigraphy, it is challenging to understand the paleolake's evolution and the driven mechanism of lake-level variation, a limitation which hinders hydrocarbon exploration. In this work, high-resolution gamma-ray logging data were analyzed to carry out the cyclostratigraphic analysis of the third member (Es_3) of the Shahejie Formation in the Dongying Depression. Significant 405-kyr eccentricity cycles were recognized based on time series analysis and statistical modeling of estimated sedimentation rates. We abstracted ~57 m cycles of the GR data in the Es_3 member, which were comparable with the long eccentricity cycles (~405-kyr) of the La2004 astronomical solution, yielding a 6.43 Myr long astronomical time scale (ATS) for the whole Es_3 member. The calibrated astronomical age of the third/fourth member of the Shahejie Formation boundary (41.21 Ma) was adopted as an anchor point for tuning our astrochronology, which provided an absolute ATS ranging from 34.78 ± 0.42 Ma to 41.21 ± 0.42 Ma in Es_3. According to the ATS, sedimentary noise modeling for the reconstruction of lake-level changes was performed through the late Eocene Es_3. The lake-level changes obtained based on sedimentary noise modeling and spectrum analysis reveal significant ~1.2 Myr cycles consistent with global sea level variations which were related to astronomical forcing. Potential driven mechanisms of marine incursion and/or groundwater table modulation were linked to explain the co-variation of global sea level changes and regional lake level changes. Our results suggest global sea level fluctuations may have played an important role in driving the hydroclimate and paleolake evolution of the late Eocene Dongying Depression.

Keywords: cyclostratigraphy; lake level; sedimentary noise modeling; Dongying Depression

Citation: Sun, Z.; Jiang, T.; Zhu, H.; Feng, X.; Wei, P. Reconstruction of Lake-Level Changes by Sedimentary Noise Modeling (Dongying Depression, Late Eocene, East China). *Energies* **2023**, *16*, 2216. https://doi.org/10.3390/en16052216

Academic Editor: Reza Rezaee

Received: 31 December 2022
Revised: 13 February 2023
Accepted: 23 February 2023
Published: 24 February 2023

1. Introduction

Cyclostratigraphy is an important branch of stratigraphy that deals with the sedimentary record of astronomically forced paleoclimate change [1–4]. The identification and deciphering of these sedimentary records improve our understanding of rates and mechanisms of biological, chemical, and physical changes, paleoclimate variations and key geological problems in Earth's distant past [4–7]. Previous cyclostratigraphy studies have mainly focused on marine environments such as deep-water facies, carbonate platforms and mixed and detrital marginal facies [8–10]. Terrestrial lake basins have recently become a hotspot for intensive petroleum exploration, with continuous successful discoveries [11–13]. Many terrestrial basins have therefore been used to conduct astronomical dating and paleoclimate reconstruction, such as Piceance Creek Basin [14], Songliao

Basin [15] and Bohaibay Basin [13,16–21]. The Bohai Bay Basin (BBB), located on the eastern part of the North China Block, is one of the largest hydrocarbon-bearing basins in China, accounting for approximately one-third of the total hydrocarbon production of the country [11]. The third member of the Shahejie Formation (Es_3) forms the most productive hydrocarbon source rock. Understanding the paleolake evolution and lake level fluctuation history is vital for clarifying the development mechanism of source rock and further enhancing hydrocarbon exploration [22]. Therefore, it is both scientifically and economically significant to fully understand the factors that influence the stratigraphic architecture and paleolake evolution in these basins. Reconstruction of high-resolution lake-level changes can better help us decipher these factors. Sedimentary noise modeling has been proven to be a powerful tool for reconstructing the sea- and lake-level changes in both marine and terrestrial lake basins [13,21,23,24]. Yet, lacking the accurate astrochronology constraints of the entire Es_3 member (previous studies only focused on lower and/or middle of the Es_3), the high-resolution lake level history has not been reconstructed [18–21].

Regarding the chronology and paleoclimate of the Paleogene Bohai Bay Basin, numerous previous studies have been employed through various methods of cyclostratigraphy and magnetostratigraphy and they have built multiple sets of chronostratigraphic frameworks [13,16–21]. The initial research postulated the boundary between the Es_3 member and the fourth member of the Shahejie Formation (Es_4) as 40,904 Ma [16]. Soon afterwards, the boundary age between the Es_3 and Es_4 was recalibrated by Liu et al. (2018) [17] according to the top age of the Chrons C18r of the GTS2012 [25], which yielded a new age of 42.47 Ma. However, updated GTS2020 determined the base age of Chron C19n as being 41.18 Ma, which resulted in the calibrated age being 42.47 Ma. This age is about 1.29 Ma older than the previously defined 41.18 Ma rising new controversy. Meanwhile, new ATS of the upper of Es_4 and lower of Es_3 was established based on high-resolution measurements of magnetic susceptibility of Well Fanye-1 in the same year [18]. However, lacking the accurate age anchor points was a constraint that caused these ATS to have various degrees of discrepancy. Subsequently, Shi et al. (2019) [19] updated the ATS using new acquired paleomagnetic data but lacked a statistically-based astronomical sedimentation rate estimation to better constrain the ATS. Given this, multiple statistical methodologies were employed to rebuild the ATS for Well Fanye-1 with an updated age model in GTS2020, using the short eccentricity cycles(~100-kyr) tuning [20]. Terrestrial strata are characterized by widely variable sedimentation rates and transient depositional or erosional hiatuses. The ~100 kyr cycle is thought to be unstable, which could lead to significant errors in the short-eccentricity tuning approach [21]. Synthesizing the above differences, the new ATS and magnetostratigraphy of the middle-late Eocene was further calibrated using 405 Kyr long eccentricity cycle tuning method developed by Ma et al. (2023) [21]. Compared with previous ATS and GTS2020, the model shows minor error uncertainties indicating the ATS of the Es_4 and Es_3^L is robust and reliable. Therefore, the ATS provided a basis for us to construct a complete astronomical time framework for the entire Es_3 member.

In this paper, a cyclostratigraphy study of the late Eocene (Bartonian to Priabonian) terrestrial succession of Well S146 in the Dongying Depression is conducted using gamma-ray series. The overall objective of the study is to (1) construct an absolute astronomical time scale (ATS) for the late Eocene Es_3 member in the Dongying Depression, Bohai Bay Basin in eastern China; (2) reconstruct lake-level changes based on sedimentary noise modeling of the tuned GR series; and (3) explore the potential driven mechanisms of million-year-scale lake level variations.

2. Geological Background

The Dongying Depression is located in the southern part of the Jiyang subbasin of the Bohai Bay Basin (BBB), which is one of the most petroliferous basins in China. The Dongying Depression covers an area of approximately 5700 km^2 which is bounded by the Qingtuozi Uplift in the east, the Qingcheng and Binxian Uplift in the west, Chenjiazhuang Uplift in the north, and Luxi and Guangrao Uplift in the south [26] (Figure 1). It is characterized by faulting in the north and overlapping in the south, forming landscape with steep slope in the north and slow slope in the south. Additionally, several subunits can be identified, including the northern steep slope zone, the central uplift zone, the Minfeng sag, the Lijin sag, the Niuzhuang sag, the Boxing sag, and the southern gentle slope zone [26,27] (Figure 1B).

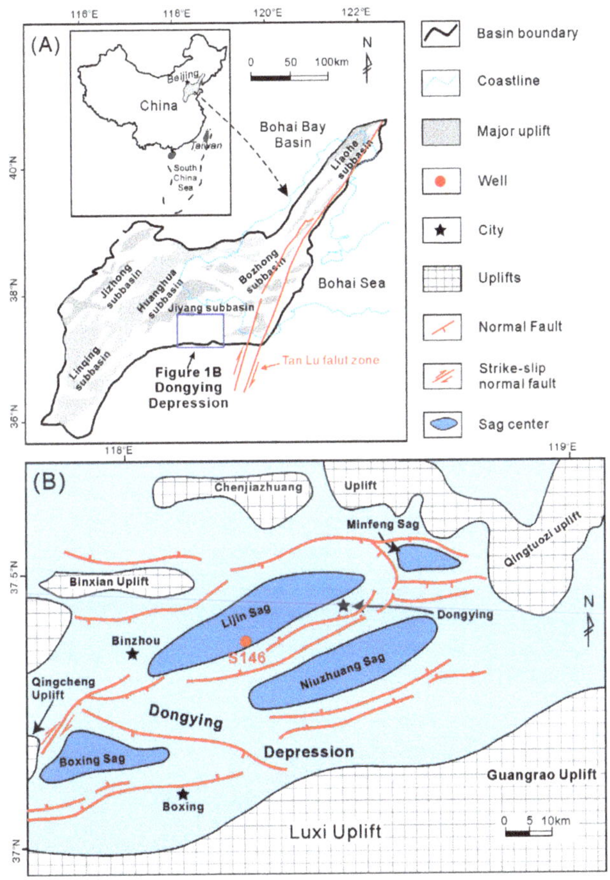

Figure 1. (**A**) Location map of the Bohai Bay Basin showing the major tectonic elements. (**B**) Tectonic map of the Dongying Depression showing locations of the studied Well S146.

The Depression, tectonically similar to the BBB, underwent several complex and distinct phases of Mesozoic and Cenozoic continental rifting, differential subsidence and down warping resulting in thick Paleogene synrift (65.0–24.6 Ma) and Neogene-Quaternary post-rift (24.6 Ma to the present) successions. From bottom to top, Paleogene synrift strata can be further divided into Kongdian (E*k*), Shahejie (Es) and Dongying (E*d*) formations. The Paleogene synrift experienced three stages: (1) initial rifting with lake division into sub-lake basin, (2) rift climax with lake expansion, and (3) weakened rifting with shrinkage of lake basin, forming a complete sedimentary cycle of shallow lake facies, deep lake facies, and successive fluvial and shallow lake facies [26]. The third member of the Shahejie Formation (Es_3) is subdivided into the lower (Es_3^L), middle (Es_3^M) and upper (Es_3^U) units, and the fourth member of the Shahejie Formation (Es_4) is subdivided into the lower (Es_4^L) and upper (Es_4^U) units (Figure 2).

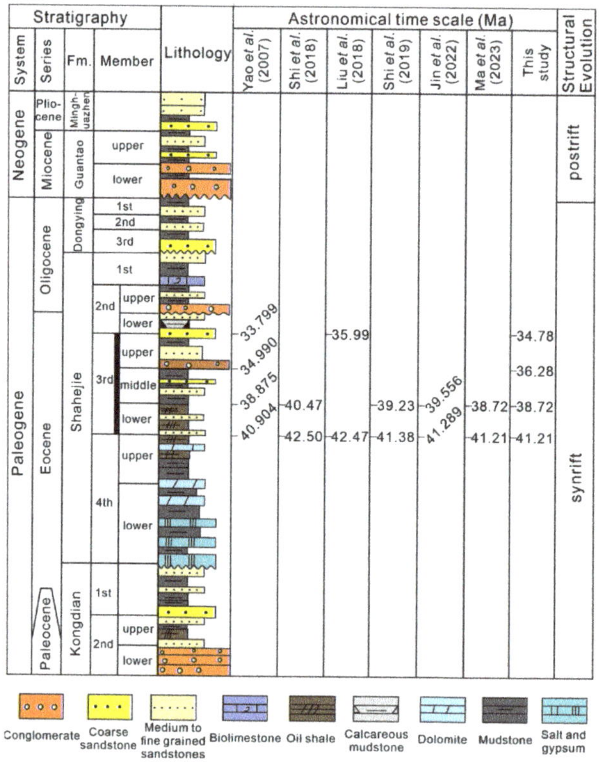

Figure 2. Generalized stratigraphic column of the Dongying Depression, modified from Feng et al. (2013) [26]. The target interval is marked in the black bold line. Fm = Formation [16–21].

The boundaries of all above members and units identified in the Shahejie Formation can be correlated and traced across the whole Depression. The Shahejie Formation was mainly deposited in lacustrine setting during the synrift period. The Es_3 and Es_4 members developed important Paleogene source rocks. The $Es_4{}^L$ is dominated by purple-red mudstone with extensively developed salt rock and gypsum. The $Es_4{}^U$ is dominated by gray and dark gray mudstone intercalated with carbonate, sandstone, and oil shale. The $Es_3{}^L$ mainly consists of dark gray mudstone and taupe oil shale and is high-quality source rock. The $Es_3{}^M$ comprises mudstone, argillaceous siltstone, and calcareous mudstone interbedded with thin sandstone. The $Es_3{}^U$ is dominated by mudstone, and contains argillaceous limestone [27,28]. Target well 146 is located in the south part of the Lijin sag, which is the sag center of the Dongying Depression. The $Es_4{}^U$ and $Es_3{}^M$ of well 146 developed massive dark mudstone, interpreted as deep and semi-deep lacustrine deposits in a stable tectonic setting. In seismic profiles, these are characterized by medium- to low-amplitude continuous parallel reflections, which are typical features of a lake depositional environment [26].

3. Materials and Methods

3.1. *Gamma Ray Logging*

In this study, cyclostratigraphic analysis was conducted on the gamma-ray (GR) logging data from the S146 well which is situated in the depocenter of the Dongying Depression where tectonic influences on sedimentation are lacking.

Gamma logging is a downhole measurement of the intensity of gamma rays emitted by the decay of naturally occurring radionuclides in sedimentary rocks. GR data record variations in the content of radioactive isotope of uranium (U), thorium (Th), and potassium (K) in the natural rocks. These three radioactive isotopes are sensitive to the variations of clay minerals and organic matter which, in abundance, are sensitive to environmental change [2]. Therefore, GR data preserve original signals related to climatic changes and can be interpreted to develop a history of the climate changes [29]. GR has proven to be one of the most efficient proxies reflecting the paleoclimatic changes in both marine and continental strata [15]. In a warm and humid environment, generally accompanied by intense chemical weathering and enhanced precipitation with higher eccentricity, more clay minerals and organic matter are input to the lake basin, corresponding to the high value of the GR logging curve. In contrast, the content of clay mineral is reduced, and the content of sandstone and the inorganic carbonate increased under cool and arid environments with lower eccentricity, which results in lower GR values. Many cyclostratigraphic studies based on the GR proxy have demonstrated the potential of GR recording the Milankovitch signals and proposed GR as a climatic proxy for identifying wet-arid climate change [15,30,31]. Meanwhile, an East Asian Lake hydrology study from four lake basins (Dongying Depression, Nanxiang Basin, Jianghan Basin, and Fushun Basin) in eastern China by Ma et al. (2023) [21] also related the hydroclimate and lake level change recorded in GR data to astronomical forcing. By reason of the foregoing, GR logging is reasonable for cyclostratigraphic analysis and lake level reconstruction in this study.

Here, we present continuous GR logging data of the Shahejie Formation in S146 well, drilling from the Lijin sag located in the middle of Dongying Depression of the BBB. The parameters for study: S146 borehole, from the base of Es_3^L to the top of Es_3^U with depth ranging from 3736 to 2912 m (a thickness of 824 m). The GR data of the borehole was acquired by continuous downhole measurements conducted by the Shengli Oilfield Company of SINOPEC with an average sampling rate of 0.125 m.

3.2. Time Series Analyses Methods

Prior to performing spectrum analysis, a series of data preprocessing were carried out. We applied linear interpolation on the GR data to ensure that the sampling intervals were equal, because the raw sample interval of the GR data collected from the S146 borehole may be unequal owing to unknown empty sampling. The raw GR data were detrended, adopting the LOWESS method by subtracting a 30% (i.e., 247.2 m) weighted average to remove low-frequency long-term trends prior to spectral analysis [32]. To preliminarily evaluate the variations of sedimentation rates, we used a sliding-window evolutionary spectrum method (fast Fourier transform spectra) to examine the changes in orbital cycle frequencies related to varied sedimentation rates [33]. We employed the correlation coefficient (COCO) approach to objectively identify the optimal sedimentation rates in order to more properly assess the variations in sedimentation rates through the studied succession. The COCO method was developed to estimate the correlation coefficient between the power spectra of the sedimentary record in the depth domain and an astronomical solution in the time domain. Based on this, various "test" sedimentation rates were computed to convert the proxy series of sedimentary records from depth to time [34] (Li et al., 2018). The optimal sedimentation rate corresponds to the one yielding the highest correlation coefficient. Meanwhile, the evolutionary correlation coefficient (eCOCO), using a sliding-window approach were performed to track variable sedimentation rates across the Gr series and to relate power spectra of GR series to its associated astronomical solution [34]. Additionally, a null hypothesis of no astronomical force is included with significance tested using Monte Carlo simulation [34] (2000 simulations). Monte Carlo simulation is a broad class of computational algorithms that relies on repeated random sampling and statistical analysis to model the probability of various outcomes. Combined with estimated sedimentation rate, a 2π multitaper (MTM) method of spectral analysis of both depth- and time-domain GR series were carried out [35]. The robust red noise model proposed by Mann and Lees (1996) [36] was performed to statistically assess the significance of peaks obtained by MTM analyses at the mean, 90%, 95%, and 99% confidence levels. To identify which of these spectral peaks corresponds to the orbital eccentricity, obliquity, and precession, those peaks detected above 95% confidence levels underwent additional analysis using the La2004 astronomical model. The MTM-identified long eccentricity cycles (405 kyr) and short eccentricity cycles (100 kyr) were separated using Gaussian bandpass filtering and visual interpretation. These isolated cycles were then counted and tuned to convert GR series from the depth domain to the time domain. The 405 kyr long eccentricity cycle produced by the gravitational effects of Jupiter and Venus on the Earth is the most stable astronomical orbital period since the past 250 Ma, providing a "pendulum" for geological time timing and serving as the main period for geological time calibration. Therefore, this paper uses ~405 kyr long eccentricity cycle to tune the astronomical age of the target interval, and directly compares the filters with the eccentricity curve of La2004 (Laskar et al., 2004) to establish a floating astronomical time scale (ATS).

3.3. *Sedimentary Noise Modelling*

The sedimentary noise model was originally proposed to reconstruct past sea level change in marginal marine environments [23]. In recent years, more and more case studies have proven that the sedimentary noise model can also be used for restoring lake level changes [13,21,24]. The sedimentary noise model comprises two complementary approaches: dynamic noise after orbital tuning analysis (DYNOT) and lag-1 autocorrelation coefficient (ρ1) analysis [23]. The DYNOT model calculated the ratio of non-orbital signal variance versus the total variance via extracting noise recorded in water-depth related proxies to indicate relative sea-level changes. The ρ1 model is an independent tool for relative water-level reconstruction which is based on the lag-1 autocorrelation coefficient of both tuned and untuned dataset. Generally, high water-level (sea- or lake-level) yields a low sedimentary noise corresponding to a low DYNOT value and a high ρ1 value. In contrast, lower water-level results in high sedimentary noise level, which yields a high DYNOT value and a low ρ1value [13,21,23]. All abovementioned analyses were carried out in *Acycle v2.5* software [37].

4. Results

4.1. *Time Series Analysis*

The GR series of studied intervals of the Well S146 exhibits significant consistency with lithological variations with high GR value corresponding to mudstone whereas low GR values are dominated by fine sandstone and sandstone (Figure 3a). The MTM power spectrum analysis of the untuned GR logging data in depth domain of the Es_3 interval of Well 146 in the Dongying Depression shows prominent periodic peaks, with confidence levels more than 95%, at wavelengths including 57, 28.8, 14.3, 7.0, 5.4, 4.7, 2.9, 2.7 and 2.2 m (Figure 4c).

The COCO analysis proposed by Li et al. (2018) was used to obtain the potential optimal deposition rate. A total of 2000 Monte Carlo simulations with tested sedimentation rates ranging from 0 to 25 cm/kyr were carried out to perform a null hypothesis test for non-astronomical orbital period signals. The COCO analysis of untuned GR data from the S146 borehole shows multiple peaks of sedimentation rate at 1.8, 4.0, 6.8, 9.4, 13.0, and 16.0 cm/kyr, with null hypothesis (H_0) significance level no more than 0.01 (Figure 3c,e). Among them, the most remarkable cluster of peaks is from 13.0 to 16.0 cm/kyr (with the highest peak at 13.0 mc/kyr), which may indicate the potential mean sedimentation rate is between 13.0 and 16.0 cm/kyr. Notably, the sedimentation rates band ranges from 13.0 to 16.0 cm/kyr is in line with a mean sedimentation rate of 16.8 (Es_3) and 10 cm/kyr (Es_3^L) in the Dongying Depression as inferred by Liu et al. (2018) and Shi et al. (2018), respectively.

The eCOCO analysis of the same untuned GR series in Well S146 was performed to track variable sedimentation rates across the studied interval. We ran 2000 Monte Carlo simulations with tested sedimentation rates ranging from 0 to 25 cm/kyr and a sliding step of 3 m and sliding window of 200 m. The results of the eCOCO reveal sedimentation rates of 13.0–16.0 cm/kyr, producing a broader frequency band than those of 4.0–6.8 cm/kyr consistent with results of COCO analyses (Figure 3d). Moreover, the correlation coefficient values for sedimentation rates of 13.0–16.0 cm/kyr are greater than 0.3 and typically substantially higher than those at sedimentation rates of 4.0–6.8 cm/kyr. A sedimentation rate of 13.0–16.0 cm/kyr is relatively stable through the entire succession. To further determine which of these two groups of peaks is the optical sedimentation rate, we boldly assumed about 5.4 cm/kyr (the average value of 4.0–6.8 cm/kyr) as the mean sedimentation rate and yielded the duration of the Es_3 at 15.8 Ma which deviate from our existing understanding [37]. Taken together, the estimated average sedimentation rate of 13.0 cm/kyr for Es_3 member derived from COCO analysis is robust and consistent with estimations of sedimentation rate produced by the eCOCO.

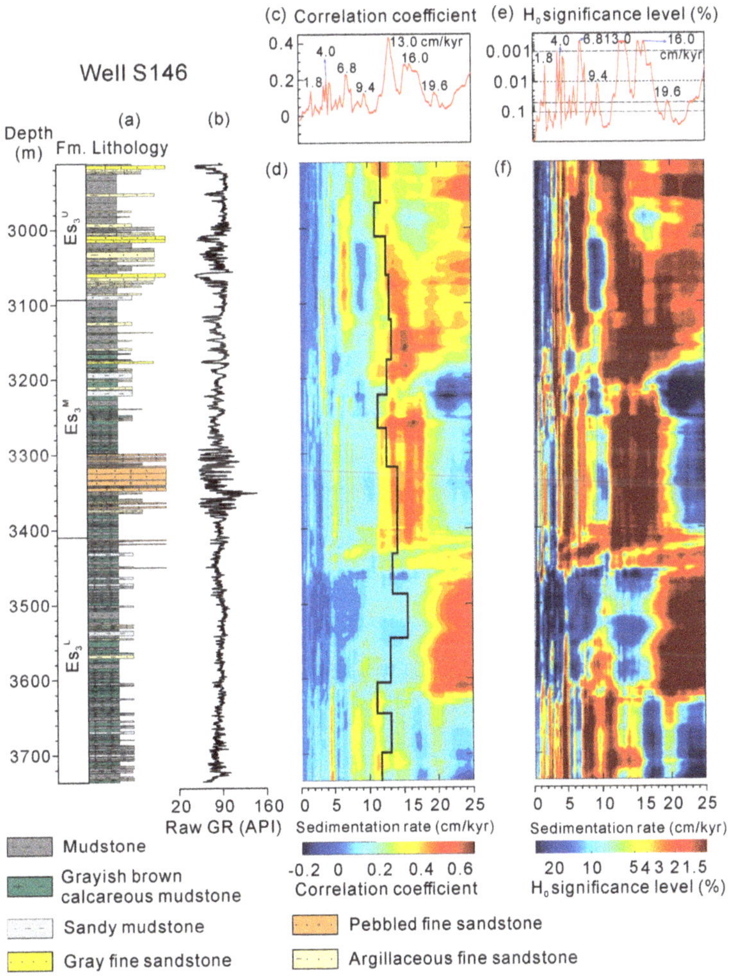

Figure 3. COCO analysis and eCOCO sedimentation rate map of the GR series in the Well S146 (2912–3736 m): (**a**) lithological column of the Well S146; (**b**) raw GR series of the Well S146; the correlation coefficient (**c**) and evolutionary correlation coefficient (**d**) shown with sedimentation rate curve (black line) based on 405-kyr tuning of the longest (~57 m) statistically significant cycles in the data; (**e**) null hypothesis test; and (**f**) evolutionary null hypothesis (H0) significance level.

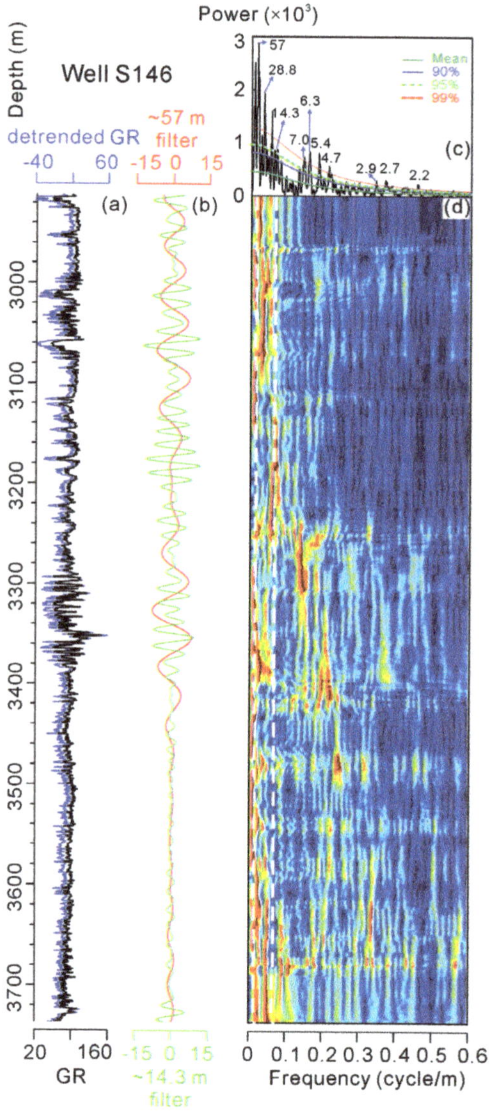

Figure 4. (**a**) Detrended and raw GR series of the Well S146 in depth domain; (**b**) filters with wavelength of ~57 m and ~14.3 m; (**c**) 2π MTM power spectrum in depth domain shown with background AR(1) model, and 90%, 95%, and 99% confidence levels; and (**d**) evolutionary power spectra calculated using a 400-kyr running window.

The wavelength ratio of the clusters of 57, 14.3, 7.0-5.4-4.7, and 2.9-2.7-2.2 m (ca. 57:14:5.4:2.7) identified in the MTM power spectrum of the untuned GR series of Well S146 is near the ~20:5:2:1. Moreover, based on our objective sedimentation rate estimate by the COCO and eCOCO analysis, the estimated average sedimentation rate is 13.0 cm/kyr. This yields tentative durations of ~438 kyr for the~57 m cycle, ~110 kyr for the ~14.3 m cycle, ~54-42-36 kyr for the ~7.0-5.4-4.7 cycles, and ~22-20-17 kyr for the 2.9-2.7-2.2 m cycles. These periodicities are broadly consistent with the expected ratio of long-eccentricity (405 kyr), short-eccentricity (~100 kyr), obliquity (41 kyr) and precession

(19–23 kyr), respectively. Therefore, the sedimentary strata can be linked to the above astronomical forcing.

4.2. Astronomical Tuning

According to the above analysis of MTM, evolutionary spectrum, and the COCO and eCOCO of S146 borehole, ~57 m sedimentary cycles were tied to the 405 kyr long-eccentricity metronome. To extract the counterpart cycles from the GR series, we applied a bandpass filter using a ~57 m wavelength (passband of 0.0175 ± 0.004375) obtaining about 16 long-eccentricity cycles (405-kyr). Meanwhile, cycles of ~14.3 m wavelength (passband of 0.07 ± 0.0175) were band-pass filtered to isolate short eccentricity cycle (100 kyr) yielding ~64 cycles. Tuning ~57 m sedimentary cycles using the 405-kyr eccentricity tuning approach allowed us to build a floating ATS for the Es_3 member with duration of ~ 6.43 Ma (Figure 5).

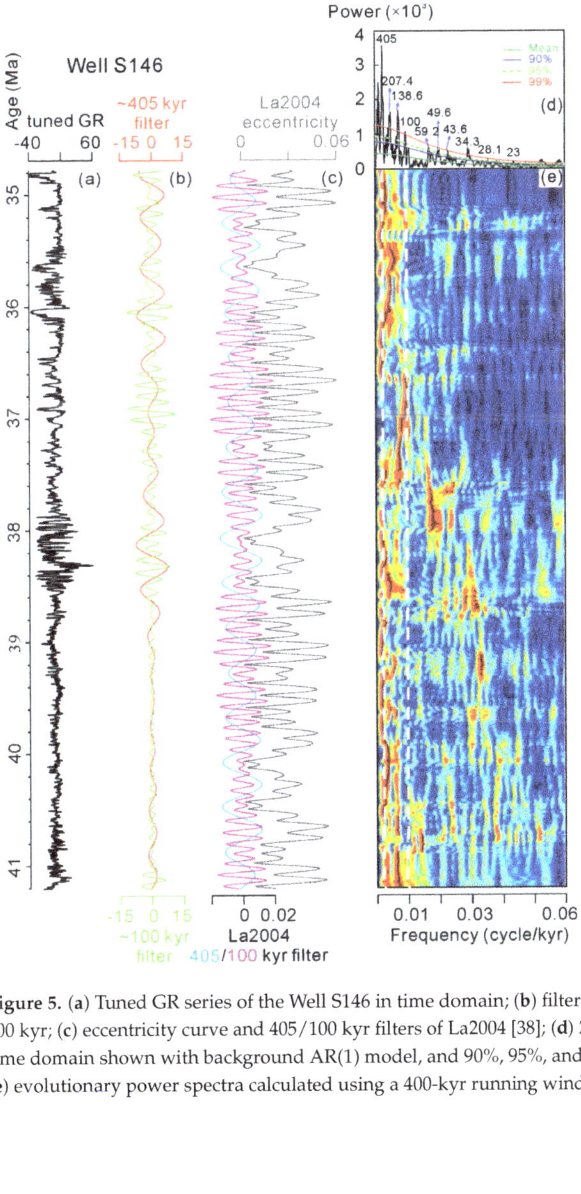

Figure 5. (**a**) Tuned GR series of the Well S146 in time domain; (**b**) filters with cycles of 405 kyr and 100 kyr; (**c**) eccentricity curve and 405/100 kyr filters of La2004 [38]; (**d**) 2π MTM power spectrum in time domain shown with background AR(1) model, and 90%, 95%, and 99% confidence levels; and (**e**) evolutionary power spectra calculated using a 400-kyr running window.

Based on this tuning method, the entire Es_3 interval yields a mean sedimentation rate of ~12.8 cm/kyr which compares well with those (13.0 cm/kyr) estimated with the COCO/eCOCO analysis. The MTM power spectrum of the tuned GR series in Well S146 shows significant cycle peaks of 405, 207.4, 138.6, 100, 59.2, 49.6, 43.6, 34.3, 28.1 and 23 kyr. The evolutionary spectrum of the tuned GR series reveals obvious 405 kyr periodic signals and 100 kyr periodic signals close to a straight line. The calculated sedimentation rate with 405 kyr tuning varies from 11.1 to 15.5 cm/kyr, with an average sedimentation rate of 12.8 cm/kyr. The age of boundary between the Es_4 and Es_3 (BES_{43}) is used as anchor point for calibrating our floating ATS. In addition, the $Es_3{}^L$ and $Es_3{}^M$ boundary ($BES_3{}^{LM}$) is adopted as checked point to further verify our ATS. Recent astrochronologic and magnetostratigraphic study conducted in the Dongying Depression has provided a robust astronomical time scale and magnetostratigraphic timescale for the Es_4 and $Es_3{}^L$ formations, with the BES_{43} age of 41.21 Ma and the $BES_3{}^{LM}$ age of 38.72 Ma, respectively [21]. According to the anchor point, our floating ATS of were able to transform to absolute ATS extending from 34.78 $\pm$ 0.42 Ma back to 41.21 $\pm$ 0.42 Ma from top of the $Es_3{}^U$ to base of $Es_3{}^L$ (Figure 5a). This period corresponds to the La2004 astronomical solution from 86th to 103rd 405-kyr eccentricity cycles (E86–E103) [38] (Figure 5).

4.3. Sedimentary Noise Modeling of Lake-Level Changes

We reconstructed sedimentary noise curves for the Dongying Depression based on DYNOT and $\rho 1$ modeling, with 5000 Monte Carlo simulations of the tuned GR data of Es_3 using a 400 kyr sliding window. Sedimentary noise curves of DYNOT and $\rho 1$ modeling display consistent patterns. Significant enhanced sedimentary noises were identified in five intervals and highlighted in gray bars (Figure 6).

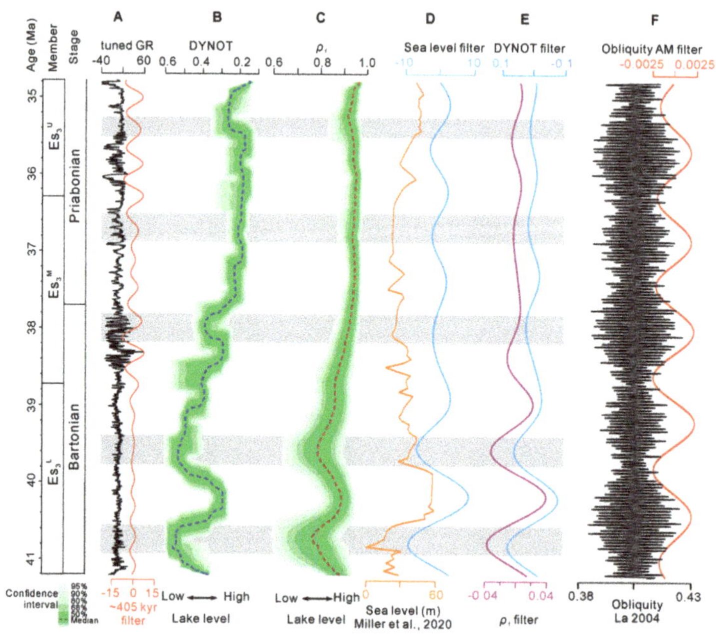

Figure 6. Sedimentary noise model interpretation of lake-level changes in Well S146. (**A**) Tuned GR series of the Well S146 (black) with 405-kyr filtered output (red); (**B**,**C**) DYNOT and $\rho 1$ models of tuned GR series; (**D**) global sea level changes modified from Miller er al., 2020 with 1.2 Myr filtered output [10]; (**E**) 1.2 Myr cycles filtered from DYNOT and $\rho 1$ models; and (**F**) earth's obliquity solution (Black, Laskar et al., 2004) [38] and its 1.2 Myr-AM cycles (red).

According to spectral analysis of the DYNOT output series (median value), we observed five cycles in sedimentary noise series with periods of 1.2 Myr (Figure 7). Similarly, spectral detection of the ρ1 output series (median value) reveals cycles that have periods of 1.2 Myr (Figure 7).

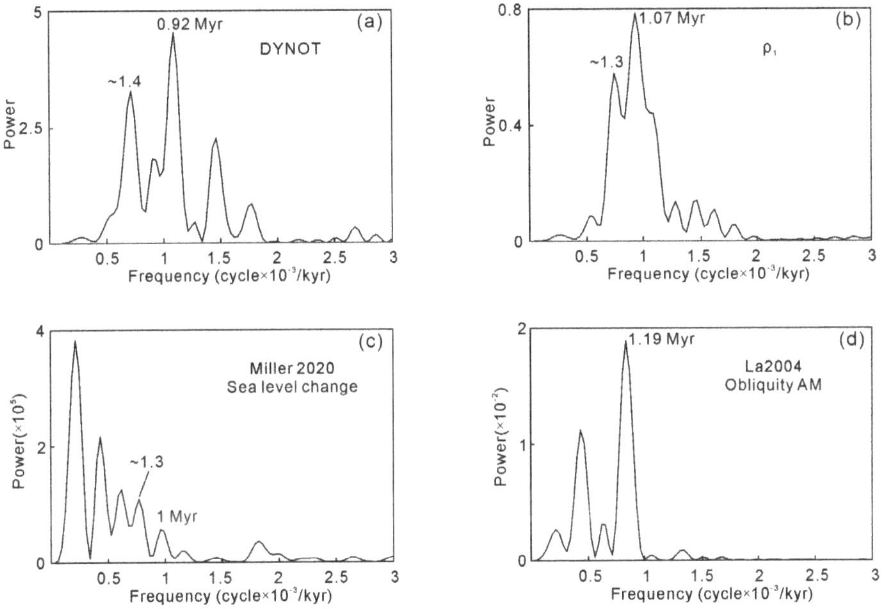

Figure 7. Periodograms of the detrended median output by DYNOT (**a**) and ρ1 (**b**) models, AM of the sea level curve by Miller 2020 [10] (**c**) and AM of the obliquity (**d**) in La2004 [38] astronomical model from 34.78 to 41.21 Ma.

5. Discussion

5.1. An Astronomical Time Scale (ATS) in the Dongying Depression

We constructed a ~6.43 Myr long floating ATS for Es_3 member in Well Shi 146 of the Dongying Depression through the late Eocene. According to our cyclostratigraphic analysis, the studied interval has a total duration of 6.43 ± 0.42 Ma, ranging from 41.21 ± 0.42 Ma to 34.78 ± 0.42 Ma. The base age of Es_3^L, Es_3^M and Es_3^U are 41.21 ± 0.42 Ma, 38.72 ± 0.42 Ma, and 36.28 ± 0.42 Ma, respectively.

The 405 kyr eccentricity bandpass filter extracted from our constructed ATS has a significant in-phase correlation with the 405 kyr filter extracted from the La2004 eccentricity solution. The prominent eccentricity, obliquity, and precession cycles were identified based on the MTM spectra analysis of the tuned GR series. These cycles match well with astronomical models of La2004 [38]. The sedimentation curve generated from 405 kyr cycles compares well with the objective sedimentation rate estimated by COCO and eCOCO (Figure 3).

The uncertainty of the tuning and anchor point may result in additional uncertainty. Therefore, it is necessary to conduct uncertainty analysis of all potential sources. Three sources of uncertainty for our ATS and reported ages were considered including: (1) the uncertainty of BES_{43} age in the Dongying Depression, as reported by Ma et al. (2023) [21]; (2) the uncertainty of contact relationship of Es_4 and Es_3; (3) the uncertainty of the floating ATS we established. The ages of base boundaries of Chrons C18n.1n through C20n with mean errors in the FY1 borehole were reported by Ma et al. (2023) [21] and compared to those of previous six studies. The base of the Es_3 roughly corresponds to the base of the C19n. Although Ma et al. (2023) [21] did not report the uncertainty of BES_{43} age in the Dongying Depression, he indeed provided seven uncertainty errors spanning the boundary between the Es_4 and Es_3 ranging from C20n to C18n.1n. These uncertainties ranged from 0.100 Ma to 0.106 Ma, with a mean uncertainty of 0.103 Ma. Thus, we suggest 0.103 Ma as the uncertainty of the BES_{43} age in the Dongying Depression. Seismic reflection profiles showing the boundary between the Es_4 and Es_3 are characterized by stable seismic conditions even with parallel reflection and conformable across the Dongying Depression [26]. There is no report of onlap or traction seismic reflection which can indicate significant tectonic activities. Our S146 well is located in the depocenter of the Dongying Depression in a stable sedimentation setting with volcanic activities reported. Taken together, these points of evidence eliminate the second source of uncertainty. An empirical uncertainty of 0.103 Myr for the BES_{43} in the Dongying Depression may be obtained from superposition calculation of the first two sources of uncertainty. The identified long eccentricity cycles in our floating ATS may cause an additional uncertainty of one 405-kyr cycle for the third source. Consequently, we can obtain a total uncertainty of 0.42 Myr (calculated form $\sqrt{0.103^2 + 0.405^2}$) [13] for the base of the $Es_3{}^L$, $Es_3{}^M$ and $Es_3{}^U$.

Our ATS of Es_3 member appears to be compatible with previous astrochronology of the same interval performed by cyclostratigraphy studies [13,17,18,20,21]. The base ages of $Es_3{}^M$ reported by Shi et al. (2018) [18], Jin et al. (2022) [20] and Ma et al. (2023) [21] are 39.23 Ma, 39.556 Ma and 38.72 Ma respectively, which have discrepancies of 0–0.836 Ma. The top age of $Es_3{}^U$ reported by Liu et al. (2018) [17] and Wang et al. (2020) [13] leads to 1.21–1.3 Ma discrepancies. The duration of the Es_3 of our ATS is 6.43 Ma, which is quite close to the 6.48 reported by Liu et al. (2018) [17]. Therefore, we argue these discrepancies could be caused by diachronous sub-member boundaries within the Es_3, age differences between anchor points or erroneous identification of long eccentricity cycles in either or both those previous studies and our study [13]. Overall, we argue that we have constructed a robust and reliable absolute ATS of the Es_3 of the Dongying Depression according to our uncertainty constraints discussed above.

5.2. *Verification of Sedimentary Noise Model for Reconstructing Lake Levels*

Sedimentary noise in data series can be influenced by multiple factors including proxy related noise (e.g., dating uncertainty), unsteady sedimentation such as post-depositional reworking, local, short term tectonic, depositional, and volcanic activities in additition to lake level changes [23]. The noise associated with paleoclimate proxy indicators is affected by three factors: index sensitivity, measurement error, and dating error. Both previous studies and this study have confirmed that GR signals can well record paleoclimate changes in this area, indicating that the sensitivity of the indicator persists [20,21]. Obtaining logging data by wireline logging is a very mature technology. At the same time, the measurements are taken in a stable borehole, so the measurement errors are relatively small. Dating in this study is based on the COCO and ECOCO sedimentation rate assessment models, which are very close to the previous age anchor point, indicating our ATS is very reliable. Unstable depositional environment and short-term tectonic activities may lead to increased depositional noise. However, the sedimentary environment of the Es_3 in Well Shi 146 was a semi-deep lake-deep lake facies with stable structural environment during the depositional period, and no faults were developed [39]. Therefore, unstable depositional environments and short-term tectonic activity had limited impact on sedimentary noise. A hydrological shift driven by volcanic activity may have increased additional sedimentary noise. However, lithostratigraphic observations showing no volcanic rocks were developed in Es_3 of Well Shi 146, implying volcanic activities did not affect the Lijin sag and related sedimentary noise [39,40]. Post-depositional reworking mainly includes bioturbation and diagenesis, which may have a greater impact on sedimentary noise. Nevertheless, there is no report of bioturbation structure in this well. Moreover, the lithology of the Es_3 is dominated by mudstone, which suffered normal compaction, being less affected by post-depositional reworking. And yet, the complexity of sedimentary noise of terrestrial strata has not yet been thoroughly understood. The similar variation patterns of the DYNOT and ρ1 models combined with the above discussion indicates that the sedimentary noise-modeling approach is nonetheless an effective method for lake-level reconstruction in lacustrine strata.

5.3. Astronomical Forcing on Lake-Level Fluctuations in the Dongying Depression

Astronomical forcing of lake-level fluctuations has been reported throughout the Phanerozoic at several basins with different latitudes across the word [13,21,24,41,42]. The robust ATS constructed in this study provides a basis for linking the astronomical cycles, lake hydraulic cycle and global sea level changes. The DYNOT modeling results shows significant anti-phase relationship with that of the ρ1 modeling. Multiple alternation of high and low sedimentary noise level was observed throughout the entire Es_3 succession. The high sedimentary noise levels were interpreted as low lake levels. In opposition, the low sedimentary noise levels were interpreted to correspond to high lake levels.

Periodogram analysis of both the DYNOT and ρ1 modeling output median values showing clear long-term cycles with periods of ~1.2 Myr (Figure 7). The 1.2 Myr cycles were identified in both the DYNOT and ρ1 modeling output series, which show similar anti-phase relationships compared to their original modeling results. This may correspond to 1.2 Myr obliquity modulation cycles, indicating a potential causal link between astronomical forcing and million-year scale lake-level change. Previous studies have reported ~1.2 Myr periodicities in the Eocene [13,43–45]. The obliquity modulation cycle of ~1.2 Myr is demonstrated to affect heat and moisture transport flux from low latitudes to high latitudes and further manage the meridional insolation gradient on Earth [23,46]. During the 1.2 Myr obliquity nodes (modulation minima), heat and moisture transport flux to high latitudes would have been reduced. Consequently, lakes in mid-to-high latitudes would become cooling and drying with low water level and vice versa [13,21,23]. Our results show a clear in-phase relationship of 1.2 Myr cycles isolated form DYNOT and ρ1 with 1.2 Myr obliquity amplitude modulation cycles of La2004 solution during the late Eocene. The global sea-level curves proposed by Miller et al. (2020) [10] were used to compare with the lake level changes of the Dongying Depression. Spectral analysis of the sea-level records shows clear ~1.2 Myr cycles which is close to 1.2 Myr obliquity cycle. The 1.2 Myr cycles of lake level in Dongying Depression are generally in-phase with those of sea-level changes and obliquity cycle. These results suggest 1.2 Myr obliquity drove the changes of both lake-level and sea-level during the late Eocene. Identifying the driven mechanism of the coevolution of the global sea level and reginal lake level remains a challenge. Multitudinous hypothesizes have been made to explain this co-variation [13,21]. We believe that one at least or more of the following mechanisms may have contributed to the regulation of the coevolution in the late Eocene.

Regional transgression events during marine transgressions impose significant effects on the rise of the lake level. Four intervals of marine incursions into the Dongying Depression have been reported based a combination of interpretations with high B/Ga ratios, high S/TOC ratios and high Sr/Ba [47]. Moreover, marine fossils including foraminifera, calcareous nanofossils and trace fossil *Paleodictyon* were observed and further confirmed the marine incursion in Dongying Depression [21,47]. Our reconstructed lake-level changes appear to be well consistent with global sea level changes implying that marine incursions play an important role in regulating the lake level changes in the Dongying Depression.

Groundwater table modulation is another driven mechanism for explaining the co-variation. By an integrated analysis of four lake basins (Dongying Depression, Nanxiang Basin, Jianghan Basin, and Fushun Basin) in east Asia, groundwater table modulation was used to clarify the mechanism for co-variation of the lake level and sea level during the middle-late Eocene [21]. Notably, the mechanism for driving the lake level changes of these four basins spans time periods of 48.51-38.52 not fully covering our studied interval with time periods of 41.21-34.78. Despite this, the groundwater table modulation is nonetheless a powerful potential driven mechanism. The shift in lake level reflects changes in groundwater storage and serves as a gauge for the continental groundwater table [21,48,49]. Deep faults can serve as efficient conduits for groundwater migration connecting lakes and open sea and further regulate the lake's level [50]. Subduction of the Pacific Plate in the late Eocene has an important impact on the evolution of the Bohai Bay Basin, forming a series of deep faults [31,51]. Moreover, some of the preexisting faults are reactivated and deepen into the crust. The Tanlu fault zone is the typical deep fault in which the kinematic direction shifts from left lateral motion to right lateral motion in this instance. Taken together, these make it possible that the BBB was connected to the open sea in the late Eocene.

6. Conclusions

The natural GR series of the S146 borehole through Es_3 member in the Dongying Depression provides an archive of paleoclimate and environmental changes for lake level reconstruction. The following conclusions can be made:

(1) We constructed a robust astronomical time scale (ATS) of the Es_3 member of the late Eocene in the Dongying Depression based on the identification of 405 kyr-long eccentricity cycles in the S146 well. The developed high-resolution and continuous ATS spans the time from 34.78 to 41.21 Ma.

(2) A quantitative assessment technique of eCOCO combined with COCO analysis generates an optimal sedimentation rate of 13.0 cm/kyr and variable sedimentation rates through the studied interval, which better supports our robust ATS.

(3) Paleolake level changes of the late Eocene were reconstructed based on sedimentary noise modeling. The lake-level changes in 1.2 Myr cycles in Dongying Depression were in phase with the same cycles of synchronous global sea-level changes. We suggest that marine incursion and/or groundwater table modulation are the main driven mechanisms of the co-variation of global sea level changes and regional lake level changes.

Author Contributions: Conceptualization, Z.S. and T.J.; methodology, X.F.; software, P.W.; validation, X.F. and P.W.; formal analysis, X.F.; investigation, Z.S.; resources, T.J.; data curation, H.Z.; writing—original draft preparation, Z.S.; writing—review and editing, H.Z. and T.J.; visualization, Z.S.; supervision, H.Z. and T.J.; project administration, H.Z. and T.J.; funding acquisition, Z.S. and H.Z. All authors have read and agreed to the published version of the manuscript.

Funding: The research presented in this paper was funded by China Postdoctoral Science Foundation (No. 2022M712950), the Open Funds for Hubei Key Laboratory of Marine Geological Resources, China University of Geosciences (No. MGR202213). This research was also funded by the National Natural Science Foundation of China (No. 42172127, No. 41872149).

Data Availability Statement: Not applicable.

Acknowledgments: The Geoscience Institute of the Shengli Oilfield, SINOPEC is thanked for providing access to their geological data and permitting publication of the results. The editor and anonymous reviewers are thanked for helpful comments and constructive suggestions that greatly improved our manuscript.

Conflicts of Interest: The authors declare no conflict of interest.

References

1. Hilgen, F.; Schwarzacher, W.; Strasser, A. Concepts and definitions in cyclostratigraphy (second report of the cyclostratigraphy working group). *SEPM. Spec. Publ.* **2004**, *81*, 303–305.
2. Strasser, A.; Hilgen, F.J.; Heckel, P.H. Cyclostratigraphy-concepts, definitions, and applications. *Newsl. Stratigr.* **2006**, *42*, 75–114. [CrossRef]
3. Hinnov, L.A. Cyclostratigraphy and astrochronology in 2018. In *Stratigraphy & Timescales*; Elsevier: Amsterdam, The Netherlands, 2018; Volume 3, pp. 1–80.
4. Meyers, S.R. Cyclostratigraphy and the problem of astrochronologic testing. *Earth Sci. Rev.* **2019**, *190*, 190–223. [CrossRef]
5. Strasser, A.; Hillgärtner, H.; Pasquier, J. Cyclostratigraphic timing of sedimentary processes: An example from the Berriasian ofthe Swiss and French Jura Mountains. *SEPM. Spec. Publ.* **2004**, *81*, 135–151.
6. De Vleeschouwer, D.; Whalen, M.T.; Day, J.E.; Claeys, P. Cyclostratigraphic calibration of the Frasnian (Late Devonian) time scale (western Alberta, Canada). *Bulletin* **2012**, *124*, 928–942. [CrossRef]

7. Huang, C.; Hesselbo, S.P. Pacing of the Toarcian Oceanic Anoxic Event (Early Jurassic) from astronomical correlation of marine sections. *Gondwana Res.* **2014**, *25*, 1348–1356. [CrossRef]
8. Argenio, B.D.; Fischer, A.G.; Silva, I.P.; Weissert, H.; Ferreri, V. Cyclostratigraphy: Approaches and case histories. *SEPM. Spec. Publ.* **2004**, *81*, 1–311.
9. Wagreich, M.; Hohenegger, J.; Neuhuber, S. Nannofossil biostratigraphy, strontium and carbon isotope stratigraphy, cyclostratigraphy and an astronomically calibrated duration of the Late Campanian Radotruncana calcarata Zone. *Cretac. Res.* **2012**, *38*, 80–96. [CrossRef]
10. Miller, K.G.; Browning, J.V.; Schmelz, W.J.; Kopp, R.E.; Mountain, G.S.; Wright, J.D. Cenozoic sea-level and cryospheric evolution from deep-sea geochemical and continental margin records. *Sci. Adv.* **2020**, *6*, eaaz1346. [CrossRef]
11. Hao, F.; Zhou, X.; Zhu, Y.; Bao, X.; Yang, Y. Charging of the Neogene Penglai 19-3 field, Bohai Bay Basin, China: Oil accumulation in a young trap in an active fault zone. *AAPG Bull.* **2009**, *93*, 155–179. [CrossRef]
12. Sun, Z.; Zhu, H.; Xu, C.; Yang, X.; Du, X. Reconstructing provenance interaction of multiple sediment sources in continental down-warped lacustrine basins: An example from the Bodong area, Bohai Bay Basin, China. *Mar. Petrol. Geol.* **2020**, *113*, 104142. [CrossRef]
13. Wang, M.; Chen, H.; Huang, C.; Kemp, D.B.; Xu, T.; Zhang, H.; Li, M. Astronomical forcing and sedimentary noise modeling of lake-level changes in the Paleogene Dongpu Depression of North China. *Earth Planet. Sci. Lett.* **2020**, *535*, 116116. [CrossRef]
14. Dyer-Pietras, K.M. Insolation forcing of sub-lacustrine debris flows–Could monsoon intensification have played a role? Eocene lacustrine Green River Formation, Piceance Creek Basin, Colorado. *Palaeogeogr. Palaeoclimatol. Palaeoecol.* **2020**, *553*, 109738. [CrossRef]
15. Wu, H.; Zhang, S.; Hinnov, L.A.; Jiang, G.; Yang, T.; Li, H.; Wan, X.; Wang, C. Cyclostratigraphy and orbital tuning of the terrestrial upper Santonian–Lower Danian in Songliao Basin, northeastern China. *Earth Planet. Sci. Lett.* **2014**, *407*, 82–95. [CrossRef]
16. Yao, Y.; Xu, D.; Zhang, H.; Han, Y.; Zhang, S.; Yin, Z.; Li, B.; He, Q.; Bian, X. A Brief Introduction To The Cenozoic Astrostratigraphic Time Scale For The Dongying Depression, Shandong. *J. Stratigr.* **2007**, *31*, 423–429. (In Chinese with English Abstract)
17. Liu, Z.; Huang, C.; Algeo, T.J.; Liu, H.; Hao, Y.; Du, X.; Lu, Y.; Chen, P.; Guo, L.; Peng, L. High-resolution astrochronological record for the Paleocene-Oligocene (66–23Ma) from the rapidly subsiding Bohai Bay Basin, northeastern China. *Palaeogeogr. Palaeoclimatol. Palaeoecol.* **2018**, *510*, 78–92. [CrossRef]
18. Shi, J.; Jin, Z.; Liu, Q.; Huang, Z.; Hao, Y. Terrestrial sedimentary responses to astronomically forced climate changes during the Early Paleogene in the Bohai Bay Basin, eastern China. *Palaeogeogr. Palaeoclimatol. Palaeoecol.* **2018**, *502*, 1–12. [CrossRef]
19. Shi, J.; Jin, Z.; Liu, Q.; Zhang, R.; Huang, Z. Cyclostratigraphy and astronomical tuning of the middle eocene terrestrial successions in the Bohai Bay Basin, Eastern China. *Glob. Planet. Chang.* **2019**, *174*, 115–126. [CrossRef]
20. Jin, S.; Liu, S.; Li, Z.; Chen, A.; Ma, C. Astrochronology of a middle Eocene lacustrine sequence and sedimentary noise modeling of lake-level changes in Dongying Depression, Bohai Bay Basin. *Palaeogeogr. Palaeoclimatol. Palaeoecol.* **2022**, *585*, 110740. [CrossRef]
21. Ma, Y.; Fan, M.; Li, M.; Ogg, J.G.; Zhang, C.; Feng, J.; Zhou, C.; Liu, X.; Lu, Y.; Liu, H.; et al. East Asian lake hydrology modulated by global sea-level variations in the Eocene warmhouse. *Earth Planet. Sci. Lett.* **2023**, *602*, 117925. [CrossRef]
22. Jia, C.; Zou, C.; Yang, Z.; Zhu, R.; Chen, Z.; Zhang, B.; Jiang, L. Significant progress of continental petroleum geological theory in basins of Central and Western China. *Pet. Explor. Dev.* **2018**, *45*, 573–588. [CrossRef]
23. Li, M.; Hinnov, L.A.; Huang, C.; Ogg, J.G. Sedimentary noise and sea levels linked to land–ocean water exchange and obliquity forcing. *Nat. Commun.* **2018**, *9*, 1004. [CrossRef] [PubMed]
24. Wang, M.; Li, M.; Kemp, D.B.; Boulila, S.; Ogg, J.G. Sedimentary noise modeling of lake-level change in the Late Triassic Newark Basin of North America. *Glob. Planet. Chang.* **2022**, *208*, 103706. [CrossRef]
25. Gradstein, F.; Ogg, J.G.; Schmitz, M.D.; Ogg, G.M. *The Geologic Time Scale*; Elsevier: Amsterdam, The Netherland, 2012; Volume 2.
26. Feng, Y.; Li, S.; Lu, Y. Sequence stratigraphy and architectural variability in Late Eocene lacustrine strata of the Dongying Depression, Bohai Bay Basin, Eastern China. *Sediment. Geol.* **2013**, *295*, 1–26. [CrossRef]
27. Lin, H.; Yan, J.; Yuan, W.; Chen, S.Y. Sedimentary Facies Types of The Third Member of The Shahejie Formation in The Paleogene of The Dongying Depression, The Bohaiwan Basion and Third Distribution Characters on Plane. *Pet. Geol. Exp.* **2005**, *27*, 55–61. (In Chinese with English Abstract)
28. Wang, J. High Frequency Sequence Stacking Mode of Dongying Delta-turbidite Fans and Lithological Trap. *Acta Sedimentol. Sin.* **2005**, *23*, 303–309.
29. Baumgarten, H.; Wonik, T. Cyclostratigraphic studies of sediments from Lake Van (Turkey) based on their uranium contents obtained from downhole logging and paleoclimatic implications. *Int. J. Earth Sci.* **2015**, *104*, 1639–1654. [CrossRef]
30. Li, M.; Ogg, J.; Zhang, Y.; Huang, C.; Hinnov, L.; Chen, Z.; Zou, Z. Astronomical tuning of the end-Permian extinction and the Early Triassic Epoch of South China and Germany. *Earth Planet. Sci. Lett.* **2016**, *441*, 10–25. [CrossRef]
31. Xu, K.; Chen, H.; Huang, C.; Ogg, J.G.; Zhu, J.; Lin, S.; Yang, D.; Zhao, P.; Kong, L. Astronomical time scale of the Paleogene lacustrine paleoclimate record from the Nanxiang Basin, central China. *Palaeogeogr. Palaeoclimatol. Palaeoecol.* **2019**, *532*, 109253. [CrossRef]
32. Cleverand, W.S. Robust locally weighted regression and smoothing scatterplots. *J. Am. Stat. Assoc.* **1979**, *74*, 829–836. [CrossRef]
33. Kodama, K.P.; Hinnov, L.A. *Rock Magnetic Cyclostratigraphy*; John Wiley & Sons: Hoboken, NJ, USA, 2014; pp. 1–147.

34. Li, M.; Kump, L.R.; Hinnov, L.A.; Mann, M.E. Tracking variable sedimentation rates and astronomical forcing in Phanerozoic paleoclimate proxy series with evolutionary correlation coefficients and hypothesis testing. *Earth Planet. Sci. Lett.* **2018**, *501*, 165–179. [CrossRef]
35. Thomson, D.J. Spectrum estimation and harmonic analysis. *Proc. IEEE* **1982**, *70*, 1055–1096. [CrossRef]
36. Mann, M.E.; Lees, J.M. Robust estimation of background noise and signal detection in climatic time series. *Clim. Chang.* **1996**, *33*, 409–445. [CrossRef]
37. Li, M.; Hinnov, L.; Kump, L. Acycle: Time-series analysis software for paleoclimate research and education. *Comput. Geosci.* **2019**, *127*, 12–22. [CrossRef]
38. Laskar, J.; Robutel, P.; Joutel, F.; Gastineau, M.; Correia, A.C.M.; Levrard, B. A long-term numerical solution for the insolation quantities of the Earth. *Astron. Astrophys.* **2004**, *428*, 261–285. [CrossRef]
39. Yao, Y.; Liang, H.; Cai, Z.; Guan, X.; Zhao, Z.; Cheng, Z.Q.; Sun, Z.C.; Yang, S.Z. *Tertiary in Petroliferous Regions of China (IV): The Bohai Gulf Basin*; Petroleum Industry Press: Beijing, China, 1994; 240p. (In Chinese)
40. Zeng, G.; Wang, F.; Zheng, H.; Fu, J. Cenozoic volcanic rocks and its relations to basin evolution and oil accumulation in dongying sag. *Earth Sci.* **1997**, *22*, 157–164. (In Chinese with English Abstract)
41. Olsen, P.E.; Kent, D.V. Milankovitch climate forcing in the tropics of Pangaea during the Late Triassic. *Palaeogeogr. Palaeoclimatol. Palaeoecol.* **1996**, *122*, 1–26. [CrossRef]
42. Trauth, M.H.; Maslin, M.A.; Deino, A.; Strecker, M.R. Late cenozoic moisture history of East Africa. *Science* **2005**, *309*, 2051–2053. [CrossRef]
43. Boulila, S.; Galbrun, B.; Miller, K.G.; Pekar, S.F.; Browning, J.V.; Laskar, J.; Wright, J.D. On the origin of Cenozoic and Mesozoic "third-order" eustatic sequences. *Earth-Sci. Rev.* **2011**, *109*, 94–112. [CrossRef]
44. Jovane, L.; Florindo, F.; Sprovieri, M.; Pälike, H. Astronomic calibration of the late Eocene/early Oligocene Massignano section (central Italy). *Geochem. Geophys. Geosyst.* **2006**, *7*. [CrossRef]
45. Westerhold, T.; Röhl, U. Orbital pacing of Eocene climate during the Middle Eocene Climate Optimum and the chron C19r event: Missing link found in the tropical western Atlantic. *Geochem. Geophys. Geosyst.* **2013**, *14*, 4811–4825. [CrossRef]
46. Wendler, J.E.; Wendler, I. What drove sea-level fluctuations during the mid-Cretaceous greenhouse climate? *Palaeogeogr. Palaeoclimatol. Palaeoecol.* **2016**, *441*, 412–419. [CrossRef]
47. Wei, W.; Algeo, T.J.; Lu, Y.; Lu, Y.; Liu, H.; Zhang, S.; Peng, L.; Zhang, J.; Chen, L. Identifying marine incursions into the Paleogene Bohai Bay Basin lake system in northeastern China. *Int. J. Coal Geol.* **2018**, *200*, 1–17. [CrossRef]
48. Wagreich, M.; Lein, R.; Sames, B. Eustasy, its controlling factors, and the limno-eustatic hypothesis-concepts inspired by Eduard Suess. *Austrian J. Earth Sci.* **2014**, *107*, 115–131.
49. Sames, B.; Wagreich, M.; Wendler, J.E.; Haq, B.U.; Conrad, C.P.; Melinte-Dobrinescu, M.C.; Hu, X.; Wendler, I.; Wolfgring, E.; Yilmaz, I.Ö. Short-term sea-level changes in a greenhouse world—A view from the Cretaceous. *Palaeogeogr. Palaeoclimatol. Palaeoecol.* **2016**, *441*, 393–411. [CrossRef]
50. Öztürk, K.; Yaltirak, C.; Alpar, B. The relationship between the tectonic setting of the Lake Iznik basin and the middle strand of the North Anatolian Fault. *Turk. J. Earth Sci.* **2009**, *18*, 209–224.
51. Zhu, R.; Xu, Y. The subduction of the west Pacific plate and the destruction of the North China Craton. *Sci. China Earth Sci.* **2019**, *62*, 1340–1350. [CrossRef]

energies MDPI

Article

Sedimentary Characteristics of Lacustrine Beach-Bars and Their Formation in the Paleogene Weixinan Sag of Beibuwan Basin, Northern South China Sea

Jie He [1,*], Hua Wang [2], Tao Jiang [1], Entao Liu [1], Si Chen [2] and Ping Jiang [3]

[1] College of Marine Science and Technology, China University of Geosciences, Wuhan 430074, China; taojiang@cug.edu.cn (T.J.); liuentao@cug.edu.cn (E.L.)
[2] School of Earth Resources, China University of Geosciences, Wuhan 430074, China; wanghua@cug.edu.cn (H.W.); sichen720@hotmail.com (S.C.)
[3] Zhanjiang Branch Company, China National Offshore Oil Corporation, Zhanjiang 524057, China; jiangp@cnooc.com.cn
* Correspondence: jiehe19920402@163.com

Abstract: Beach-bar reservoirs have been promising hydrocarbon-bearing exploration advances in the Beibuwan Basin, especially in the WZ12-2 area within the Weixinan sag. The sedimentary characteristics, distribution and formation mechanisms of beach-bar sand bodies in Mbr2 (Member 2) of the Paleogene Liushagang Fm. in the WZ12-2 area within the Weixinan sag were analyzed based on well-log, seismic and core data on thin section and heavy mineral data. Mbr2 in the WZ12-2 area comprises a third-order sequence, which consists of three systems tracts (lowstand systems tract, transgressive systems tract and a locally developed highstand systems tract). Thick beach-bar sand bodies are developed in the WZ12-2 area during the lowstand systems tract stage. The formation of sandy beach-bar sand bodies can be divided into five stages. By integrating lithology, mineral composition, sedimentary structures and geophysical characteristics, it can be concluded that the beach-bar sand bodies in the study area were controlled by paleotopography, hydrodynamic environment, sediment provenance and lake-level variation. The gentle slope of the Qixi uplift and relatively stable passive tectonic background during the deposition of Mbr2 of the Liushagang Fm. laid a solid paleogeomorphological foundation for beach-bar deposition. Strong hydrodynamic forces and shallow water further contributed to beach-bar sand bodies formation. In addition, the sands in the fan delta in the northwestern part of the area served as point provenance and the deposits in the southeast acted as linear provenance in providing sediments to the beach-bars. High-frequency variations of the lake level drove vertical stacking of the beach-bar sand bodies and considerable lateral extension over a large area. The sedimentary characteristics and formation mechanism of lacustrine beach-bars in this study may provide a reference for hydrocarbon exploration in other similar basins in the world.

Keywords: beach-bars; Beibuwan Basin; Weixinan sag; sedimentary characteristics; formation mechanism

Citation: He, J.; Wang, H.; Jiang, T.; Liu, E.; Chen, S.; Jiang, P. Sedimentary Characteristics of Lacustrine Beach-Bars and Their Formation in the Paleogene Weixinan Sag of Beibuwan Basin, Northern South China Sea. *Energies* **2022**, *15*, 3391. https://doi.org/10.3390/en15093391

Academic Editor: Reza Rezaee

Received: 23 March 2022
Accepted: 29 April 2022
Published: 6 May 2022

1. Introduction

Beach-bar refers to the beach and bar sand bodies in lacustrine or marine basins with a shallow water environment [1–6]. In the gently sloping area of a rifted lacustrine basin, beach-bar sand bodies tend to develop in shallow littoral lakes [4,7,8]. Beach-bars can usually be subdivided into bioclastic and terrigenous clastic types [2,4,9]. Bioclastic beach-bars usually develop in quiet and clear lake water under warm and humid condition with carbonate provenance [10]. Terrigenous clastic beach-bar sand bodies form from the redeposition and reworking of deltas and other related near-shore sedimentary systems in a shallow water environment under the effects of fluctuations in energy conditions from coastal currents and waves [11–14]. Generally, lacustrine beach-bars are mostly characterized by thin interbedded sand bodies mainly composed of fine sandstone and siltstone with restricted distributions [2].

The Beibuwan Basin, which is an extensional Cenozoic sedimentary basin, is characterized by lacustrine deposits in Paleogene and littoral-neritic sea sediments in Neogene-Quaternary [15]. It is a productive basin in the northern South China Sea (Figure 1A) [15,16]. The northern part of the Beibuwan Basin has been proven to be a prospective area for further petroleum exploration [17–20]. Recently, important breakthroughs in the Weixinan sag have been achieved (Figure 1B) [21]. The Southeast Slope Belt is one of the largest hydrocarbon-producing targets in the Weixinan sag (Figure 1C) [19]. Among the oil fields in the Southeast Slope Belt, the WZ12 oil field produces abundant oil and gas. The beach-bars developed in the WZ12-2 area of Mbr2 of the Liushagang Fm. have now become important exploration targets in the Beibuwan Basin [20,22,23]. A beach-bar sand reservoir has grain size dominated by fine sand and sand, good sorting, well-developed primary and secondary porosity, and relatively high permeability. Beach-bar sand bodies change frequently on a lateral scale, easily forming lithologic lenticular traps and sandstone pinch-out traps [4,14,24]. The dark mudstone at the bottom provides hydrocarbons for the beach-bar reservoirs in the Mbr2 stage of Liushagang Formation [16]. The overlying thick dark mudstone can act as a superior caprock for beach-bar reservoirs underneath [25–27]. With superior hydrocarbon generation and sealing conditions, beach-bar sand bodies in the WZ12-2 area are evaluated as prospective exploration targets. During the depositional period of Mbr2 in the Weixinan sag, especially during the lowstand systems tract (LST) stage, beach-bar sand bodies formed on a large scale [23,28]. In the past few years, considerable attention has been given to the petroleum source of WZ12 oils in Mbr2 of the Liushagang Fm. in the Weixinan sag [16,17,21,29]. However, studies on the reservoir are scarce.

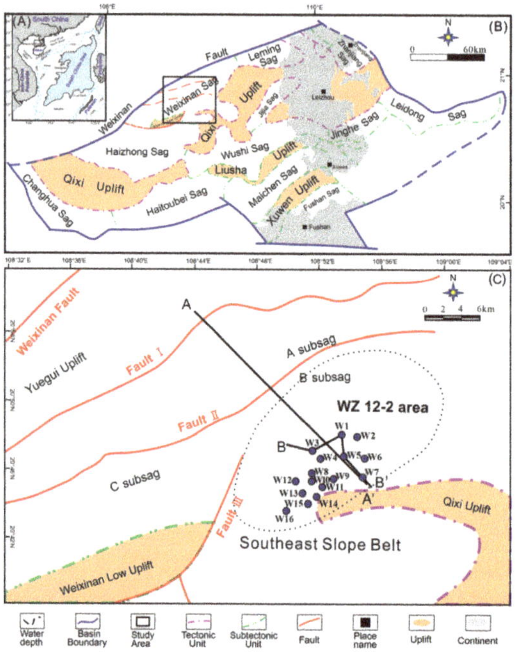

Figure 1. (**A**) Location of sedimentary basins in the South China Sea. The black box indicates the location of Figure 1B; (**B**) simplified geological map of the Beibuwan Basin [16,20]. The Weixinan sag was located in the northwestern Beibuwan Basin. The black box refer to the study area in Figure 1C; (**C**) simplified geological map of the Weixinan sag. The black line AA′ and BB′ mark the locations of seismic section and sequence stratigraphic correlation sections in Figures 3 and 9, respectively.

In this regard, good hydrocarbon source rocks and thick top seals are present in the Mbr2 stage in the WZ 12-2 area. Further steps are to find superior hydrocarbon "container"—reservoirs, to form potential source-reservoir-cap assemblages, which are of great significance to hydrocarbon exploration. In this study, the sedimentary characteristics, i.e., lithology, mineral composition, sedimentary structures and geophysical characteristics of the beach-bar sand bodies, are analyzed concretely to investigate the formation mechanism, on the basis of thin section, heavy mineral, core, well logging and 3D seismic data. This study can be useful for further hydrocarbon prospecting and exploration in the Weixinan sag and has great significance for further application to lacustrine rift basins with similar backgrounds in the world.

2. Geological Background

The Beibuwan Basin has an area of approximately 1.9×10^4 km^2 onshore and 1.6×10^4 km^2 offshore [15,30]. The development of the Beibuwan Basin includes two main stages: the Paleogene syn-rifting stage (65–23 Ma) and the post-rifting stage (23 Ma to present) (Figure 2) [20,31]. The Weixinan sag located in the northwestern Beibuwan Basin is a half-graben rift lake basin bounded by the Qixi uplift to the southeast, the Weixinan fault to the northwest, and the Haizhong sag and Weixinan low uplift to the southwest (Figure 1B) [15].

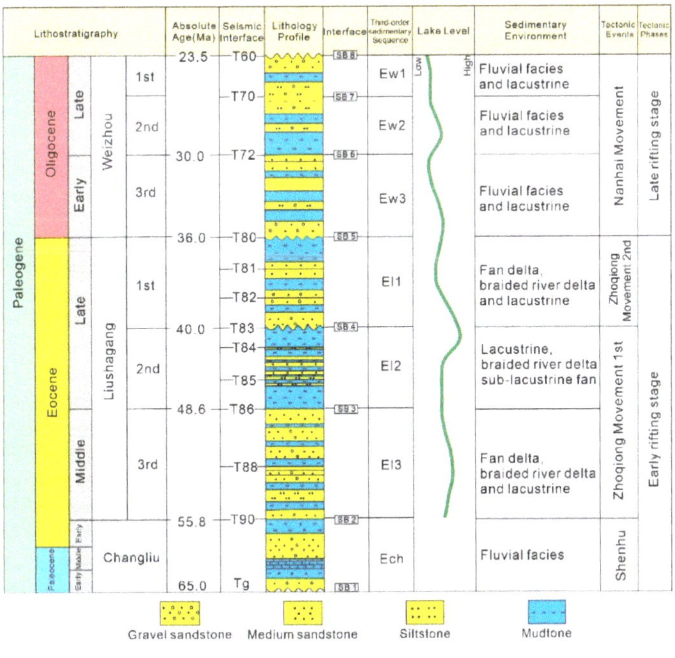

Figure 2. The Paleogene sedimentary sequence and stratigraphic framework of the Beibuwan Basin [31].

The A, B and C sub-sags formed in response to three extensional events in Paleocene, Eocene and middle-late Oligocene (Figure 1C). The three growth normal faults (Fault I, Fault II and Fault III) in the NE-SW direction have large fault throws up to several thousand meters and long extents with lengths of more than 40 km (Figure 3) [16]. In the Paleogene, the depositional environments in the Weixinan sag were dominated by fan delta, shore-shallow lake delta and lacustrine facies [31].

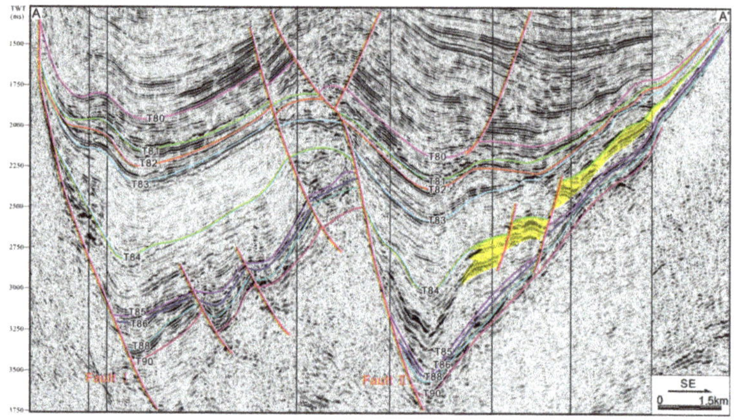

Figure 3. Seismic section (northwest to southeast) across the Weixinan sag showing the sequence boundaries, stratigraphic units and major faults. The targeted Mbr2 of the Liushagang Fm. is from T85 to T84 reflectance surface. The light yellow color in the figure refers to the beach-bar deposition.

Based on logging core, drilling, 3D seismic, paleontological and geochemical data, the stratigraphy of the Paleogene in the Weixinan depression has been reconstructed. The Paleogene sediments in the Weixinan depression are composed of the Changliu Fm., Liushagang Fm. and Weizhou Fm. and seven third-order sequences (Ech, El3, El2, El1, Ew3, Ew2 and Ew1) from bottom to top [31]. As the primary and crucial oil and gas-bearing reservoir, the Liushagang Fm. (including Mbr3, Mbr2, and Mbr1) is the main depositional unit in the Weixinan sag (Figure 2). The main sedimentary environments in Mbr3 and Mbr1 are shallow lacustrine and shore-shallow deltas, and lacustrine environments prevailed during Mbr2 deposition [20]. The beach-bars formed in Mbr2 of the Liushagang Fm. in WZ12-2 area of the Weixinan sag.

3. Materials and Methods

This study integrated logging, core and 3D seismic data with thin section, scanning electron microscopy (SEM) and heavy mineral analyses. Wireline log data from 16 wells and more than 2000 km^2 3D seismic data were analyzed. Approximately 40 m of cores from 3 wells were studied. This study also analyzed heavy minerals from 16 wells, grain-size analysis from 51 samples, and porosity and permeability data from 75 samples. All the data were acquired from the CNOOC Zhanjiang Branch Company.

Depositional systems tracts and sequences were first recognized in the WZ12-2 area of the Weixinan sag, mainly on the basis of termination patterns and seismic reflections [32]. They were further confined by core data and wireline logs along the main seismic sections. Depositional and stratigraphic cross-sections throughout the Weixinan sag were established to depict the lateral and vertical distributions of the lacustrine beach-bar facies. The logging, core, grain-size, porosity, permeability and SEM data were used to identify the sedimentary characteristics of the beach-bars. The controlling factors of lacustrine beach-bar formation in the study area were investigated by the reconstruction of paleotopography, sediment source systems and lake levels. Then, a distribution model of lacustrine beach-bar sand bodies was established within the Weixinan sag.

On the basis of well log, drilling core and 3D seismic data, Mbr2 in the Weixinan sag was identified as a third-order sequence, including a lowstand systems tract (LST) and a transgressive systems tract (TST). The highstand systems tract (HST) was not well developed and seen only as thin layers in several wells. In seismic sections, the T83 reflector marking the top sequence boundary separates Mbr1 and Mbr2. The bottom sequence boundary is marked by the T86 reflector that divides Mbr2 and Mbr3. T83 and T86 are regional unconformable surfaces along the basin edges. On seismic profiles, onlap and erosional truncation terminations are observed on the T83 and T86 surfaces (Figure 3).

Scanning electron microscopy (SEM) was conducted with TESCAN-VEGA equipped with an SDD detector, pulse processor and Esprit software workstation. Samples for heavy mineral analysis were first processed to remove the weathered rims, carbonate materials and clay minerals. Then the 63–125 μm fraction were separated by magnetic and electrostatic filters. About 700–1000 heavy mineral grains were identified under the binocular microscope for each sample. The porosity and permeability were measured following Boyle's and Darcy's laws, respectively, by using 2.5 cm diameter cylinders drilled from the cores.

4. Results

4.1. Sedimentary Characteristics

4.1.1. Lithology

The lithologies of lacustrine beach-bar sands are usually composed of siltstone, fine-grained sandstone and pelitic siltstone [4,33]. In the WZ12-2 area, the sandstones of the beach-bars are mainly grey and greyish-brown siltstone and fine- to medium-grained sandstone (Figure 4). The logging and core data indicate that there are few layers of sandstone but that the single layer thickness is large, usually thicker than 3 m. An individual sandy bar layer in the LST can be thicker than 10 m and is interbedded with thin mudstones, showing coarsening trends in the vertical direction. The sandy bar sediments in well W5 are dominated by sand, which makes up 92% of the total content (Figure 4).

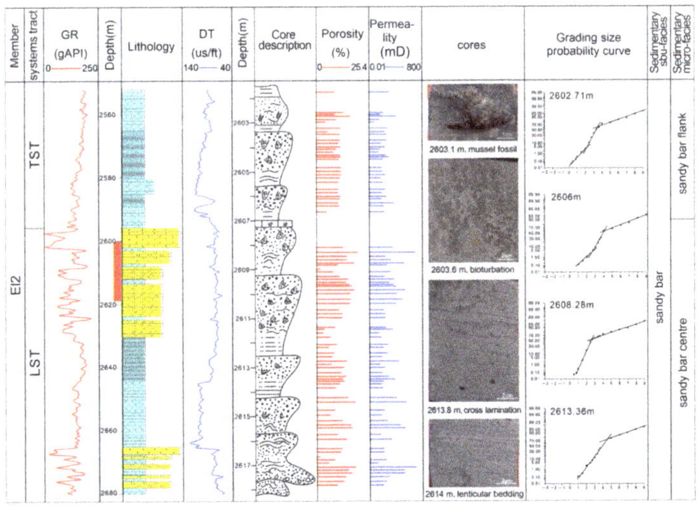

Figure 4. Single well integrated diagram of the beach-bar sand bodies of well W5 in the Weixinan sag. The red rectangle emphasized on the left logging lithologic column refers to the core description in the right.

4.1.2. Mineral Composition

Quartz is the most frequent and abundant mineral in beach-bar sandstones. The quartz content in the WZ12-2 beach-bar reservoirs ranges between 52% and 77%, with an average of 67% (Figure 5A,B). Feldspars in the study area include potassium feldspar and plagioclase (Figure 5H,L), and the percentage of plagioclase is higher than that of potassium feldspar. The average feldspar content is 10%. The lithic fragment content is approximately 5%, and fragments are predominantly argillaceous clasts. The detrital compositions of beach-bar sandstones are mainly litho-quartzose and feldspatho-litho-quartzose sandstones, with subordinate feldspatho-quartzose sandstones and minor litho-feldspatho-quartzose, quartzo-lithic and quartzose sandstones (Figure 6) [34]. The compositional maturity of the beach-bar sand bodies is high (average 2.2), ranging from 1.1 to 3.7. Thin section observations indicate that there are intraclasts of partly dissolved ooids and muddy and silty limestones with plastic deformation in sandstones of the beach-bar (Figure 5C–E,H). The carbonate sediments are redeposited in shallow lacustrine environments after reworking by waves. The intraclasts, which consist of freshly deposited and loosely compacted carbonate sediments, suggest that the carbonate sediments originated internally from the basin. Both the high compositional maturity and the occurrence of oolitic particles with thin crusts reflect the high frequency oscillation of lake water and long-term reworking of the sediments.

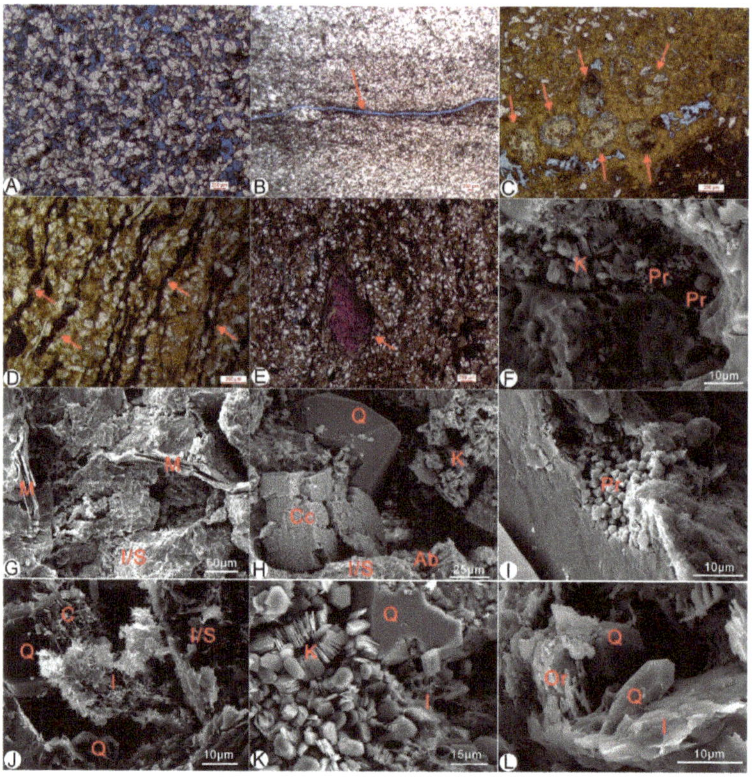

Figure 5. Thin section and SEM images providing information about the mineral composition of beach-bar sandstones in Weixinan sag. (**A**) Well W5, 2610.20 m, high content of quartz and porosity (polarized light). (**B**) Well W5, 2604.38 m, micro-fractures in beach-bar reservoir (polarized light). (**C**) Well W5, 2603.20 m, carbonate oolites (polarized light). (**D**) Well W5, 2612.56 m, black banded carbonaceous fragments (polarized light). (**E**) Well W5, 2605.71 m, carbonaceous fragments and dissolution of detrital grains (polarized light). (**F**) Well W6, 2435.90 m, booklet-like kaolinite and pyrite framboids (SEM). (**G**) Well W1, 3327.04 m, deformation of mica and hairy illite-smectite mixed layers (SEM). (**H**) Well W13, 2266.80 m, authigenic carbonate cement, quartz, booklet-like kaolinite and partly dissolved albite (SEM). (**I**) Well W1, 3331.35 m, octahedral pyrites (SEM). (**J**) Well W1, 3334.50 m, fibrous illite, quartz and flaky illite-smectite mixed layers (SEM). (**K**) Well W12, 2755.82 m, booklet-like kaolinite, quartz and honeycombed illite (SEM). (**L**) Well W12, 2699.52 m, quartz and partly dissolved orthoclase (SEM). Ab—albite, Cc—carbonate cement, I—illite, I/S-illite and smectite mixed layer, K—kaolinite, M—mica, Or—orthoclase. Pr—pyrite, Q—quartz.

Microscopic data show that quartz is common in beach-bar sand bodies (Figure 5H,J–L). Orthoclase and albite are partly dissolved (Figure 5H,L). Under SEM, micas have a flaky morphology and are deformed by compaction (Figure 5G). Carbonate cements principally include calcite, dolomite and ankerite, and the most abundant type is calcite. The calcite cements developed in the WZ 12-2 beach-bar sandstones are rhombic crystals filling in intergranular pores (Figure 5E,H). Pyrite is present as framboids (Figure 5F) and octahedrons (Figure 5I) in beach-bar sandstones. Clay minerals in the WZ 12-2 beach-bar sandstone reservoirs are mainly authigenic clay minerals and matrix materials. The clay minerals observed in beach-bar reservoirs are kaolinite (Figure 5F,H,K), illite (Figure 5J–L), illite-smectite mixed layer minerals (Figure 5G,H,J) and chlorite (Figure 5J). Kaolinite forms as pore-filling cement, presenting booklet or vermicular pseudohexagonal morphologies. Illite occurs with fibrous and sheet textures. The illite-smectite mixed-layer clays have honeycomb and flake textures. Chlorite grows as rims along detrital grain surfaces.

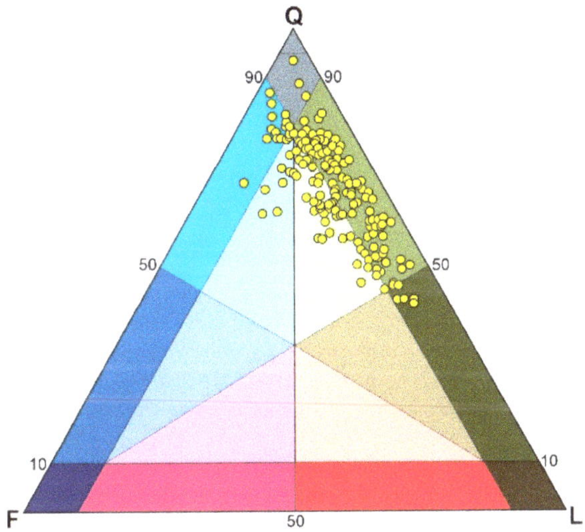

Figure 6. Detrital composition triangular diagram of the beach-bar sandstones in the Weixinan sag [34]. The detrital composition of beach-bar sandstones are mainly litho-quartzose and feldspatho-litho-quartzose sandstones.

Heavy minerals in the study area are mainly zircon, tourmaline, rutile, garnet, hematite, limonite, magnetite and titanomorphite. The zircon contents are between 0.2% and 20%, with an average value of 5.2%. The percentage of tourmaline varies from 0.2% to 27.6%, and the average value is 4.5%. Rutile has an average content of 0.82%, ranging from 0.2% to 2.9%. The garnet content varies from 0.2% to 19.4% with an average of 3.2%. Hematite and limonite have the highest contents. They account for an average of 61.2% of the total heavy minerals, varying from 1% to 97.8%. Magnetite and titanomorphite have average values of 5.7% and 19.9%, fluctuating between 0 and 30.8%, and between 0.2% and 91.6%, respectively (Figure 7).

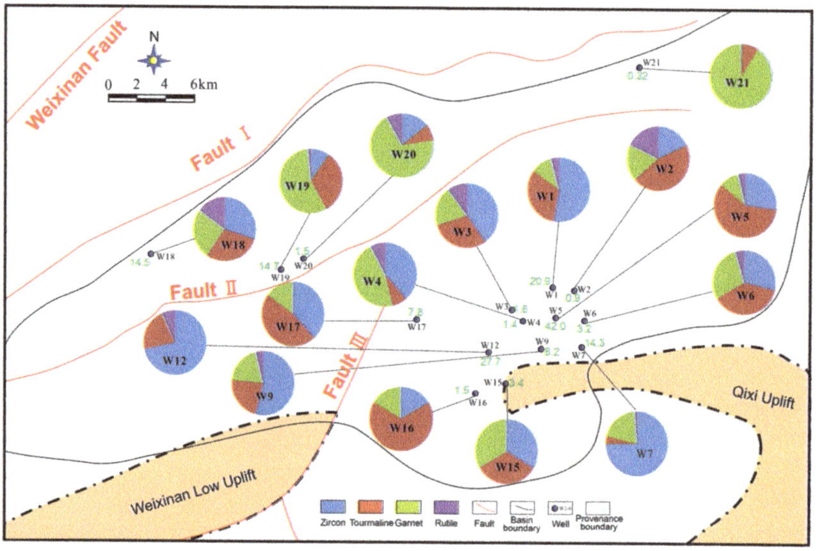

Figure 7. The pie chart of heavy mineral assemblage (zircon + tourmaline + garnet + rutile) percentage and ZTR index of the Mbr2 in the Weixinan sag. The numbers beside the wells are the ZTR values of the wells.

4.1.3. Sedimentary Structures

The sedimentary structures observed in the WZ12-2 beach-bar sandstones have obvious characteristics implying a fluctuating hydrodynamic environment in the lake. Combined-flow ripple cross-lamination or current ripple with lenticular bedding and claystone drapes are observed in the cores (Figure 4). The sandstone beds are massive or horizontally stratified and covered by combined-flow (or current) and interlaminated and interbedded with laminated siltstones or mudstones. These sedimentary structures suggest scouring, reworking and multi-directional flows by waves during deposition [2,35,36]. Teichichnus and fugichnia bioturbation structures are common with various morphologies under different depositional hydrodynamic energies. The presence of mussel bivalve fossils indicates shallow water depths during beach-bar deposition.

4.1.4. Geophysical Characteristics

As the sand bodies of sandy beach-bars in the WZ12-2 area are thick and interbedded with muds, the wireline logging curves are dagger-like or broad-amplitude tooth-shaped boxes and funnel-like shapes with relatively high amplitude (Figure 8). The beach-bars in the WZ12-2 area formed between regional unconformity interface T84 and fourth-order sequence interface T85. The beach-bar reservoirs deposited during the Mbr2 stage have distinctive seismic reflection characteristics with two strong and continuous seismic reflectors, which are also the top interfaces of beach-bar sand bodies. Two obvious inverse superimposed cycles are observed in the wireline logs of wells W1 and W5. The sand thickness increases upward in each cycle, and the mud layers become thinner, indicating that the base level was falling when the beach-bars formed [2]. The grain-size cumulative probability curves show that the core samples of beach-bar sands from well W5 formed mainly from saltating and suspended particles. The saltation types correspond to 60–70% of the whole grain-size curve. A saltation-type curve section is composed of 2–3 segments. The intersection point between the suspended and saltation types is at 3–4Φ (Figure 4). The porosity of well W5 ranges from 2.32% to 23.4% with an average of 13.7%. The permeability of well W5 has an average value of 11.38 mD, varying from 0.01 mD to 357 mD.

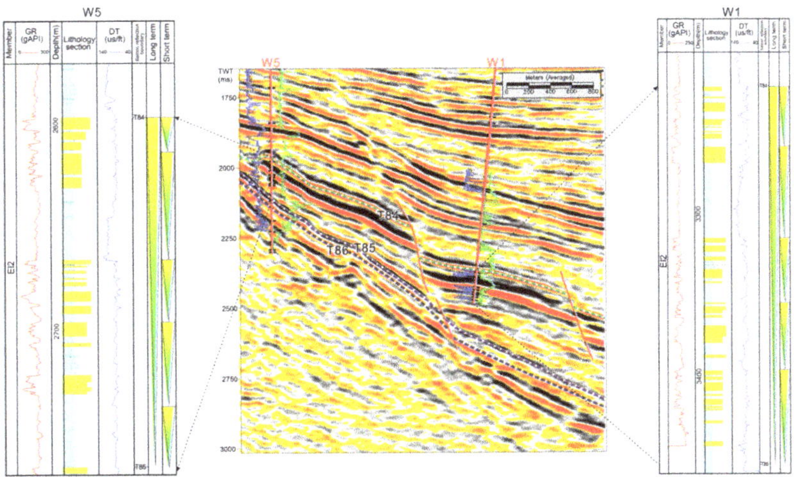

Figure 8. Base level division combined with seismic section of beach-bar sand bodies of Mbr2 stage in the Weixinan sag. The beach-bar sand bodies can be divided into five stages, and two strong and continuous seismic reflections are observed in beach-bar seismic section.

4.2. Distribution Pattern of Beach-Bar Sand Bodies

The beach-bars developed in LST within Mbr2 in the WZ12-2 area of the Weixinan sag and can be further divided into five parasequences (Figure 8). Five stages of sand bodies are identified according to the five parasequences. From bottom to top, the percentage of sand decreases and the mud color becomes darker, indicating a rising lake level and a trend of retrogradational packages. The same phenomenon was also reported in beach-bar syndeposition in the Boxing and Banqiao sags within the Bohai Bay Basin [2,4,14,37]. The five stages of beach-bar sand bodies gradually retrograde towards the uplift (Figure 9).

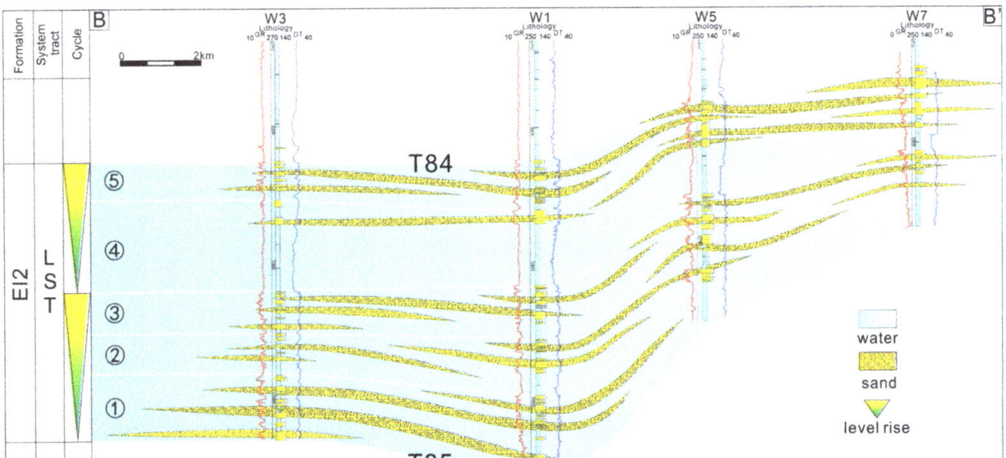

Figure 9. Sequence stratigraphic correlation section of beach-bar sand bodies of Mbr2 stage in the Weixinan sag. The five sand bodies in the LST show an obvious retrogradation upward.

Sand thickness contour maps were constructed to depict the distribution regularity of the five stages of beach-bar sand bodies in the WZ12-2 area of the Weixinan sag (Figure 10).

The shapes of five sand thickness contour maps vary with different depocenters. Sand ① has three depocenters, which are located near wells W3, W15 and W5. The directions of the long axis are northeast. Sand ② also has three depocenters near wells W5, W12 and W16. The long axis direction of two depocenters is northeast, and that of the third is north. Sand ③ has two depocenters in the vicinity of wells W13 and W5 with long axis directions oriented northeast. Compared with sand ③, sand ④ has one long axis direction oriented NNW. Sand ⑤, again, has three depocenters, but the long axes of the three depocenters change NNW and west-east. Although the long axis directions of the depocenters differ in each stage, the long axis direction of the whole beach-bar sand body remains to be northeast, which is constrained by the uplift geomorphology of Fault III in the northwest and the Qixi uplift in the northeast (Figure 11). Among these five stages of sands, the thickness of sand ③ is greatest, and the distribution area of sand ② is broadest. The distribution of the five stages of sands also shows retrogradation towards the uplift from sand ① to sand ⑤ (Figure 9). Additionally, changes in the depocenters, which are closely related to shoreline trajectory in the lake environment, show landward trend as transgression.

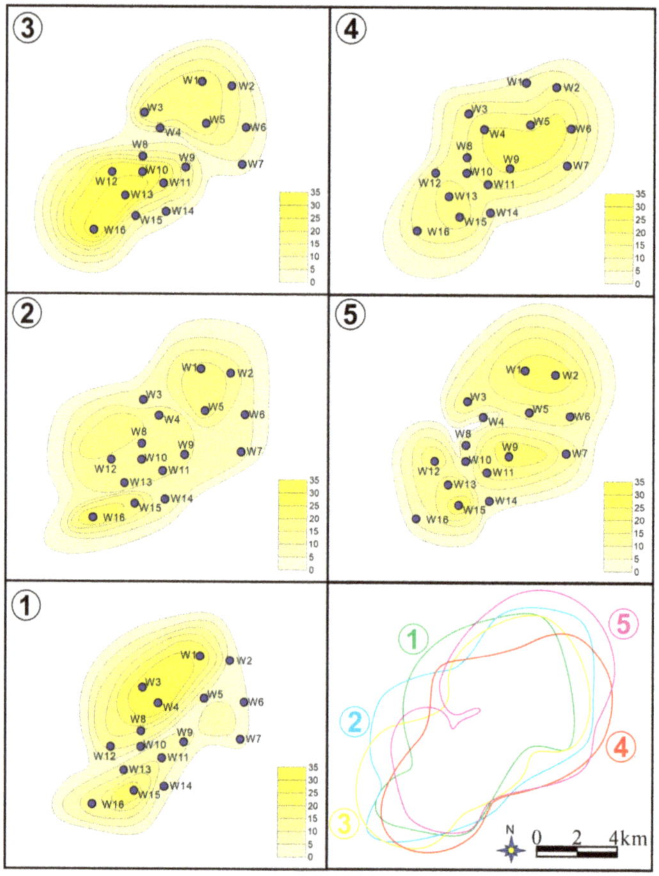

Figure 10. Isopach maps of sandstone thickness and distributions of the five stage sands in the LST of Mbr2 stage in the Weixinan sag. Note that the thickness of sand ③ is greatest, and the distribution area of sand ② is largest. Horizontal distribution of five stage sands also show a retrograding trend towards the uplift. The numbers ①–⑤ indicate the five stage sands in the LST of Mbr2 stage in the Weixinan sag from the bottom to top.

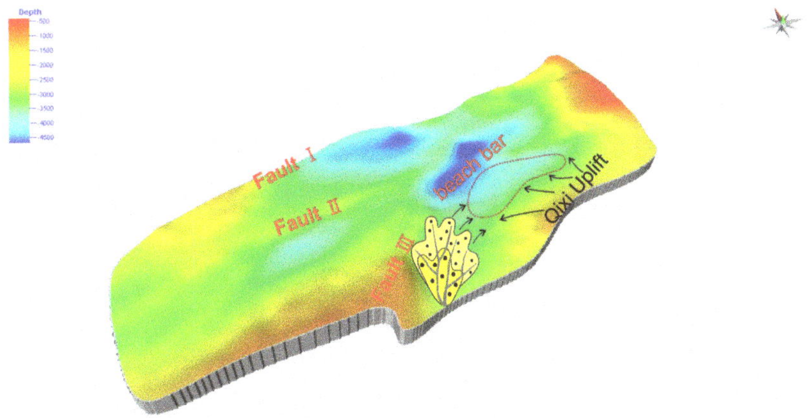

Figure 11. Paleotopography of depositional stage of Mbr2 showing the sediment provenance from the fan deltaic front and the Qixi uplift.

5. Discussion

5.1. Formation Mechanism

Comparing the HST and TST of the Mbr2 stage in the Weixinan sag, large-scale thick beach-bar sand bodies developed during the LST period. The formation of well-developed beach-bar sand bodies in the LST stage in the Weixinan sag was controlled by the paleotopography, hydrodynamic environment, sediment provenance and lake level variation.

5.1.1. Paleotopography

Beach-bars usually form in the early rifting or fault-depression transitional stage, during which tectonic movements is relatively stable [38]. Tectonic movements are closely related to paleotopography. Paleotopography is one of the most crucial factors controlling sedimentation [2,39–42]. Beach-bars may develop on large-scale gentle slopes of basins by studying the modern beach-bars in a faulted lacustrine basin [5]. It has been proposed that beach-bars may be well-developed on gentle slopes of lacustrine basins [2,4]. The favorable places for beach-bar deposition are subaqueous low uplifts, the flanks of nose structures and structural slope-break zones [43]. In this study, the beach-bar sand bodies that formed in the LST stage of Mbr2 in the WZ 12-2 area are situated on the southern slope of the Southeast Slope Belt, where a nose structure formed [28] and the slope angle was less than 1.2° (Figure 11) [23]. Therefore, the paleotopography in the WZ 12-2 area of the Weixinan sag provided good formation conditions for lacustrine beach-bar deposits.

5.1.2. Hydrodynamic Environment

The hydrodynamic environment is an important factor in beach-bar sand body formation and distribution [2,44–46]. Sediments from onshore continental areas are strongly scoured and reformed by lake waves and finally form beach-bar sand reservoirs in the gentle slope zone [4,47,48]. Different from rip and tide currents in marine coastal zones, lake currents are not very strong, and the primary source of hydrodynamic forces is mainly from shore wind waves [49]. On the windward side of the basin, the water in the lake is dragged to move, and friction is generated by wind on the water surface, finally producing waves [4]. Numerous factors, e.g., wind direction, wind force, wind duration and fetch length, can affect wave actions. The directions of the prevailing modelled surface currents (upper 100 m) during the summer and winter monsoons in the Beibuwan Basin during the Mbr2 period were southwest and northeast, respectively [50,51]. Therefore, the waves moved back and forth between the southwest and the northeast. Thus, waves affected a broad area in the southern Southeast Slope Belt. In the northwest, the sands of the fan delta developed along the Fault III were scoured and reworked by lake waves. The waves transported the sands to the gentle slope area and formed beach-bar deposits. The hydrodynamic conditions during the LST stage of the Weixinan sag were relatively strong. The paleowater water depth during the LST stage of Mbr2 was relatively shallow. The paleowater depth of the northwestern footwall of Fault III was deeper than that of the southern Southeast Slope Belt. In general, the Weixinan sag in the Beibuwan Basin during the LST period had strong hydrodynamic conditions with shallow lake water, which laid a good foundation for beach-bar sand body deposition.

5.1.3. Sediment Provenance

It has been proposed that the sediment provenance for lacustrine beach-bar sand bodies may be fan or braided deltas [2,48,52]. During the Mbr2 period in the Weixinan sag, the main sediment sources for beach-bars were the northwestern fan delta and the southeastern Qixi uplift [23,28]. The northwestern fan deltaic sediments acted as a point source (Figure 11). Sand bodies from the deltaic front could have been scoured and reworked by wave currents, producing the five stages of retrograded beach-bar sand bodies. Meanwhile, the nose structure that developed in the Qixi uplift in the southeast served as a large linear source, generating a persistent and abundant sediment supply. The nose structure in the northern Qixi uplift had already been denuded since the Mbr3 period [53]. The denuded sediments were transported to the Weixinan sag and frequently reworked, scoured and redeposited by lake waves and currents.

The distribution range of the lacustrine clastic beach-bar sand reservoirs was significantly affected by the supply and position of the sediment provenance. As a point source, the fan deltaic front in the northwest provided relatively limited amounts of sediment to the Weixinan sag. The water depth of the fan delta was relatively deep, and most of the fan deltaic sand bodies were below the wave base. Only few deposits could be reworked by lake waves. Sediments from the fan delta could reach the locations of wells W15 and W16 (Figure 12), which were the southwestern part of the beach-bar sand reservoirs. However, the linear source, the Qixi uplift, provided large quantities of sediments to the southern of Weixinan sag (Figure 12). The water depth in the southern Weixinan sag was shallow. Sediments from the Qixi uplift were continuously reworked by lake waves and currents, so the Qixi uplift was the main sediment source for the WZ12-2 beach-bar sand reservoirs in the Weixinan sag.

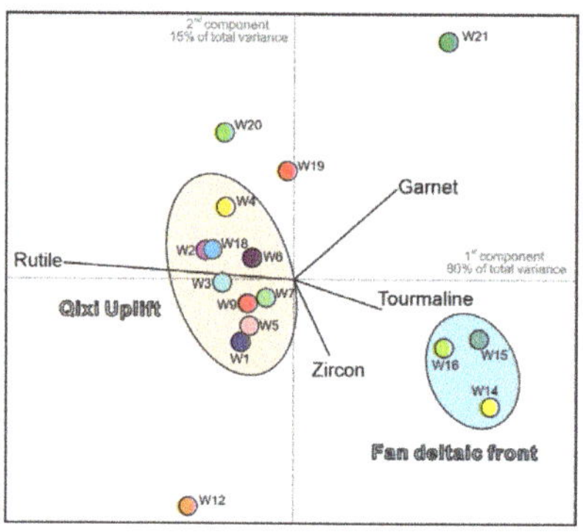

Figure 12. The rutile, garnet, tourmaline and zircon contents of beach-bar sandstones of Mbr2 stage in the Weixinan sag. The concentration of heavy minerals is controlled by sediment provenance. The beach-bar sediments were from the fan deltaic front and the Qixi uplift. The compositional biplot [54], drawn using CoDaPack software [55], displays multivariate observations (points) and variables (rays). The length of each ray is proportional to the variance of the corresponding element in the data set. If the angle between two rays is close to 0°, 90°, or 180°, then the corresponding elements are directly correlated, uncorrelated, or inversely correlated, respectively. Data for the biplot are provided in supplementary material.

5.1.4. Lake Level Oscillations

Lacustrine shallow littoral sand bodies usually spread parallel to shorelines, where lake level oscillations trigger back and forth movements of the shorelines, thus, controlling the shallow littoral sand body lateral distributions [2,4,53–59]. Previous research has proposed that repeated variations in lake level probably had a close relationship with Milankovitch cycles [46,60,61]. In this study, the lakeshore moved frequently and the water depth rose and fell rapidly during each parasequence set, most likely indicating the overfilled status of the lake [62,63]. Under the circumstances of low-relief paleogeomorphology, lake level changes can lead to the lateral migration of shorelines. The high-frequency variation of the lake level in the Weixinan sag contributed to the five stages of vertically stacked sands and to the formation of a wide lateral distribution of beach-bar sand reservoirs.

During the Mbr2 stage, lacustrine clastic beach-bar sand reservoirs formed during the LST period in the WZ12-2 area of the Weixinan sag. However, beach-bar formation was limited in the TST and HST stages. During the TST period, the quick rise in lake water level, together with the low sediment supply, led to starvation of the basin and formation of dark grey mudstones. In the LST stage, the lake level rise in the Weixinan sag was slow, and the source supply was limited [16], thus, the distribution range of beach-bars was largest in the LST period in the WZ12-2 area of the Weixinan sag.

5.2. Depositional Model of Lacustrine Beach-Bars

On the basis of the discussion above, a conceptual depositional model of lacustrine beach-bars in the WZ12-2 area of the Weixinan sag has been constructed to show the formation mechanism of beach-bar sand bodies (Figure 13). The deposition of beach-bar sand bodies in the WZ12-2 area of the Weixinan sag was favored by multiple factors: paleotopography, hydrodynamic environment, sediment provenance and lake level variation. During the deposition of Mbr2, the prevailing southwest summer and northeast winter monsoons and the paleotopography of the Weixinan sag, subjected the fan deltaic and gentle slope zones of the Southeast Slope Belt to lake wave impacts. Lake waves repeatedly washed and reworked the sand bodies sourced from the northwestern fan deltaic front and the southeastern Qixi uplift. Given the lake level oscillations, the shoreline migrated back and forth during each parasequence set. The frequent oscillations in lake level contributed to a wide distribution of beach-bars. The five stages of beach-bar sand bodies retrograded towards the shore with rising lake water.

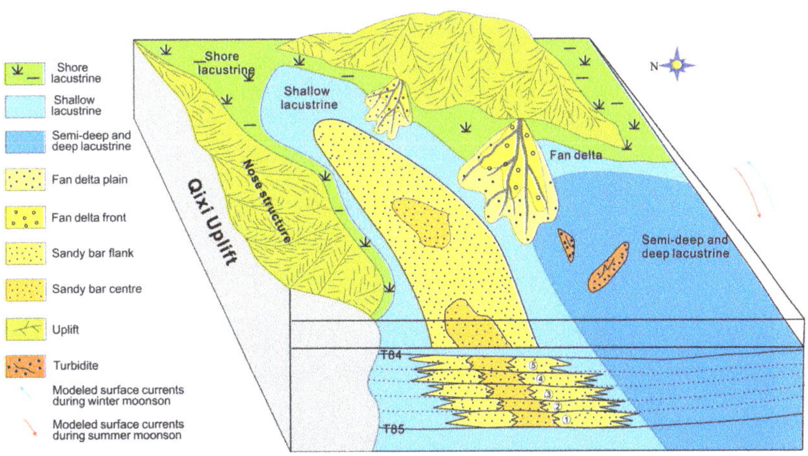

Figure 13. Depositional model for lacustrine beach-bar sand bodies in the LST stage of the Mbr2 period in the Weixinan sag.

In a sandy bar, the sandy bar center has thicker sandstones and higher porosity and permeability than the sandy bar flanks (Figure 4) and, thus, better reservoir quality. With thick source rocks at the bottom and good cap rock on top, a series of lithologic lenticular traps and sandstone pinch-out traps developed in the WZ12-2 area of the Weixinan sag. Oil exploration of lacustrine beach-bars in South China has rarely been reported [23,28,64], while much research on lacustrine beach-bars in East China have been published [2,4,11,14,65], especially in the Bohai Bay Basin. Important breakthroughs have been made in the exploration for oil in beach-bar deposits in the Bohai Bay Basin. The beach-bar sand bodies developed in South China and East China have similar characteristics. Beach-bars in both regions were located in gentle slope belts of a faulted basin during the middle stage of the Eocene, and controlled by the hydrodynamic environment, sediment source and lake level oscillations. Lacustrine beach-bar studies in East China basins can serve as a reference for further exploration of good lacustrine beach-bar sand body reservoirs in South China basins. The depositional model proposed in this study provides guidance for further prospecting and exploration of large-scale and thick beach-bar sand bodies during the LST periods in similar basins.

6. Conclusions

Beach-bar sand reservoirs formed in the Mbr2 of the Liushagang Fm. in the Weixinan sag. Five stages of beach-bar sand bodies developed vertically during the LST period of the Mbr2 sequence. The gently sloping paleotopography, strong hydrodynamic conditions in shallow water, abundant sediment sources and frequent lake water variations were the main factors that controlled beach-bar deposition. The northwestern fan deltaic front and the southeastern Qixi uplift were provenances for the beach-bar depositions. The beach-bar sand bodies that developed in the gently sloping belt within the Weixinan sag were subjected to strong hydrodynamic forces under shallow water conditions. Given these forces, together with repeated variations in lake level, the beach-bars retrograded vertically towards the shore and were distributed laterally over a large area. This study provides guidance for comparable beach-bar reservoirs in rift lacustrine basins with similar tectonic and depositional settings in the world.

Supplementary Materials: The following supporting information can be downloaded at: https://www.mdpi.com/article/10.3390/en15093391/s1, Table S1: The heavy mineral contents of beach-bar sandstones of Mbr2 stage in the Weixinan sag.

Author Contributions: Conceptualization, J.H. and H.W.; methodology, J.H.; software, J.H.; validation, E.L., S.C. and T.J.; formal analysis, S.C.; investigation, J.H.; resources, P.J.; data curation, T.J.; writing—original draft preparation, J.H.; writing—review and editing, T.J.; visualization, S.C.; supervision, H.W.; project administration, E.L.; funding acquisition, H.W. All authors have read and agreed to the published version of the manuscript.

Funding: This research was funded by the National Natural Science Foundation of China (U19B2007, 41972117, 41702121) and by the Major National Science and Technology Programs in the "Thirteenth Five-Year" Plan period (2016ZX05024-006-002).

Institutional Review Board Statement: Not applicable.

Informed Consent Statement: Not applicable.

Data Availability Statement: Not applicable.

Acknowledgments: We appreciate Zhanjiang Branch Company of China Offshore Oil Corporation for providing the data and permission to publish this paper. We are also very grateful to the reviewers and editors for their contributions to improving this paper.

Conflicts of Interest: The authors declared that they have no conflict of interest to this work. We declare that we do not have any commercial or associative interest that represents a conflict of interest in connection with the work submitted.

References

1. Basilici, G. Sedimentary facies in an extensional and deep-lacustrine depositional system: The Pliocene Tiberino Basin, Central Italy. *Sediment. Geol.* **1997**, *109*, 73–94. [CrossRef]
2. Jiang, Z.; Liu, H.; Zhang, S.; Su, X.; Jiang, Z. Sedimentary characteristics of large-scale lacustrine beach-bars and their formation in the Eocene Boxing Sag of Bohai Bay Basin, East China. *Sedimentology* **2011**, *58*, 1087–1112. [CrossRef]
3. Monroe, S. Late Oligocene-early Miocene facies and lacustrine sedimentation, upper Ruby River basin, Southwestern Montana. *J. Sediment. Res.* **1981**, *51*, 939–951.
4. Song, F.; Su, N.N.; Yang, S.C.; Yao, R.X.; Chu, S.M. Sedimentary characteristics of thick layer lacustrine beach-bars in the Cenozoic Banqiao Sag of the Bohai Bay Basin, East China. *J. Asian Earth Sci.* **2018**, *151*, 73–89. [CrossRef]
5. Soreghan, M.J.; Cohen, A.S. Textural and compositional variability across littoral segments of Lake Tanganyika: The effect of asymmetric basin structure on sedimentation in large rift lakes. *AAPG Bull.* **1996**, *80*, 382–409.
6. Zhu, X.M.; Xin, Q.L.; Zhang, J.R. Sedimentary characteristics and models of the beach-bar reservoirs in faulted down lacustrine basins. *Acta Sedimentol. Sin.* **1994**, *12*, 20–27, (In Chinese with English Abstract).
7. Otvos, E.G. Beach ridges definitions and significance. *Geomorphology* **2000**, *32*, 83–108. [CrossRef]
8. Tamura, T. Beach riges and prograded beach deposits as palaeoenvironment records. *Earth-Sci. Rev.* **2012**, *144*, 279–297. [CrossRef]
9. Reid, I.; Frostick, L.E. Beach orientation, bar morphology and the concentration of metalliferous placer deposits: A case study, Lake Turkana, N Kenya. *J. Geol. Soc.* **1985**, *142*, 837–848. [CrossRef]
10. Williamson, C.R.; Picard, M.D. Petrology of carbonate rocks of the Green River Formation (Eocene). *J. Sediment. Res.* **1974**, *44*, 738–759.
11. Deng, H.; Xiao, Y.; Ma, L.; Jiang, Z. Genetic type, distribution patterns and controlling factors of beach and bars in the second member of Shahejie Formation in the dawanbei sag, Bohai Bay, China. *Geol. J.* **2011**, *46*, 380–389.
12. Gao, X.; Deng, H.; Lin, H. Stratigraphic controls on the evolution and distribution of lacustrine beach and bar Sands-Bohai Bay basin in China. *Manuf. Eng. Environ. Eng.* **2014**, *84*, 1513–1519.
13. Goulart, E.S.; Calliari, L.J. Medium-term morphodynamic behavior of a multiple sand bar beach. *J. Coast. Res.* **2013**, *65*, 1774–1779. [CrossRef]
14. Wang, J.; Cao, Y.; Liu, K.; Wang, X.; Xiao, J.; Xie, N. Mass transfer between mudstone-sandstone interbeds during diagenesis as revealed from the type and distribution of carbonate cements in the Eocene beach-bar sandstones, Bohai Bay Basin. *Mar. Petrol. Geol.* **2019**, *110*, 21–34. [CrossRef]
15. Zhao, Y.; Wang, H.; Yan, D.; Jiang, P.; Chen, S.; Zhou, J.; Zhao, Y. Sedimentary characteristics and model of gravity flows in the Eocene Liushagang Formation in Weixi'nan depression, South China Sea. *J. Petrol. Sci. Eng.* **2020**, *190*, 107082. [CrossRef]
16. Cao, L.; Zhang, Z.; Li, H.; Zhong, N.; Xiao, L.; Jin, X.; Li, H. Mechanism for the enrichment of organic matter in the Liushagang Formation of the Weixinan Sag, Beibuwan Basin, China. *Mar. Petrol. Geol.* **2020**, *122*, 104649. [CrossRef]
17. Huang, B.J.; Tian, H.; Wilkins, R.W.T.; Xiao, X.M.; Li, L. Geochemical characteristics, palaeoenvironment and formation modelof Eocene organic-rich shales in the Beibuwan Basin, South China Sea. *Mar. Petrol. Geol.* **2013**, *48*, 77–89. [CrossRef]
18. Xie, J.Y.; Li, J.; Mai, W.; Zhang, H.L.; Cai, K.L.; Liu, X.Y. Palynofloras and age of the Liushagang and Weizhou formations in the Beibuwan Basin, south China sea. *Acta Palaeontol. Sin.* **2012**, *51*, 385–394.
19. Xu, X.D.; Wang, B.W.; Li, X.H.; Liu, M.Q.; Zhang, Y.Z.; Hang, B.J. Oil sources of concealed reservoirs in Liushagang formations of the Weixi'nan sag and accumulation feature, Beibuwan Basin. *Nat. Gas Geosci.* **2012**, *23*, 92–98.
20. Zhou, X.; Gao, G.; Lü, X.; Zhao, L.; Dong, Y.; Xu, X.; Gan, J. Petroleum source and accumulation of WZ12 oils in the Weixi'nan sag, South China Sea, China. *J. Petrol. Sci. Eng.* **2019**, *177*, 681–698. [CrossRef]
21. Huang, B.J.; Xiao, X.M.; Cai, D.S.; Wilkins, R.W.T.; Liu, M.Q. Oil families and their source rocks in the weixinan sub-basin, Beibuwan basin, South China Sea. *Org. Geochem.* **2011**, *42*, 134–145. [CrossRef]
22. Gao, G.; Zhou, X.X.; Dong, Y. *Detailed research of petroleum generation and accumulation in Liushagang Formation of Weixi'nan Sag, Beibuwan Basin (Internal Report of Zhanjiang Branch of the China National Offshore Oil Company)*; China National Offshore Oil Company: Zhanjiang, China, 2014.
23. Luo, B.H.; He, J.; Jiang, P.; Qin, C.Y.; Yang, X.B.; Lu, J.; Zhao, Y.P. Sedimentary Characteristics of Lacustrine Beach-bars and Their Formation of the Second Member of Liushagang Formation in the Paleogene Weixinan Sag of Beibuwan Gulf Basin, Northern South China Sea. *Geol. Sci. Technol. Inf.* **2019**, *38*, 186–196, (In Chinses with English Abstract).
24. He, J.; Wang, H.; Tang, J.; Shangguan, Y.; Yang, X.; Gong, Y.; Zhao, R.; Zhou, X. Evaluation of lacustrine mudstone top sealing capacity in the Lenghu-5 area, Qaidam Basin, Tibetan Plateau. *J. Petrol. Sci. Eng.* **2020**, *188*, 106973. [CrossRef]
25. Gao, Z.; Yang, X.; Hu, C.; Wei, L.; Jiang, Z.; Yang, S.; Fan, Y.; Xue, Z.; Yu, H. Characterizing the pore structure of low permeability Eocene Liushagang Formation reservoir rocks from Beibuwan Basin in northern South China Sea. *Mar. Petrol. Geol.* **2019**, *99*, 107–121.
26. Hao, F.; Zhou, X.H.; Zhu, Y.M.; Bao, X.H.; Yang, Y.Y. Mechanisms for oil depletion and enrichment on the Shijiutuo uplift, Bohai Bay Basin, China. *AAPG Bull.* **2009**, *93*, 1015–1037. [CrossRef]
27. Wang, J.; Cao, Y.C.; Li, J.L. Sequence structure and non-structural traps of the Paleogene in the Weixi'nan Sag, Beibuwan Basin. *Pet. Explor. Dev.* **2012**, *39*, 325–334. [CrossRef]

28. Wang, J.; Cao, Y.C.; Li, J.L.; Yue, Y.X. Distribution of beach-bar sandbodies of the second member of Liushagang Formation of Paleogene of south slope in Weixinan Depression, Beibuwan Basin. *Acta Sedimentol. Sin.* **2013**, *31*, 536–544, (In Chinese with English Abstract).
29. Liu, P.; Xia, B.; Tang, Z.Q.; Wang, X.G.; Zhang, Y. Fluid inclusions in reservoirs of Weixinan Sag, Beibuwan Basin. *Petrol. Explor. Dev.* **2008**, *35*, 164–200. [CrossRef]
30. Liu, Y.; Wu, Z.; Liu, L.; Yan, S.; Hu, L.; Ping, M.; Zhang, M. Cenozoic structure and tectonics of North subbasins in Beibu Gulf Basin, northern South China Sea. *Tectonophysics* **2021**, *812*, 228912. [CrossRef]
31. Liu, E.; Wang, H.; Li, Y.; Zhou, W.; Leonard, N.D.; Lin, Z.; Ma, Q. Sedimentary characteristics and tectonic setting of sublacustrine fans in a half-graben rift depression, Beibuwan Basin, South China Sea. *Mar. Petrol. Geol.* **2014**, *52*, 9–21. [CrossRef]
32. Vail, P.R.; Audemard, F.; Bowman, S.A. The stratigraphic signatures of tectonics, eustasy and sedimentology—An overview. In *Cycles and Events in Stratigraphy*; Springer: Berlin/Heidelberg, Germany, 1991; Volume 6, pp. 7–659.
33. Scherer, C.M.; Lavina, E.L.; Dias Filho, D.C.; Oliveira, F.M.; Bongiolo, D.E.; Aguiar, E.S. Stratigraphy and facies architecture of the fluvial-Aeolian-lacustrine Sergi Formation (Upper Jurassic), Reconcavo Basin, Brazil. *Sediment. Geol.* **2007**, *194*, 169–193. [CrossRef]
34. Garzanti, E. Petrographic classification of sand and sandstone. *Earth-Sci. Rev.* **2019**, *192*, 545–563. [CrossRef]
35. Schwartz, R.K. Bedform, texture, and longshore bar development in response to combined storm wave and current dynamics in a nearshore helical flow system. *J. Coast. Res.* **2012**, *28*, 1512–1535. [CrossRef]
36. Short, A.D. Multiple offshore bars and standing waves. *J. Geophys. Res.* **1975**, *80*, 3838–3840. [CrossRef]
37. Yang, Y.Q.; Qiu, L.W.; Jiang, Z.X.; Yin, X.L. A depositional pattern of beach bar in continental rift lake basins: A case study on the upper part of the fourth member of the Shahejie Formation in the Dongying Sag. *Acta Petrol. Sin.* **2011**, *32*, 417–423.
38. Lin, H.X.; Deng, H.W.; Qin, Y.Q.; Gao, X.P.; Long, G.Q.; Zhao, N.; Yan, H.; Pan, F.Y.; Di, Y.X. Control of sequence stratigraphic evolution on the distribution and hydrocarbon accumulation of beach and bar reservoirs. *Petrol. Expl. Dev.* **2010**, *37*, 680–689, (In Chinese with English Abstract).
39. Distefano, S.; Gamberi, F.; Baldassini, N.; Di Stefano, A. Neogene stratigraphic evolution of a tectonically controlled continental shelf: The example of the Lampedusa island. *Ital. J. Geosci.* **2019**, *138*, 418–431. [CrossRef]
40. Guo, S. The influence of the pre-existing topography on the depositionary systems, the development of the Lower Jurassic reservoirs and hydrocarbon accumulation in Central Western Ordos Basin. *J. Petrol. Sci. Eng.* **2010**, *75*, 129–134. [CrossRef]
41. Paul, J.M.; Peter, M.I.; Dale, A.L. Use of micromorphology for palaeoenvironmental interpretation of complex alluvial palaeosols: An example from the MillCreek Formation (Albian), southwestern Alberta, Canada. *Palaeogeogr. Palaeoclimatol. Palaeoecol.* **1998**, *143*, 87–110.
42. Zhu, X.M.; Liu, Q.H.; Ge, J.W.; Dong, Y.L.; Zhu, S.F.; Tan, M.X.; Yang, Y. Reconstruction of sediment-dispersal patterns using seismic sedimentology in the southeastern Zhanhua Sag, Bohai Bay Basin, China. *J. Petrol. Sci. Eng.* **2019**, *182*, 106335.
43. Jiang, Z.X.; Wang, J.H.; Zhang, Y.F. Advances in beach-bar research: A review. *J. Palaeogeogr.* **2015**, *17*, 427–440, (In Chinese with English Abstract).
44. Graham, J.P. Revised stratigraphy, depositional systems, and hydrocarbon exploration potential for the lower Cretaceous muddy sandstone. Northern Denver Basin. *AAPG Bull.* **2000**, *84*, 183–209.
45. Wang, J.; Xu, S.; Yu, J.; Han, W.; Cui, H.; Zhai, Y. Prediction of beach-bar sand reservoirs using waveform analysis: A case study on Es4s in the west area of the Dongying sag. *Earth Sci. J. Chin. Univ. Geosci.* **2008**, *33*, 627–633.
46. Zhang, Y.; Hu, C.; Wang, M.; Ma, M.; Wang, X.; Jiang, Z. A quantitative sedimentary model for the modern lacustrine beach bar (Qinghai Lake, Northwest China). *J. Paleolimnol.* **2018**, *59*, 279–296. [CrossRef]
47. Gao, Y.F.; Peng, S.P.; He, H.; Kong, W. Reservoir characters and evaluation of the Tertiary clastic sediments in the Northern Qaidam basin, Northwest. *China. Petrol. Expl. Dev.* **2003**, *30*, 40–42.
48. Keighley, D.; Flint, S.; Howell, J.; Moscariello, A. Sequence stratigraphy in lacustrine basin: A model for part of the Green River Formation (Eocene), Southwest Uinta Basin, Utah, USA. *J. Sediment. Petrol.* **2003**, *73*, 987–1006. [CrossRef]
49. Cook, D.O. The occurrence and geologic work of rip currents of southern California. *Mar. Geol.* **1970**, *9*, 173–186. [CrossRef]
50. Huang, E.; Tian, J.; Steinke, S. Millennial-scale dynamics of the winter cold tongue in the southern South China Sea over the past 26ka and the East Asian winter monsoon. *Quat. Res.* **2011**, *75*, 196–204. [CrossRef]
51. Schroeder, A.; Wiesner, M.G.; Liu, Z.F. Fluxes of clay minerals in the South China Sea. *Earth Planet. Sci. Lett.* **2015**, *430*, 30–42. [CrossRef]
52. Zhu, X.M.; Zhong, D.K.; Yuan, X.J.; Zhang, H.L.; Zhu, S.F.; Sun, H.T.; Gao, Z.Y.; Xian, B.Z. Development of sedimentary geology of petroliferous basins in China. *Petrol. Expl. Dev.* **2016**, *43*, 890–901. [CrossRef]
53. Xie, N.; Cao, Y.C.; Wang, J.; Jin, J.H.; Zhang, W.J.; Zhong, Z.H. Diagenesis and its control on physical property of the reservoirs in the 3rd member of the Paleogene Liushagang Formation in Weixinan Depression, Beibuwan Basin. *Nat. Gas Geosci.* **2019**, *30*, 1743–1754, (In Chinese with English Abstract).
54. Gabriel, K.R. The biplot graphic display of matrices with application to principal component analysis. *Biometrika* **1971**, *58*, 453–467. [CrossRef]
55. Comas Cufí, M.; Thió Fernández de Henestrosa, S. CoDaPack 2.0: A stand-alone, multi-platform compositional software. In *CoDAWork'11: 4th International Workshop on Compositional Data Analysis, Sant Feliu De Guíxols*; CoDAWork'11: Girona, Spain, 2011.

56. Distefano, S.; Gamberi, F.; Baldassini, N.; Di Stefano, A. Quaternary evolution of coastal plain in response to sea-level changes: Example from south-east Sicily (Southern Italy). *Water* **2021**, *13*, 1524. [CrossRef]
57. Carroll, A.R.; Bohacs, K.M. Lake type control on hydrocarbon source potential in nonmarine basins. *AAPG Bull.* **2001**, *85*, 1033–1053.
58. Hou, F.H.; Jiang, Y.Q.; Fang, S.X.; Guo, G.A.; Yang, J.L. Sedimentary model of sandstone in second and fourth members of Xiangxi formation in the upper Triassic of Sichuan basin. *Acta Petrol. Sin.* **2005**, *26*, 30–37.
59. Yuretich, R.F.; Hickey, L.J.; Gregson, B.P.; Hsia, Y.L. Lacustrine deposits in the Paleocene Fort Union Formation, Northern Bighorn Basin. Montana. *J. Sediment. Petrol.* **1984**, *54*, 836–852.
60. Xia, S.; Lin, C.; Du, X.; Jia, D.; Ahmad, N.; Zhang, Z.; Gao, L.; Zhu, Y. Correspondences among lacustrine fluctuations, climate changes and the Milankovitch cycles in the Paleogene through tracking onlap points and correlating palaeontology in Liaozhong Depression, Bohai Bay Basin, NE China. *Geol. J.* **2020**, *55*, 6527–6543. [CrossRef]
61. Milankovitch, M. Kanon der Erdbestrahlung und seine Anwendung auf das Eiszeitenproblem. *Beogr. Königlich Serbische Akad.* **1941**, *32*, 633.
62. Bohacs, K.M.; Carroll, A.R.; Neal, J.E.; Mankiewicz, P.J. Lake-Basin type, source potential, and hydrocarbon character: An integrated sequence-stratigraphic-geochemical framework. In *Lake Basins through Space and Time*; Gierlowski-Kordesch, E.H., Kelts, K.R., Eds.; American Association of Petroleum Geologists: Tulsa, OK, USA, 2000; Volume 46, pp. 3–34.
63. Cong, F.; Zhu, F.; Cai, Z.; Chen, H.; Li, J.; Wang, Y.; Wang, L. Orbitally forced glacio-eustatic origin of third-order sequences and parasequences in the Middle Permian Maokou Formation, South China. *Mar. Petrol. Geol.* **2019**, *99*, 237–251. [CrossRef]
64. Liu, Q.; Zhu, H.; Shu, Y.; Zhu, X.; Yang, X.; Chen, L.; Tan, M.; Geng, M. Provenance identification and sedimentary analysis of the beach and bar systems in the Palaeogene of the Enping Sag, Pearl River Mouth Basin, South China Sea. *Mar. Petrol. Geol.* **2016**, *70*, 251–272. [CrossRef]
65. Wang, Y.; Yang, S.; Zhang, Y.; Lu, Y.; Wang, Y.; Zhao, Y. Investigation of pore structure and reservoir quality of Eocene beach-bar sandstones in Dongying Depression, Bohai Bay Basin, East China. *J. Petrol. Sci. Eng.* **2020**, *189*, 106854. [CrossRef]

energies

Article

Effects of Depositional Processes in Submarine Canyons and Distribution of Gas Chimneys on Gas Hydrate Accumulation in the Shenhu Sea Area, Northern South China Sea

Yunlong He [1,*], Zenggui Kuang [2], Cong Cheng [1], Tao Jiang [1], Cheng Zhang [1], Biyu Lu [3], Chengzhi Yang [2], Jiayu Liu [1] and Changlong Xiang [1]

1 Hubei Key Laboratory of Marine Geological Resources, China University of Geosciences, Wuhan 430074, China
2 National Engineering Research Center of Gas Hydrate Exploration and Development, Guangzhou Marine Geological Survey, Guangzhou 510075, China
3 Guangxi Zhuang Autonomous Region Marine Geological Survey Institute, Beihai 536000, China
* Correspondence: ylhe@cug.edu.cn

Abstract: Previous gas hydrate production tests conducted by the Guangzhou Marine Geological Survey (GSGM) in 2017 and 2020 indicated the great potential of gas hydrates in the Shenhu Sea area in the Pearl River Mouth Basin (PRMB), China. In this study, the effects of deposition processes in submarine canyons and the distribution of gas chimneys on gas hydrate accumulation were investigated using high-resolution two- dimensional (2D) and three-dimensional (3D) seismic data. Four intact submarine canyons were identified in the study area. Five deepwater depositional elements are closely related to submarine canyons: lateral accretion packages (LAPs), basal lags, slides, mass transport deposits (MTDs), and turbidity lobes. MTDs and lobes with multiple stages outside the distal canyon mouth reveal that the sedimentary evolution of the canyon was accompanied by frequent sediment gravity flows. Gas chimneys originating from Eocene strata are generally up to 3 km wide and distributed in a lumpy or banded pattern. The analysis of seismic attributes confirmed fluid activity in these gas chimneys. Gas hydrates are mainly distributed in ridges among different canyons. Based on the gas sources of gas hydrates and depositional evolution of submarine canyons, depositional processes of sediment gravity flows in submarine canyons and the distribution of gas chimneys significantly affect the accumulation of gas hydrates. Based on these findings, this study establishes a conceptional model for the accumulation of gas hydrate, which can provide guidance in the prediction for favorable gas hydrates zones in the area and nearby.

Keywords: submarine canyon; depositional process; gas chimney; accumulation; gas hydrate; South China Sea

Citation: He, Y.; Kuang, Z.; Cheng, C.; Jiang, T.; Zhang, C.; Lu, B.; Yang, C.; Liu, J.; Xiang, C. Effects of Depositional Processes in Submarine Canyons and Distribution of Gas Chimneys on Gas Hydrate Accumulation in the Shenhu Sea Area, Northern South China Sea. *Energies* **2023**, *16*, 234. https://doi.org/10.3390/en16010234

Academic Editor: Ingo Pecher

Received: 30 November 2022
Revised: 20 December 2022
Accepted: 22 December 2022
Published: 25 December 2022

1. Introduction

Gas hydrates, which are ice-like solid compounds formed by water and methane, are considered important unconventional energy sources [1,2]. They are widely distributed in marine sediments and permafrost and the estimation of the total resources of gas hydrates can be up to 3000×10^{12} m^3 [3,4]. Because of their great resource potential, they have received widespread attention from industry and academia [1,5–8]. To determine the distribution of gas hydrates in marine sediments, a series of exploration projects have been carried out worldwide, and many gas hydrates discoveries have been made on continental margins [9–13]. These projects worldwide have helped scientists to find that the accumulation of gas hydrates is controlled by several factors, such as the origin of the gas, pathway of gas migration, reservoir characteristic, stability condition, water, and time [14]. The geological characteristics of the reservoir, such as grain size, hydraulic conductivity and/or tectonics, have a great influence on hydrate accumulation. Based on numerous studies, researchers found that grain size of sediments hosting the gas hydrates are variable. For instance, in the Mallik of Canada and the Nankai Trough of Japan, the gas hydrates were found in coarse-grained sediment [15,16], while those found in South China Sea were recovered from fine sediments with silt content ranging from 70% to 80% [13,17]. Furthermore, on the continental margin, the evolution of sediment gravity flows within the submarine canyon can also result in the dissolution and subsequent leakage of gas hydrates [18,19].

Hydrate exploration in the northern South China Sea began at the end of the last century. In the early stage, scientists studied the characteristics of seismic facies and distribution of gas hydrates by using seismic data and established a preliminary accumulation model [20–25]. Owing to the tremendous potential of gas hydrates of the South China Sea, the GSGM has conducted several hydrate drilling voyages (GMGS1-5) since 2007, and gas hydrates have been found in different basins along the northern margin of the South China Sea [26–32]. By using a series of well logs, core data, and seismic data, researchers have studied the lithological characteristics and types of depositional elements of sediments containing gas hydrate in detail. Significant differences were observed with respect to the occurrence and gas sources of gas hydrates in different basins, which implied that there are significant differences in the factors controlling hydrate accumulation in different areas [20,24,29,30,33–35]. For instance, gas hydrates have been found in different voyages in the Shenhu sea area in PRMB [28,32,34]. However, the distribution and saturation of gas hydrates here is heterogeneous, even within two close sites which have a very similar morphological and geological background [32,36,37]. For example, in the GMGS1, there were 8 sites located on the canyon ridges besides both sides of a submarine canyon, but only SH2, SH3, and SH7 have recovered the gas hydrates on the canyon ridges of western side and there are no gas hydrates in SH5, SH6 and SH9 on canyon ridge of eastern side [20,38,39]. Therefore, it is critical to find out the factors controlling the distribution of gas hydrates and clarify their accumulation mechanism, from the origin and migration of gases to the accumulation processes and conceptional model of accumulation of gas hydrates.

By using high-resolution 2D and 3D seismic data covering the Shenhu Sea area in the Pearl River Mouth Basin (PRMB), the deepwater depositional system of the hydrate development area was investigated in this study. Based on the theory of sequence stratigraphy, sedimentology and gas hydrate accumulation systems, gas hydrate sources, and migration pathway, the accumulation processes of gas hydrates were clarified. This paper presents a conceptual model for gas hydrate accumulation that can help predict the distribution of gas hydrates in different areas of the SCS, which can be of great importance in guiding the exploration of gas hydrate in this area.

2. Geological Background

The Shenhu Sea area is located on the slope of the Baiyun Sag in the PRMB, which is characterized by abundant oil and gas resources in the northern South China Sea (Figure 1) [40–42]. Since the Eocene, the PRMB has undergone five major tectonic movements, and extremely thick sediments have been deposited. Tectonic activity was intense in the Eocene. Since the Oligocene, the basin has entered the post-rifting stage and the fault activity has decreased. Since the Miocene, only a few faults have remained active [43]. Many diapirs with heights reaching >8 km developed in the central part of the Baiyun Sag [44].

In the early stage of the basin development during the Eocene, the PRMB was dominated by a lacustrine sedimentary environment. After the Oligocene, the basin was characterized by a marine sedimentary environment. Three sets of source rocks developed in the basin during this process, that is, the Wenchang and Enping formations in the Eocene and the Zhuhai Formation in the Oligocene [45–47]. Paleontological data indicate that a deepwater environment began to appear and typical progradational continental shelf margin and deepwater fan systems began to develop in the PRMB during the Oligocene [48,49]. Since then, numerous submarine canyons, which show characteristics of unidirectional migration under the influence of contour currents, developed at the margin of the northern shelf of the South China Sea, especially during the Late Miocene [49–52]. In addition, many mass transport deposits (MTDs) developed from the shelf margin to the deepwater basin, which led to the reshaping of the geomorphology of the shelf margins [53–56]. Complex interaction processes between turbidity currents, contour currents, and MTDs led to large differences in the sediment properties in different areas [13].

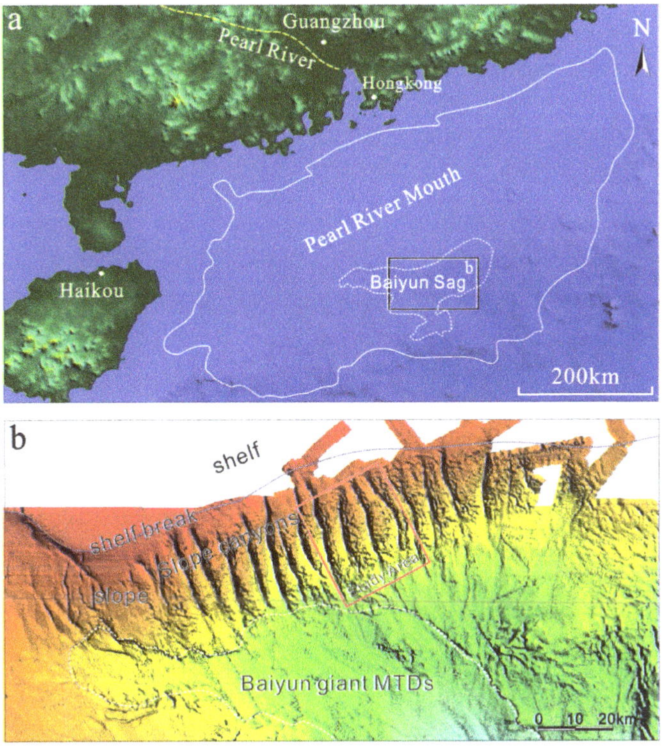

Figure 1. (**a**) Location of Pearl River Mouth Basin; (**b**) characteristics of seafloor morphology of shelf margin in Pearl River Mouth Basin, red rectangular represents the location of study area.

3. Methodology

High-resolution two-dimensional (2D) and 3D seismic data used in this study were obtained from the GMGS. The sequence stratigraphy framework covering PRMB was established by China National Offshore Oil Company (CNOOC), combined with well data obtained by CNOOC in the PRMB. In this study, the 2D seismic data with a total length of more than 1350 km cover a wider area and were used to determine the correlation with seismic data from CNOOC nearby (Figure 2). Nine sequence boundaries were identified in this study: T80, T70, T60, T50, T40, T32, T30, T20, and T10. The 3D seismic data obtained in the submarine canyon developed area on the shelf margin cover an area of ~800 km^2. The major frequency and vertical resolution of 3D seismic data can reach 40 Hz and 20 m, respectively. By using high-resolution 3D seismic data and seismic facies, the distribution of deepwater depositional elements, gas hydrates, and typical gas chimney structures can be determined. The results will provide insights into the accumulation of gas hydrates in this area. In this study, seismic attributes also were extracted based on the 3D seismic data, including instantaneous frequency and coherence slices. Generally, the instantaneous frequency can be used to identify the activity of fluid, while the coherence slices are used to meticulously describe the geological bodies or the structures of different scales in the basin.

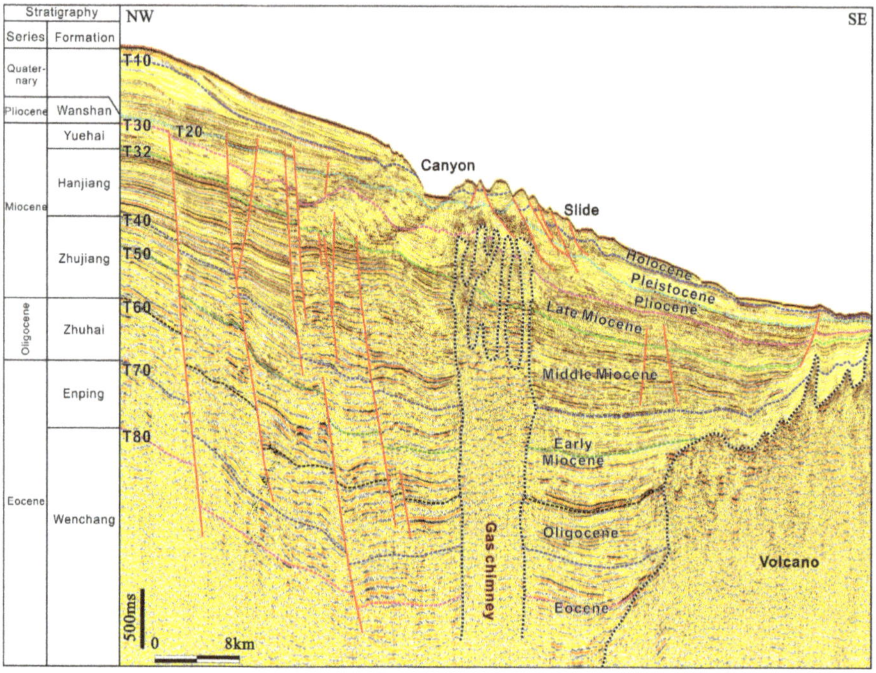

Figure 2. Framework of sequence stratigraphy and typical seismic profile of the Pearl River Mouth Basin.

4. Results

4.1. Depositional System Related to Submarine Canyons

Based on the bathymetric chart, many submarine canyons are distributed along the shelf margin of the Baiyun Sag in the PRMB. Beyond the distal mouth of the submarine canyons, giant mass transport deposits developed, which are known as Baiyun giant landslides. The Shenhu Sea area is located between the shelf margin and Baiyun giant landslides; it is characterized by well-developed submarine canyons (Figure 1).

4.1.1. General Features of Submarine Canyons

Based on the morphology of the seabed or the isobathic chart of T20, four relatively complete submarine canyons are developed in the study area (Figures 1b and 3a). The canyons are generally ~25 km long and ~2–3 km wide (Figures 1b and 3a,b). The middle of the canyon, with a depth of around 400 m (from the canyon bottom to the ridge top; with an average velocity of 1.5 km/s of acoustic waves), has experienced the strongest erosion. The incision depth gradually decreases from the middle to both ends of the canyon (Figure 3b). The concave pattern of erosion of T32 in the crossing profile of the canyon shows that submarine canyons have been initiated since the Late Miocene (Figures 2 and 3b). Along with the development of the canyons, the thalweg of submarine canyons shows typical unidirectional migration to the northeast. The horizontal distance of the migration from the beginning of the canyon formation to the seabed can exceed 3 km (Figure 3b). The characteristics of these submarine canyons are similar at different stages; however, the incision depth is relatively small in the early stages of the canyon development.

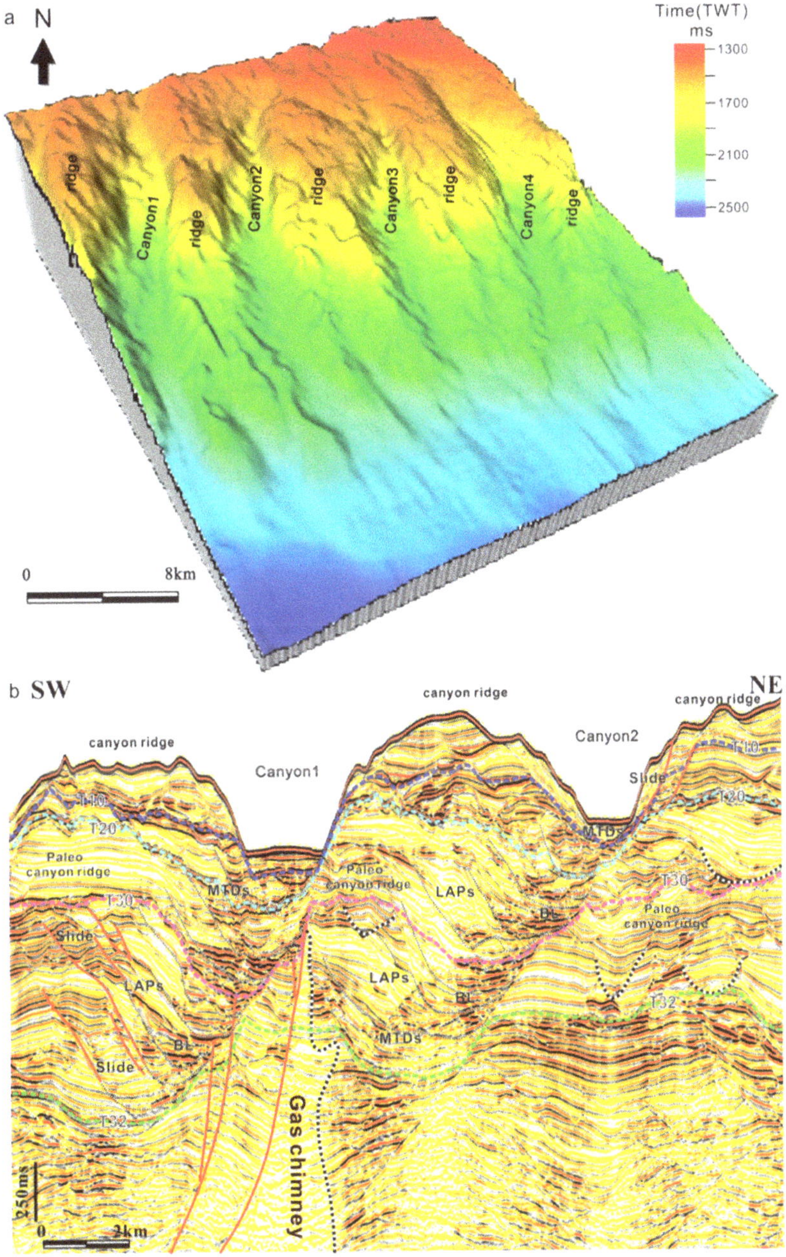

Figure 3. (**a**) 3D view of time depth of sequence boundary of T20(boundary between Pliocene and Pleistoncene); (**b**) typical features of submarine canyons on seismic profile perpendicular to canyons.

4.1.2. Elements and Distribution of Depositional Systems

Lateral Accretion Packages

As mentioned above, the migration of submarine canyons since the Late Miocene can be clearly observed in the crossing seismic profiles. On the eastern margin of the canyon, an erosional feature is revealed by the truncation of continuous seismic reflections of canyon ridges (Figures 3b and 4a). On the western side of the canyon, a succession of strata inclined toward the canyon axis developed, which are named lateral accretion packages (LAPs). Their reflections are characterized by a moderate amplitude and fair continuity. These LAPs developed from the early stage of the canyon till today (Figure 3b). Considering the paleoceanography background, researchers believe that the unidirectional migration of these LAPs is due to bottom-current or contour-current activities in this area [49,50,52].

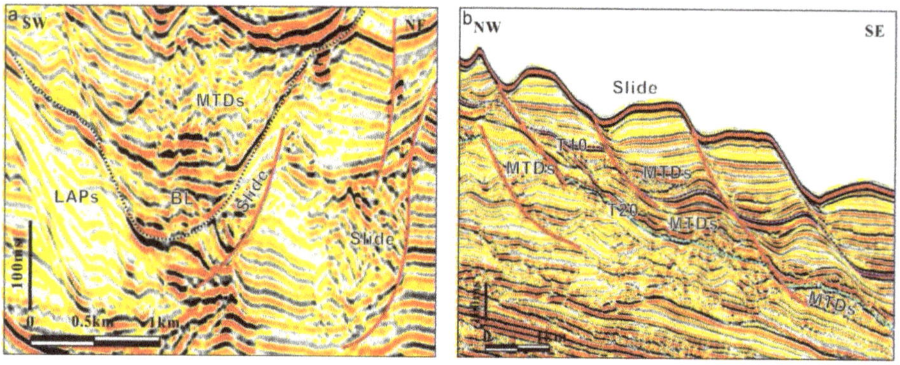

Figure 4. (**a**) Slides developed on margin of canyon and MTDs infilling the canyon; (**b**) slides developed on the canyon ridges.

Basal Lags

At the bottom of the submarine canyons, relatively short seismic reflections with high amplitudes were observed near the thalweg. Their assembly presents a lenticular geometry, which differs from the adjacent reflections infilling the canyon with low to moderate amplitude and fair continuity (Figures 3b and 4a). However, reflections with such characteristics have not continuously developed throughout the evolutionary stages of the canyons but mostly developed at the canyon bottom with distinct erosion (Figure 3b). Compared with other submarine canyons, these assemblies of reflections are interpreted as basal lags caused by irregular turbidity currents in canyons.

Slides

Slides are a common depositional element in the study area. Reflections within the slides show continuous and parallel characteristics, separated by faults between different blocks in the slides. Slides near the seafloor have a typical ladder shape (Figures 2 and 4b). Slides are occasionally developed on canyon margins and ridges between the canyons. The majority of them maintain continuous, parallel features, but a few exhibit more significant deformation on their edges due to sliding and the rotation of blocks (Figure 4a). In the early stage of canyon evolution, several slides with large thicknesses developed on the canyon margin. Faults among these blocks within the slides almost combined with the canyon bottom, leading to the widening of the canyon (Figure 3b).

MTDs

MTDs are another important depositional element in and around the canyon. In seismic profiles, MTDs are characterized by chaotic, discontinuous, and low-amplitude reflections. Within canyons, MTDs generally cover the basal lag at the bottom or deposit aside of the canyons. Confined by a concave morphology, MTDs in the canyon are generally thick but have a smaller lateral extension, which results in a lenticular geometry (Figures 3b and 4a). However, MTDs outside the distal mouth of the canyon are relatively thin but have a greater horizontal extension (>3 km) because of the absence of a confining canyon topography. Multiple depositions of MTDs indicate that the canyon has experienced several MTD events (Figure 5).

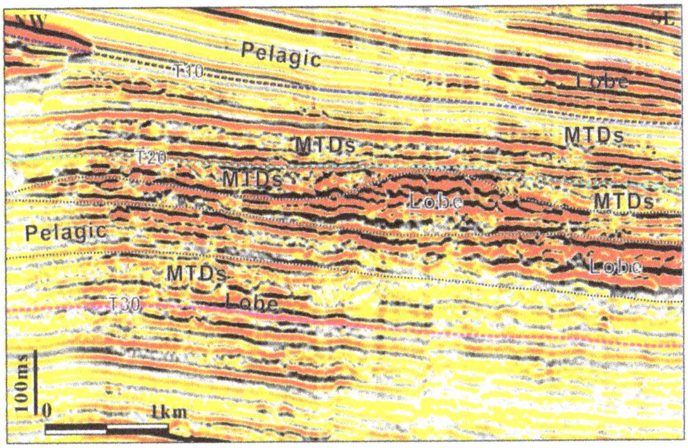

Figure 5. Successive MTDs and turbidity lobes developed outside the distal mouth of canyons.

Turbidity Lobe

Several reflections outside the distal canyon mouth are characterized by a continuous, high amplitude. In contrast to the basal lag, the lateral extension of these reflections is greater than that of the basal lag deposited within canyons (Figure 5). Based on their geometry and depositional environment, they represent the turbidity lobes that formed by turbidity flows from the canyons. Several lobes were identified outside the distal mouth of the canyon. Lobes that developed in the Pliocene are small and thin in the early stage, indicated by a short single reflection with high amplitude; however, they become wider in the lateral extension and thicker with a dome-like shape later on (Figure 5).

The comparison of the characteristics of seismic facies in canyons shows that the sedimentary characteristics at different stages of the canyon development are similar. LAPs are mainly developed on the western side of the canyons. Since the initiation of the canyon, the stacking of LAPs gradually prograded eastward and led to the movement of the thalweg of canyons in the same direction (Figures 3b and 6). A basal lag always develops at the bottom near the thalweg of the canyon and shows discontinuous features in the plan view and a strike similar to that of the canyons. MTDs are well-developed at the edge and outside of the distal mouth of the canyon, whereas slides are mainly developed on the ridges among different canyons. The turbidity lobe is another important depositional element outside the canyons. Lobes from different canyons merge into one greater one in the area farther away from the canyons at the edge of the study area (Figure 6).

4.2. Distribution of Gas Chimneys

Diapirs are a very important and widely distributed structural type in the study area. Two types of diapirs can be observed in the study area, that is, gas chimneys and diapirs associated with volcanic activity, which exhibit different seismic characteristics (Figures 2 and 7). Generally, fluids in sediment significantly affect the high-frequency seismic energy in seismic data and consequently result in a reduction in the instantaneous frequency [57]. Gas chimneys are characterized by blank or low-amplitude seismic reflections. Compared with volcanic diapirs, gas chimneys are small in scale but large in number. Their widths are generally less than 3 km. The tops of gas chimneys are relatively flat or slightly convex and the lateral extension is greater than their body, showing a mushroom-like shape. Their bottom is generally located in Eocene or Oligocene strata (Figure 7). In instantaneous frequency attribute profiles, the gas chimneys show a noticeable zone with low frequency. Diapirs associated with volcanoes are large and have a base width exceeding 20 km. These diapirs originate from the deep earth crest or mantle and intrude through the basement and thick strata within the basin. Inside, diapirs are characterized by chaotic reflections, but high-amplitude reflections with irregular distribution and poor continuity can be observed locally (Figure 2). At the top of volcanic diapirs, a series of sharp small bulges can be generally observed, which is caused by the differential intrusion of volcanic rocks.

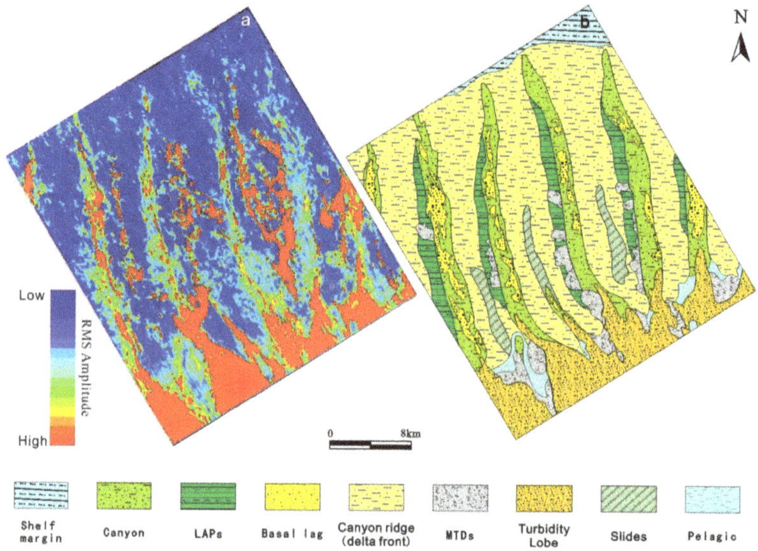

Figure 6. (**a**) RMS attribution of horizontal slice following T20; (**b**) Interpretation of depositional system in the Pliocene due to RMS attribution and typical seismic profiles.

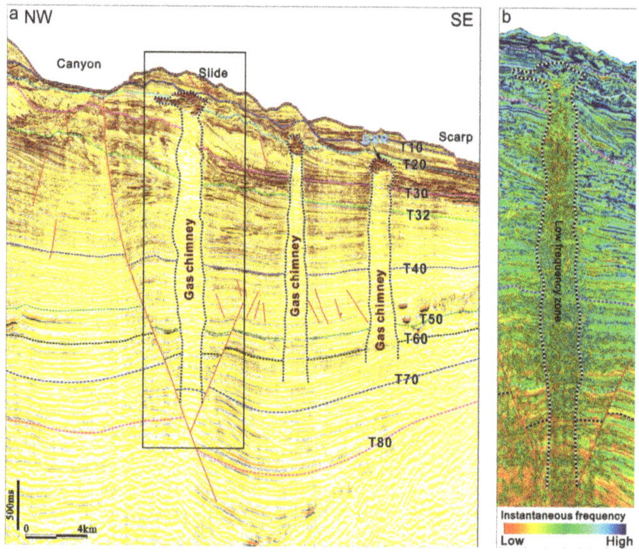

Figure 7. (**a**) Typical features of gas chimneys in the general seismic profile; (**b**) Features of gas chimney on instantaneous frequency attribute profile, in which a low frequency zone (especially at the upper part of the gas chimney) could be found.

Based on the analysis of the study area, gas chimneys are well developed and volcanic diapirs are only distributed in the lower left corner of the study area. The results of detailed studies showed that most of the gas chimneys are located in Pliocene strata (between T20 and T30; Figures 2, 3 and 7). In the plane view, gas chimneys show lumpy or banded shapes and can be located at canyons or ridges among different canyons, but most of them are located east of Canyon 1. Furthermore, larger gas chimneys with lengths up to 10 km developed on two ridges to the east of Canyons 1 and 3 (Figure 8).

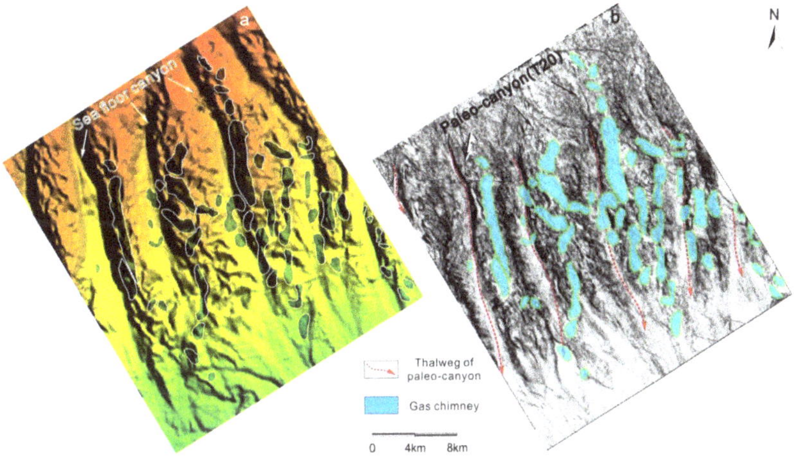

Figure 8. Distribution of gas chimneys. (**a**) Bathymatry of sea floor; (**b**) the background is a coherence slice which has been extracted following the Horizon T20 in the 3D seismic data. In the slice, eastern margins of the paleo-canyons can be easily identified and the red dashed lines (approximate parallel to the canyon margin) with arrows pointing toward downstream standing for the thalweg of paleo-canyons.

4.3. Gas Hydrate Distribution

The occurrence of bottom-simulating reflector (BSR) is often an indication of the accumulation of free gas beneath the gas hydrate zone [6]. Owing to the lack of well data, BSRs were used in this study to infer gas hydrates. In profiles perpendicular to the canyon, BSRs are distributed on the ridge among the canyons with different widths, but most of them are less than 2 km wide. The characteristics of BSRs on different ridges differ (Figure 9). The BSR with a flat shape on the ridge between Canyons 1 and 2 is clear and continuous and significantly differs from the attitude of nearby strata. However, the BSR on the ridge between Canyons 2 and 3 has a relatively poor continuity, making it difficult to identify it. The outline of high-amplitude reflections, which coincide with the irregular topography of the seafloor, can be used to locate the BSR nearby (Figure 9). In the profile along the strike of the canyons, reflections with high amplitude have a lateral extension reaching up to 8 km, indicating a wider distribution of BSRs in this direction. Owing to the irregular topography of the seafloor, the overall continuity of BSRs is bad (Figure 10).

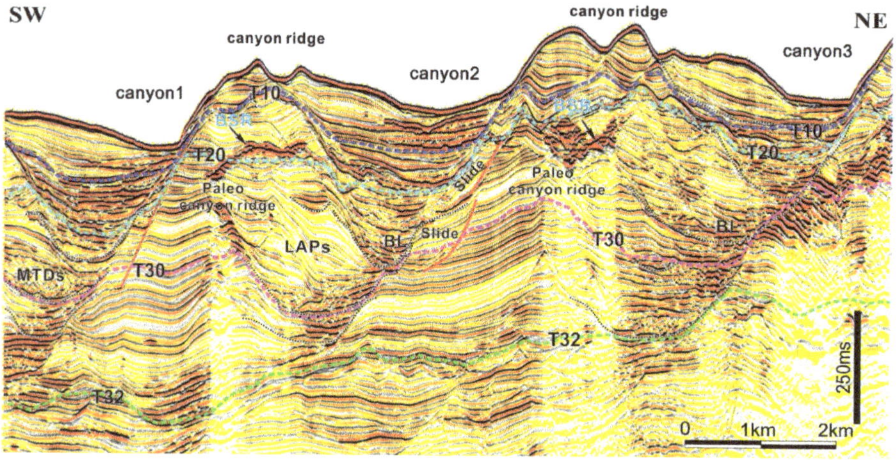

Figure 9. Characteristics and distribution of BSRs in the profile perpendicular to canyons.

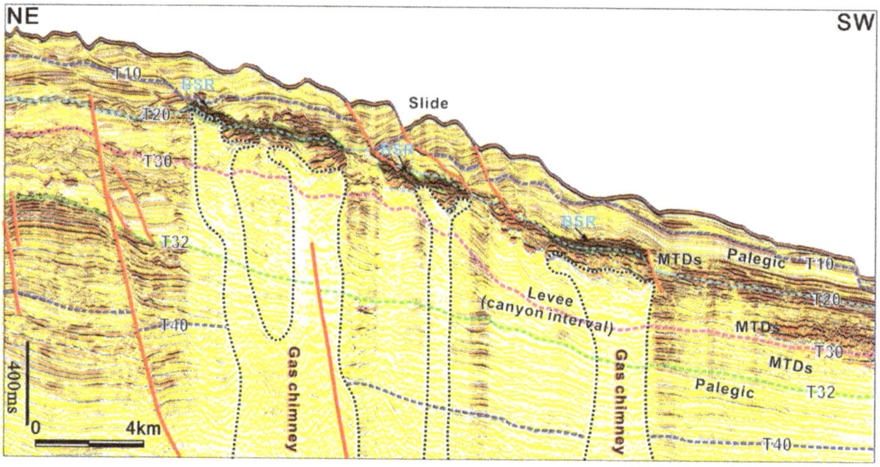

Figure 10. Characteristics and distribution of BSRs in the profile on the canyon ridge.

The distribution characteristics of BSRs were determined based on the interpretation of BSRs in the study area. Most BSRs are located in Pliocene strata and few were identified in Quaternary layers (Figures 7, 9 and 10). In the plane view, BSRs represent lumpy and banded shapes. The lumpy ones are smaller, with lengths ranging from 1 km and diameters of 4 m, but larger in number than the banded ones. The regulation of the distribution of BSRs with a lumpy outline is inconspicuous, but banded BSRs that develop on canyon ridges follow a notable distribution rule. The long axial direction of banded canyons with lengths up to 15 km coincides with the strike of the canyons (Figure 11).

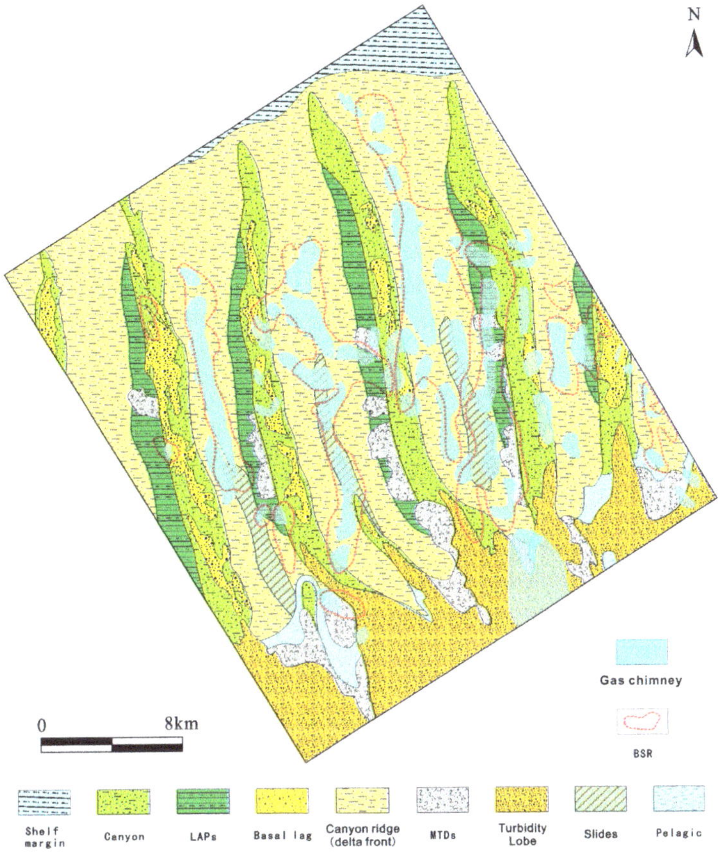

Figure 11. Spatial location relationship of BSRs, gas chimneys and sedimentary facies.

5. Discussion

5.1. Origin of Gas in Gas Hydrates

The accumulation of gas hydrates is a complex process that requires not only an environment with low temperature and high pressure, but also stable gas sources [58–61]. Generally, three gas sources of gas hydrates can be identified: microbiogenic gas, thermogenic gas, and a mixture of the two [59,62].

To identify the resource potential of gas hydrates in the Northern South China Sea, the GMGS has conducted several hydrate drilling voyages in the Shenhu area (GMGS1, GMGS3, GMGS4, GMGS5) and samples of gas hydrates have been obtained during all voyages [62–64]. However, there is a difference in understanding the source of gas in the hydrate in Shenhu sea area. In GMGS1, the methane content of the gas hydrate samples reaches 99.66%. Based on the molecular compositional ratio of C1/(C2 + C3) and $\delta^{13}C$ values, the methane in the gas hydrates in the Shenhu area is mainly of microbial origin and mixed with a limited amount of thermogenic gas [22,62,65]. However, the molecular composition of gases in the hydrate from GMGS3 and GMGS4 shows that the proportion of C2+ hydrocarbons in the gas can reach ~3%. Paragenetic relationships were also identified between shallow gas hydrate and deep conventional reservoirs, which are supplied by the hydrocarbon kitchens in the Baiyun Sag. Combined with isotopic signatures, it has been considered that thermogenic gas supplies part of the gas for hydrates in the Shenhu area [28,32,34,64]. Hence, it is reasonable to conclude that the gas sources of the hydrates are biogenic gas and a mixture of biogenic and thermogenic gases [24,28,33,34,60,64].

5.2. Gas Hydrate Accumulation

The results of previous studies showed that the Baiyun Sag is a hydrocarbon-rich sag in the PRMB. Three main sets of source rocks have been identified: lacustrine source rocks from the Wenchang and Enping formations and marine source rocks from the Zhuhai Formation [45–47]. The source rocks from the Wenchang and Enping formations are currently in the over- and high-maturity stages and have generated thermogenic gas since the Middle Miocene. In contrast, the source rocks from the Zhuhai Formation are still in a low-maturity stage and cannot generate a large amount of thermogenic gas. As mentioned above, gas chimneys originate from the Eocene strata, including the Wenchang and Enping formations. The comparison of the planimetric positions of the gas chimneys and BSRs shows that their plane distributions strongly correlate (Figure 11). Combined with the low-frequency zone indicating the activities of fluids or gases in gas chimneys, it is reasonable to believe that these gas chimneys are important pathways for the migration of deep thermogenic gas to the shallow region (Figure 7). Hence, the results of this study show that the components of thermogenic gases of gas hydrates originate from source rocks of the Wenchang and Enping formations, whereas microbiogenic gases are derived from source rocks of the Zhuhai Formation.

In addition, note that several gas chimneys are developed in Canyon 4, but no BSRs exist. There are several reasons for this phenomenon. Many researchers have studied the evolutionary history of submarine canyons in the northern South China Sea and concluded that submarine canyons are affected by the interaction of contour and turbidity currents [49,50,52]. The LAPs that are developed on the western margin of the canyons also indicate the significant effects of contour currents on the development and evolution of the canyon (Figures 3b and 9). However, depositional processes within the canyons seem to be more important for submarine canyons. Generally, the incision of the canyon is closely related to the activity of turbidity currents, which play a significant role in deepening the canyon and sustaining its V-shaped incision pattern [50,66–68]. It also has been demonstrated that MTDs and slides can lead to strong erosion of the canyon bottom [69,70]. The canyon is completely filled with sediments without this great incision. In these successive incising processes caused by various sediment gravity flows, a large amount of sediment at the bottom of the canyons is eroded and transported down the slope, which can change the pressure condition for gas hydrates beneath the canyon. The gas hydrate will dissociate, gas leakage will occur, and the gas hydrate beneath the canyon cannot be preserved.

Based on the depositional processes, distribution characteristics of gas chimneys, and gas hydrate source, a model of the gas hydrate accumulation in the Shenhu sea area was established (Figure 12). Since the Late Miocene, the source rocks in the Eocene in the Baiyun Depression have been in a high- or overmaturity stage. Gas chimneys are formed by the release of overpressure within Eocene source rocks under the influence of external factors (e.g., thermal fluid activity, regional tectonic activity), which provides a fair migration pathway for the vertical migration of thermogenic gas from source rocks during the Eocene [26,29,44]. Together with shallow microbiogenic gas from shallow source rock, a mixture of thermogenic and microbiogenic gases migrates from the gas chimneys to positions above with suitable pressure and temperature to form gas hydrates. Most of the gas hydrates distributed in the canyon ridges are preserved where less erosion was caused by different types of gravity flow, whereas gas hydrates beneath the canyon bottom for which the conditions are changing frequently due to subsequent successive sediment gravity flows within the canyons have dissociated and could not be preserved.

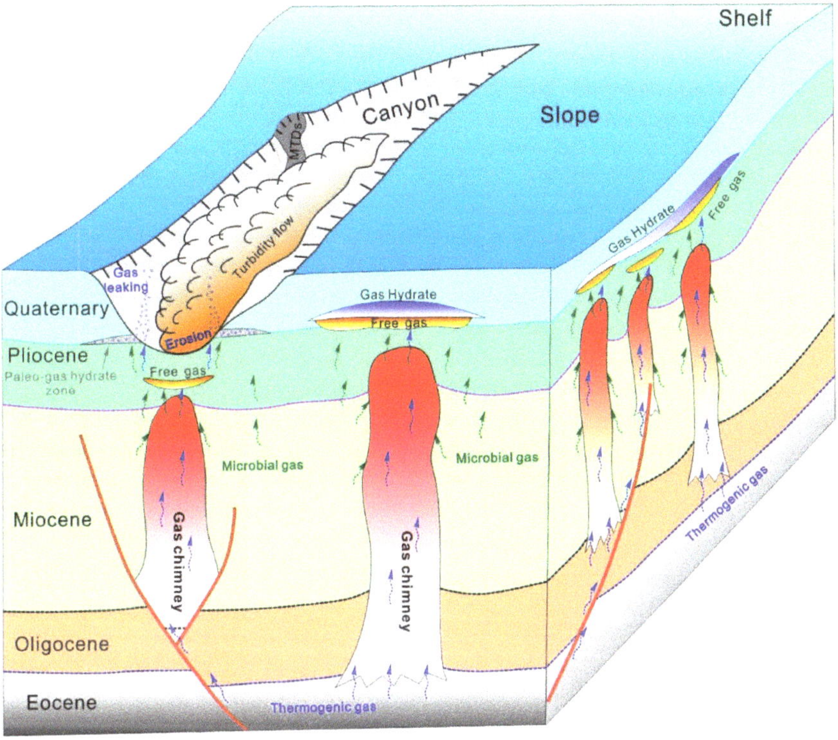

Figure 12. Accumulation model (conceptual model not to scale) of gas hydrate in a canyon developed area, northern South China Sea.

6. Conclusions

In this study, the characteristics of depositional processes within canyons and gas chimneys in the Shenhu Sea area were analyzed using high-resolution 2D and 3D seismic data. Submarine canyons exhibit the most distinct geomorphological features. Five depositional elements were identified within and adjacent to submarine canyons: LAPs, basal lags, slides, MTDs, and turbidity lobes. As an important conduit for sediment transport from shallow to deep water, the canyons have experienced frequent sediment gravity flows during their evolutionary history, which is indicated by successive MTDs and turbidity lobes that are developed outside the distal canyon mouth. Gas chimneys are important structures in the study area. The seismic profiles show that most of them originated from Eocene strata and terminated in or nearby Pliocene strata at the top. In the plane view, gas chimneys exhibit a lumpy and banded shape. Lumpy ones are small and less than 3 km wide, whereas banded ones can be up to 10 km long, indicating that the long axis coincides with the canyon strike. The distributions of BSRs and gas hydrates in the plane view are consistent. Based on the depositional processes within canyons, distribution of gas chimneys and BSRs, and gas sources, an accumulation model of gas hydrate was proposed. Thermogenic gas from Eocene source rocks migrates vertically through gas chimneys, mixes with microbiogenic gases from shallow source rocks, and then forms gas hydrates above the gas chimneys under suitable pressure and temperature conditions. Because of the strong erosion caused by sediment gravity flows within canyons and subsequent changes in the conditions, gas hydrates dissociate, followed by gas leakage. Most of the preserved hydrates can be observed in the ridges of canyons, with very few hydrates beneath the bottom of canyons. Therefore, it is of great importance to study the depositional processes, which can not only form the gas hydrate reservoir, but also destroy the preservation of gas hydrates. The conceptional model established in this study can give great guidance for the prediction of gas hydrates in the South China Sea.

Author Contributions: Methodology, Y.H.; Formal analysis, Y.H., Z.K., C.C., T.J., C.Z., B.L., C.Y., J.L. and C.X.; Investigation, Y.H., Z.K., C.C., T.J., C.Z., B.L. and C.Y.; Data curation, Y.H., C.C., C.Z., B.L., C.Y., J.L. and C.X.; Writing—original draft, Y.H.; Writing—review & editing, Y.H. and Z.K.; Project administration, Z.K.; Funding acquisition, Y.H. and Z.K. All authors have read and agreed to the published version of the manuscript.

Funding: This study was sponsored by the National Natural Science Foundation of China (grant nos. 42130408 and 41502102).

Data Availability Statement: Not applicable.

Acknowledgments: The authors wish to thank the Guangzhou Marine Geological Survey for providing the seismic data and permission to publish this work. The authors thank three anonymous reviewers for their useful and constructive comments which allowed us to improve this paper and had a significant contribution to the manuscript.

Conflicts of Interest: The authors declared that they have no conflict of interest in this work. We declare that we do not have any commercial or associative interest that represents a conflict of interest in connection with the work submitted.

References

1. Dickens, G.R. The potential volume of oceanic methane hydrates with variable external conditions. *Org. Geochem* **2001**, *32*, 1179–1193. [CrossRef]
2. Collett, T.S. Energy resource potential of natural gas hydrates. *Aapg. Bull.* **2002**, *86*, 1971–1992.
3. Boswell, R.; Collett, T.S. Current perspectives on gas hydrate resources. *Energ. Environ. Sci.* **2011**, *4*, 1206–1215. [CrossRef]
4. Milkov, A.V. Global estimates of hydrate-bound gas in marine sediments: How much is really out there? *Earth Sci. Rev.* **2004**, *66*, 183–197. [CrossRef]
5. Klauda, J.B.; Sandler, S.I. Global distribution of methane hydrate in ocean sediment. *Energ. Fuel* **2005**, *19*, 459–470. [CrossRef]
6. Boswell, R.; Shipp, C.; Reichel, T.; Shelander, D.; Saeki, T.; Frye, M.; Shedd, W.; Collett, T.S.; McConnell, D.R. Prospecting for marine gas hydrate resources. *Interpret. J. Sub.* **2016**, *4*, SA13–SA24. [CrossRef]
7. Wang, X.J.; Jin, J.P.; Guo, Y.Q.; Li, J.; Li, Y.P.; Qian, J.; Wang, B.; Zhou, J.L. Enrichment characteristics and quantitative evaluation of gas hydrate in northern South China Sea. *Earth Sci.* **2021**, *46*, 1038–1057. (In Chinese with English Abstract)
8. He, J.X.; Ning, Z.J.; Zhao, B.; Wan, Z.F.; Meng, D.J. Preliminary Analysis and Prediction of Strategic Replacement Area for Natural Gas Hydrate Exploration in the South China Sea. *Earth Sci.* **2022**, *47*, 1549–1568. (In Chinese with English Abstract)
9. Boswell, R.; Shelander, D.; Lee, M.; Latham, T.; Collett, T.; Guerin, G.; Moridis, G.; Reagan, M.; Goldberg, D. Occurrence of gas hydrate in Oligocene Frio sand: Alaminos Canyon Block 818: Northern Gulf of Mexico. *Mar. Petrol. Geol.* **2009**, *26*, 1499–1512. [CrossRef]
10. Barnes, P.M.; Lamarche, G.; Bialas, J.; Henrys, S.; Pecher, I.; Netzeband, G.L.; Greinert, J.; Mountjoy, J.J.; Pedley, K.; Crutchley, G. Tectonic and geological framework for gas hydrates and cold seeps on the Hikurangi subduction margin, New Zealand. *Mar. Geol.* **2010**, *272*, 26–48. [CrossRef]
11. Collett, T.S.; Lee, M.W.; Agena, W.F.; Miller, J.J.; Lewis, K.A.; Zyrianova, M.V.; Boswell, R.; Inks, T.L. Permafrost-associated natural gas hydrate occurrences on the Alaska North Slope. *Mar. Petrol. Geol.* **2011**, *28*, 279–294. [CrossRef]
12. Riedel, M.; Collett, T.S.; Shankar, U. Documenting channel features associated with gas hydrates in the Krishna-Godavari Basin, offshore India. *Mar. Geol.* **2011**, *279*, 1–11. [CrossRef]
13. Su, M.; Luo, K.; Fang, Y.; Kuang, Z.; Yang, C.; Liang, J.; Liang, C.; Chen, H.; Lin, Z.; Wang, C.; et al. Grain-size characteristics of fine-grained sediments and association with gas hydrate saturation in Shenhu Area, northern South China Sea. *Ore Geol. Rev.* **2021**, *129*, 103889. [CrossRef]
14. Collett, T.S.; Johnson, A.H.; Knapp, C.C.; Boswell, R. *Natural Gas Hydrates-Energy Resource Potential and Associated Geologic Hazards, AAPG Memoir 89*; AAPG: Tulsa, AK, USA, 2009.
15. Dallimore, S.R.; Collett, T.S. Summary and Implications of the Mallik 2002 Gas Hydrate Production Research Well Program. 2005. Available online: http://geopub.nrcan.gc.ca/index_e.php (accessed on 14 August 2022).
16. Uchida, T.; Tsuji, T. Petrophysical properties of natural gas hydrates-bearing sands and their sedimentology in the Nankai Trough. *Resour. Geol.* **2004**, *54*, 79–87. [CrossRef]
17. Liang, J.Q.; Wei, J.G.; Bigalke, N.; Roberts, J.; Schultheiss, P.; Holland, M.; Triaxial, P. In Laboratory Quantification of Geomechanical Properties of Hydrate-Bearing Sediments in the Shenhu Area of the South China Sea at In-Situ Conditions. 2017, pp. 1–14. Available online: https://www.geotekcoring.com/wp-content/uploads/2017/10/LiangBigalkIeICGH9.pdf (accessed on 25 November 2022).

18. Davies, R.J.; Thatcher, K.E.; Mathias, S.A.; Yang, J. Deepwater canyons: An escape route for methane sealed by methane hydrate. *Earth Planet. Sci. Lett.* **2012**, *323*, 72–78. [CrossRef]
19. Sauter, E.J.; Muyakshin, S.I.; Charlou, J.; Schlüter, M.; Boetius, A.; Jerosch, K.; Damm, E.; Foucher, J.; Klages, M. Methane discharge from a deep-sea submarine mud volcano into the upper water column by gas hydrate-coated methane bubbles. *Earth Planet. Sci. Lett.* **2006**, *243*, 354–365. [CrossRef]
20. Wang, X.; Collett, T.S.; Lee, M.W.; Yang, S.; Guo, Y.; Wu, S. Geological controls on the occurrence of gas hydrate from core, downhole log, and seismic data in the Shenhu area, South China Sea. *Mar. Geol.* **2014**, *357*, 272–292. [CrossRef]
21. Chen, D.F.; Li, X.X.; Xia, B. Distribution characteristics and resource prediction of natural gas hydrate stability domain in Qiongdongnan Basin, South China Sea. *Chin. J. Geophys.* **2004**, *47*, 483–489. (In Chinese with English Abstract) [CrossRef]
22. Liu, C.; Ye, Y.; Meng, Q.; He, X.; Lu, H.; Zhang, J.; Liu, J.; Yang, S. The Characteristics of Gas Hydrates Recovered from Shenhu Area in the South China Sea. *Mar. Geol.* **2012**, *307*, 22–27. [CrossRef]
23. Wang, X.; Qian, J.; Collett, T.S.; Shi, H.; Yang, S.; Yan, C.; Li, Y.; Wang, Z.; Chen, D. Characterization of gas hydrate distribution using conventional 3D seismic data in the Pearl River Mouth Basin, South China Sea. *Interpret. J. Sub.* **2016**, *4*, SA25–SA37. [CrossRef]
24. Su, P.B.; Liang, J.Q.; Zhang, W.; Liu, F.; Wang, F.F.; Li, T.W.; Wang, X.X.; Wang, L.F. Gas Hydrate Accumulation System in Shenhu Sea Area, Northern South China Sea. *Nat. Gas Ind.* **2020**, *40*, 77–89. (In Chinese with English Abstract)
25. Yuan, H.; Wang, Y.; Wang, X. Seismic Methods for Exploration and Exploitation of Gas Hydrate. *J. Earth Sci. China* **2021**, *32*, 839–849. [CrossRef]
26. Zhang, W.; Liang, J.; Yang, X.; Su, P.; Wan, Z. The formation mechanism of mud diapirs and gas chimneys and their relationship with natural gas hydrates: Insights from the deep-water area of Qiongdongnan Basin, northern South China Sea. *Int. Geol. Rev.* **2020**, *62*, 789–810. [CrossRef]
27. Jinqiang, L.; Wei, Z.; Jing, A.L.; Jiangong, W.; Zenggui, K.; Yulin, H. Geological occurrence and accumulation mechanism of natural gas hydrates in the eastern Qiongdongnan Basin of the South China Sea: Insights from site GMGS5-W9-2018. *Mar. Geol.* **2019**, *418*, 106042.
28. Zhang, W.; Liang, J.Q.; Lu, J.A.; Yu, J.G.; Su, P.B.; Fang, Y.X.; Guo, Y.Q.; Yang, S.X.; Zhang, G.X. Characteristics and mechanism of high-saturation natural gas hydrate accumulation in Shenhu area, northern South China Sea. *Pet. Explor. Dev.* **2017**, *44*, 670–680. (In Chinese with English Abstract) [CrossRef]
29. Cheng, C.; Jiang, T.; Kuang, Z.G.; Yang, C.Z.; Zhang, C.; He, Y.L.; Cheng, Z.; Tian, D.M.; Xiong, P.F. Characteristics of gas chimneys and their implications on gas hydrate accumulation in the Shenhu area, northern south China sea. *J. Nat. Gas. Sci. Eng.* **2020**, *84*, 103629. [CrossRef]
30. Chen, Z.; Jiang, T.; Kuang, Z.; Cheng, C.; Xiong, P.; Chen, Y. Accumulation Characteristics of Gas Hydrate-Shallow Gas Symbiotic System in Qiongdongnan Basin. *Earth Sci.* **2022**, *47*, 1619–1634. (In Chinese with English Abstract)
31. Liang, C.; Liu, C.Y.; Xie, X.N.; Yu, X.H.; He, Y.L.; Su, M.; Chen, H.; Zhou, Z.; Tian, D.M.; Mi, H.G.; et al. Basal shear zones of recurrent mass transport deposits serve as potential reservoirs for gas hydrates in the Central Canyon area, South China Sea. *Mar. Geol.* **2021**, *441*, 106631. [CrossRef]
32. Jiangong, W.; Yunxin, F.; Hailong, L.; Hongfeng, L.; Jingan, L.; Jinqiang, L.; Shengxiong, Y. Distribution and characteristics of natural gas hydrates in the Shenhu Sea Area, South China Sea. *Mar. Petrol. Geol.* **2018**, *98*, 622–628.
33. Yang, S.X.; Liang, J.Q.; Lu, J.A.; Qu, C.W.; Liu, B. New understanding of gas hydrate accumulation characteristics and main controlling factors in Shenhu area of northern South China Sea. *Earth Sci. Front.* **2017**, *24*, 1–14. (In Chinese with English Abstract)
34. Zhang, W.; Liang, J.Q.; Wei, J.G.; Su, P.B.; Lin, L.; Huang, W. Origin of natural gases and associated gas hydrates in the Shenhu area, northern South China Sea: Results from the China gas hydrate drilling expeditions. *J. Asian Earth Sci.* **2019**, *183*, 103953. [CrossRef]
35. Jin, J.; Wang, X.; Guo, Y.; Li, J.; Li, Y.; Zhang, X.; Qian, J.; Sun, L. Geological controls on the occurrence of recently formed highly concentrated gas hydrate accumulations in the Shenhu area, South China Sea. *Mar. Petrol. Geol.* **2020**, *116*, 104294. [CrossRef]
36. Yang, S.; Zhang, M.; Liang, J.Q.; Lu, J.; Zhang, Z.J.; Holland, M.; Schultheiss, P.; Fu, S.Y.; Sha, Z.B. Preliminary results of China's third gas hydrate drilling expedition: A critical step from discovery to development in the South China Sea. *Cent. Nat. Gas Oil* **2015**, *412*, 386–7614.
37. Su, M.; Yang, R.; Wang, H.; Sha, Z.; Liang, J.; Wu, N.; Qiao, S.H.; Cong, X. Gas hydrates distribution in the Shenhu Area, northern South China Sea: Comparisons between the eight drilling sites with gas-hydrate petroleum system. *Geol. Acta Int. Earth Sci. J.* **2016**, *14*, 79–100.
38. Wang, X.J.; Wu, S.G.; Lee, M.; Guo, Y.Q.; Yang, S.X.; Liang, J.Q. Gas hydrate saturation from acoustic impedance and resistivity logs in the Shenhu area, South China Sea. *Mar. Petrol. Geol.* **2011**, *28*, 1625–1633. [CrossRef]
39. Wu, N.; Zhang, H.; Yang, S.; Zhang, G.; Liang, J.; Su, X.; Schultheiss, P.; Holland, M.; Zhu, Y. Gas Hydrate System of Shenhu Area, Northern South China Sea: Geochemical Results. *J. Geol. Res.* **2011**, *2011*, 370298. [CrossRef]
40. Liu, B.J.; Pang, X.; Yan, C.Z.; Liu, J.; Lian, S.Y.; He, M.; Shen, J. Evolution of Oligocene-Miocene shelf slope-break zone in Baiyun deep water area of Pearl River Mouth Basin and its significance for oil and gas exploration. *Acta Pet. Sin.* **2011**, *32*, 234–242. (In Chinese with English Abstract)

41. Liu, B.J.; Pang, X.; Xie, S.W.; Mei, L.F.; Zheng, J.Y.; Sun, H.; Yan, H.; Wu, Y.X.; Xiang, X.H.; Feng, X. Control of crust-mantle detachment fault activity on deep large delta sedimentary system in baiyun sag, pearl river estuary basin. *Earth Sci.* **2022**, *47*, 2354–2373. (In Chinese with English Abstract)
42. Liao, J.H.; Wu, K.Q.; Er, C. Deep Reservoir Characteristics and Effective Reservoir Controlling Factors in Baiyun Sag, Pearl River Mouth Basin. *Earth Sci.* **2022**, *47*, 2454–2467. (In Chinese with English Abstract)
43. Chao, H.; Jianye, R.; Yan, W.; Yanghui, Z. Sequence Stratigraphic Framework and its Formation Mechanism for the Break-up Sequence in the Baiyun Sag, Northern South China Sea. *J. Earth Sci.* **2022**. Available online: https://kns.cnki.net/kcms/detail/42.1788.P.20220512.1022.004.html (accessed on 21 December 2022).
44. Wang, J.H.; Pang, X.; Wang, C.W.; He, M.; Lian, S.Y. Discovery and identification of central diapir zone in Baiyun Sag, Pearl River Mouth Basin. *Earth Sci.* **2006**, *31*, 209–213. (In Chinese with English Abstract)
45. Pang, X.; Shi, H.S.; Zhu, M.; Yan, C.Z.; Liu, J.; Zhu, J.Z.; Liu, B.J. The oil and gas exploration prospect in Baiyun deep water area is discussed again. *China Offshore Oil Gas* **2014**, *26*, 23–29. (In Chinese with English Abstract)
46. Mi, L.; Zhang, Z.; Pang, X.; Liu, J.; Zhang, B.; Zhao, Q.; Feng, X. Main controlling factors of hydrocarbon accumulation in Baiyun Sag at northern continental margin of South China Sea. *Pet. Explor. Dev.* **2018**, *45*, 963–973. [CrossRef]
47. Shi, H.S.; Liu, B.J.; Yan, C.Z.; Zhu, M.; Pang, X.; Qin, C.G. Hydrocarbon Accumulation Conditions and Exploration Potential in Baiyun-Liwan Deepwater Area of Pearl River Mouth Basin. *China Offshore Oil Gas* **2010**, *22*, 369–374. (In Chinese with English Abstract)
48. Zhang, L.L. Early Oligocene-Miocene Paleontological Strata and Sedimentary Environment in Jieyang Sag, Pearl River Mouth Basin. *Acta Micropalaeontolo. Sin.* **2020**, *37*, 266–277. (In Chinese with English Abstract)
49. Gong, C.L.; Wang, Y.M.; Zhu, W.L.; Li, W.G.; Xu, Q. Upper Miocene to Quaternary unidirectionally migrating deep-water channels in the Pearl River Mouth Basin, northern South China Sea. *Aapg. Bull.* **2013**, *97*, 285–308. [CrossRef]
50. He, Y.; Xie, X.; Kneller, B.C.; Wang, Z.; Li, X. Architecture and controlling factors of canyon fills on the shelf margin in the Qiongdongnan Basin, northern South China Sea. *Mar. Petrol. Geol.* **2013**, *41*, 264–276. [CrossRef]
51. Wang, X.X.; Zhuo, H.T.; Wang, Y.M.; Mao, P.X.; He, M.; Chen, W.T.; Zhou, J.W.; Gao, S.M.; Wang, M.H. Controls of contour currents on intra-canyon mixed sedimentary processes: Insights from the Pearl River Canyon, northern South China Sea. *Mar. Geol.* **2018**, *406*, 193–213. [CrossRef]
52. Zhu, M.; Graham, S.; Pang, X.; McHargue, T. Characteristics of migrating submarine canyons from the middle Miocene to present: Implications for paleoceanographic circulation, northern South China Sea. *Mar. Petrol. Geol.* **2010**, *27*, 307–319. [CrossRef]
53. Sun, Q.L.; Xie, X.N.; Piper, D.; Wu, J.; Wu, S.G. Three dimensional seismic anatomy of multi-stage mass transport deposits in the Pearl River Mouth Basin, northern South China Sea: Their ages and kinematics. *Mar. Geol.* **2017**, *393*, 93–108. [CrossRef]
54. Li, W.; Wu, S.G.; Volker, D.; Zhao, F.; Mi, L.J.; Kopf, A. Morphology, seismic characterization and sediment dynamics of the Baiyun Slide Complex on the northern South China Sea margin. *J. Geol. Soc.* **2014**, *171*, 865–877. [CrossRef]
55. Sun, Q.L.; Cartwright, J.; Xie, X.N.; Lu, X.Y.; Yuan, S.Q.; Chen, C.X. Reconstruction of repeated Quaternary slope failures in the northern South China Sea. *Mar. Geol.* **2018**, *401*, 17–35. [CrossRef]
56. Su, M.; Alves, T.M.; Li, W.; Sha, Z.B.; Hsiung, K.H.; Liang, J.Q.; Kuang, Z.G.; Wu, N.Y.; Zhang, B.D.; Chiang, C.S. Reassessing two contrasting Late Miocene-Holocene stratigraphic frameworks for the Pearl River Mouth Basin, northern South China Sea. *Mar. Petrol. Geol.* **2019**, *102*, 899–913. [CrossRef]
57. Coren, F.; Volpi, V.; Tinivella, U. Gas hydrate physical properties imaging by multi-attribute analysis—Blake Ridge BSR case history. *Mar. Geol.* **2001**, *178*, 197–210. [CrossRef]
58. Wei, Z.; Jinqiang, L.; Zhifeng, W.; Pibo, S.; Wei, H.; Lifeng, W.; Lin, L. Dynamic accumulation of gas hydrates associated with the channel-levee system in the Shenhu area, northern South China Sea. *Mar. Petrol. Geol.* **2020**, *117*, 104354.
59. Kvenvolden, K.A. A review of the geochemistry of methane in natural gas hydrate. *Org Geochem* **1995**, *23*, 997–1008. [CrossRef]
60. Su, M.; Sha, Z.; Zhang, C.; Wang, H.; Wu, N.; Yang, R.; Liang, J.; Qiao, S.; Cong, X.; Liu, J. Types, Characteristics and Significances of Migrating Pathways of Gas-bearing Fluids in the Shenhu Area, Northern Continental Slope of the South China Sea. *Acta Geol. Sin. Engl.* **2017**, *91*, 219–231. [CrossRef]
61. Lai, H.; Fang, Y.; Kuang, Z.; Ren, J.; Liang, J.; Lu, J.; Wang, G.; Xing, C. Geochemistry, origin and accumulation of natural gas hydrates in the Qiongdongnan Basin, South China Sea: Implications from site GMGS5-W08. *Mar. Petrol. Geol.* **2021**, *123*, 104774. [CrossRef]
62. He, J.X.; Yan, W.; Zhu, Y.H.; Zhang, W.; Gong, F.X.; Liu, S.L.; Zhang, J.R.; Gong, X.F. Biogas/Subbiogas Resources and Gas Hydrate Accumulation in the Northern Marginal Basin of the South China Sea. *Nat. Gas Ind.* **2013**, *33*, 121–134. (In Chinese with English Abstract)
63. Sha, Z.B.; Liang, J.Q.; Su, P.B.; Zhang, G.X.; Lu, J.A.; Wang, J.L. Study on Drilling Results and Accumulation Factors of Natural Gas Hydrate in Eastern Sea Area of Pearl River Mouth Basin. *Earth Sci. Front.* **2015**, *22*, 125–135. (In Chinese with English Abstract)
64. Su, P.B.; Liang, J.Q.; Peng, J.; Zhang, W.; Xu, J.H. Petroleum systems modeling on gas hydrate of the first experimental exploitation region in the Shenhu area, northern South China sea. *J. Asian Earth Sci.* **2018**, *168*, 57–76. [CrossRef]
65. Zhu, Y.; Huang, X.; Fu, S.; Su, P. Gas Sources of Natural Gas Hydrates in the Shenhu Drilling Area, South China Sea: Geochemical Evidence and Geological Analysis. *Acta Geologica Sinica* **2013**, *87*, 767–776.
66. Hui, C.; Xinong, X.; Kainan, M.; Yunlong, H.; Ming, S.; Wenyan, Z. Depositional Characteristics and Formation Mechanisms of Deep-Water Canyon Systems along the Northern South China Sea Margin. *J. Earth Sci. China* **2020**, *31*, 808–819.

67. Baztan, J.; Berne, S.; Olivet, J.L.; Rabineau, M.; Aslanian, D.; Gaudin, A.; Rehault, J.P.; Canals, M. Axial incision: The key to understand submarine canyon evolution (in the western Gulf of Lion). *Mar. Petrol. Geol.* **2005**, *22*, 805–826. [CrossRef]
68. Su, M.; Lin, Z.X.; Wang, C.; Kuang, Z.G.; Liang, J.Q.; Chen, H.; Liu, S.; Zhang, B.D.; Luo, K.W.; Huang, S.Q.; et al. Geomorphologic and infilling characteristics of the slope-confined submarine canyons in the Pearl River Mouth Basin, northern South China Sea. *Mar. Geol.* **2020**, *424*, 106166. [CrossRef]
69. Nugraha, H.D.; Jackson, C.A.L.; Johnson, H.D.; Hodgson, D.M.; Clare, M.A. Extreme erosion by submarine slides. *Geology* **2022**, *50*, 1130–1134. [CrossRef]
70. Moscardelli, L.; Wood, L. New classification system for mass transport complexes in offshore Trinidad. *Basin Res.* **2008**, *20*, 73–98. [CrossRef]

energies

Article

Research on Diagenetic Evolution and Hydrocarbon Accumulation Periods of Chang 8 Reservoir in Zhenjing Area of Ordos Basin

Guilin Yang, Zhanli Ren * and Kai Qi

Department of Geology, Northwest University, Xi'an 710069, China; yangguilin612724@163.com (G.Y.); kaiqi0913@163.com (K.Q.)
* Correspondence: renzhanl@nwu.edu.cn

Abstract: The Mesozoic Chang 8 Section in the Zhenjing area is a typical low permeability-tight sand reservoir and is regarded as the most important set of paybeds in the study area. Guided by the principles of basic geological theory, the diagenetic evolution process and hydrocarbon accumulation periods of the Chang 8 reservoir in the study area were determined through various techniques. More specifically, core observation, scanning electron microscopy (SEM), X-ray diffraction (XRD), and vitrinite reflectance experiments were performed in combination with systematic studies on rock pyrolysis and the thermal evolutionary history of basins, the illite-dating method, and so on. The Chang 8 reservoir is dominated by feldspar lithic and lithic feldspar sandstones. Quartz, feldspar, and lithic fragments are the major clastic constituents. In clay minerals, the chlorite content is the highest, followed by illite/smectite formation and kaolinite, while the illite content is the lowest. The major diagenesis effect of the Chang 8 reservoir includes compaction, cementation, dissolution, metasomatism, and rupturing. The assumed diagenetic sequence is the following: mechanical composition → early sedimentation of chlorite clay mineral membrane → early cementation of sparry calcite → authigenic kaolinite precipitation → secondary production and amplification of quartz → dissolution of carbonate cement → dissolution of feldspar → late cementation of minerals such as ferrocalcite. Now, the study area is in Stage A in the middle diagenetic period. Through the inclusion of temperature measurements, in conjunction with illite dating and thermal evolutionary history analysis technology in basins, the Chang 8 reservoir of this study was determined as the phase-I continuous accumulation process and the reservoir formation epoch was 105~125 Ma, which was assigned to the Middle Early Cretaceous Epoch.

Keywords: diagenesis; hydrocarbon accumulation periods; Chang 8 reservoir; Zhenjing area; Ordos Basin

Citation: Yang, G.; Ren, Z.; Qi, K. Research on Diagenetic Evolution and Hydrocarbon Accumulation Periods of Chang 8 Reservoir in Zhenjing Area of Ordos Basin. *Energies* **2022**, *15*, 3846. https://doi.org/10.3390/en15103846

Academic Editors: Dameng Liu and Mofazzal Hossain

Received: 25 March 2022
Accepted: 19 May 2022
Published: 23 May 2022

1. Introduction

Zhenjing Block is located at the intersection of the Tianhuan Depression, Northern Shaanxi Slope, Weibei Hump and thrust belt at the west edge of the Ordos Basin, which is in a unique geological position (Figure 1) [1–4]. In the study area, the Mesozoic Chang 8 reservoir forms the major paybed, which is rich in oil and gas resources [3–6]. With continuous developments of the oil field, high yields have been difficult to maintain and may even worsen, largely due to insufficient understanding of the reservoir quality. The formation of the reservoir is a complicated and time-consuming process, with the three essential geological processes—sedimentation, diagenesis, and tectonism—requiring thorough examination [7]. Of these processes, diagenesis plays an important role in reservoir reformation and, as a result, has been widely examined by the scientific community in the field [8–19]. The analysis of the accumulation period is an important part of the accumulation system study and key in the analysis of the accumulation process [20–23].

Previous studies on the Chang 8 Member in the Zhenjing area of the Ordos Basin mainly focused on sedimentary facies and reservoir characteristics. However, studies on the diagenesis and accumulation stages were relatively weak. It is therefore of great significance to clarify the diagenetic evolution and hydrocarbon accumulation stage of the Chang 8 Member—the main oil-producing layer—for deepening the theoretical understanding of reservoir evaluation and hydrocarbon accumulation [9–15,20–23]. In this paper, both the diagenetic sequence and hydrocarbon accumulation periods of the Chang 8 reservoir section, which is a low-permeability-tight sandstone reservoir, are discussed systematically. First, a brief analysis of the petrology and physical characteristics of the Chang 8 reservoir was carried out based on insights from the borehole core observation, scanning electron microscopy (SEM), and slice observation. On this basis, the diagenesis of the Chang 8 reservoir section was investigated by combining X-ray diffraction (XRD) measurements, vitrinite reflectance, and rock pyrolysis. Thus, it was divided into various evolutionary sequences. Meanwhile, the hydrocarbon accumulation periods of the Chang 8 reservoir were analyzed comprehensively by conducting a thermal history analysis of basins and applying the illite dating method. The hydrocarbon accumulation times and the specific time were determined both indirectly and directly. Our work provides reliable references to the follow-up exploration and exploitation of the study area [5].

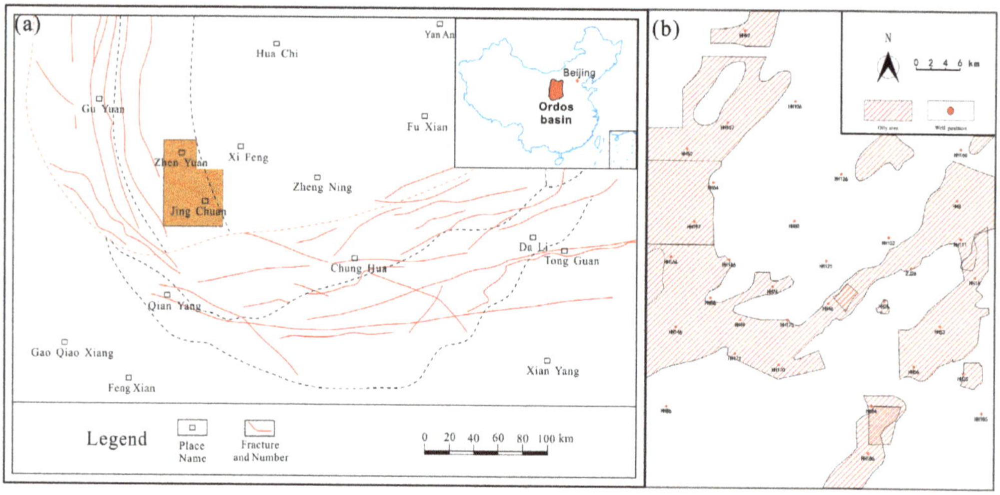

Figure 1. Location map of the study area: (**a**) Structural location map of the study area; (**b**) exploration and development area map of the study area.

2. Lithological Characteristics and Physical Characteristics of the Reservoir

Based on the statistical studies of borehole core observation, SEM and slice observation data were analyzed. At the same time, the analysis of lithologic triangulation diagram was carried out (Figure 2a). Region I represents quartz sandstone, Region II denotes feldspathic quartz sandstone, Region III stands for the rock debris quartz sandstone, Region IV signifies arkose, Region V is the rock debris arkose, Region VI represents feldspar rock debris sandstone, and Region VII is the rock debris sandstone. It was found that feldspar rock debris sandstone and rock debris arkose are the dominant lithologies in the Chang 8 reservoir (Figure 2a). With respect to the clastic constituents, quartz accounts for the highest proportion (29%), followed by feldspar (26%), which is mainly composed of potash feldspar and plagioclase. The content of rock debris is the lowest, averaging at 24%. Among them, the magmatic rock debris mainly consists of neural acidity and the metamorphic rock debris is mainly composed of quartzite, followed by phyllite. The sedimentary rock debris is mainly siltstone and silty mudstone, followed by mudstones and flint. In addition, the content of mica is the lowest (Table 1) [24].

According to SEM and XRD analysis, the illite content in sandstones of Chang 8 reservoir in the study area is the lowest, ranging from 4% to 11% and averaging at 6.85%. Under the microscope, the thin-film, schistose, hair, and fiber modes are the major adhesion modes on the particle surfaces. Chlorite content is the highest (21~45%), averaging at 36.3%. Insights from SEM analysis indicate that chlorite mainly presents as thin-film mode and foliated mode, while the cementation mode is mainly presented as the looped lining mode and pore-lining mode, followed by illite/smectite formation and kaolinite. It is interesting to notice that the illite/smectite formation looks like a honeycomb under SEM imaging and the content of kaolinite ranges from 21% to 45%, averaging at 36.3%. It was developed on a large scale as filling in the pores and is presented with good crystal form. Under SEM imaging, the crystals look like book pages and worms (Figures 2b and 4c–g).

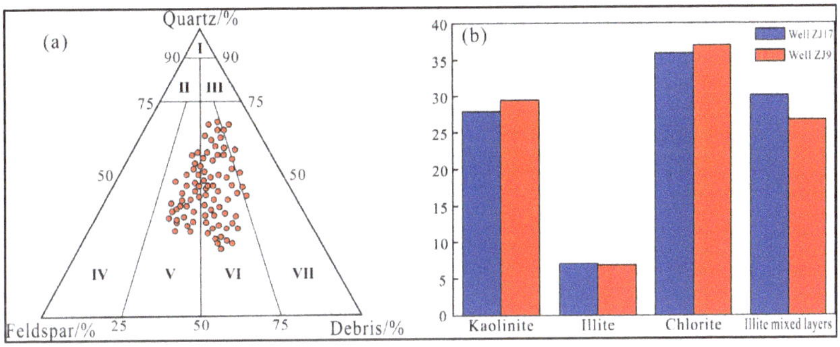

Figure 2. Comprehensive analysis of rock characteristics of the Chang 8 reservoir: (**a**) Triangular map of rock classification; (**b**) distribution histogram of clay minerals.

According to the core data test and analysis, the porosity of the Chang 8 reservoir distributes between 1.8~17.9%, averaging at 10.9%. The permeability ranges between 0.037~0.79 mD, averaging at 0.45 mD (Figure 3). Thus, it belongs to a low-porosity and low-permeability reservoir.

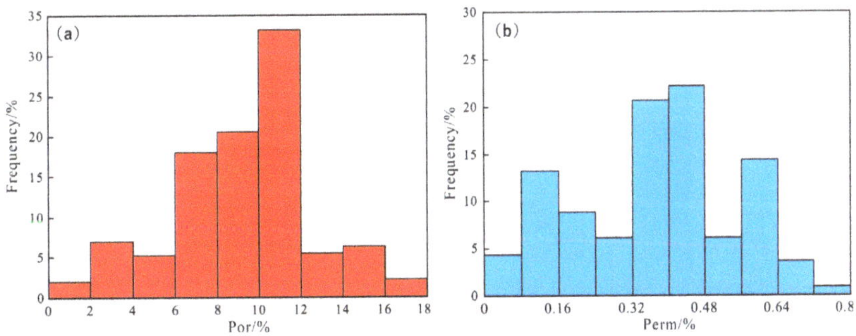

Figure 3. (**a**,**b**) Histogram of the porosity–permeability frequency distribution of the Chang 8 reservoir.

Table 1. Reservoir rock composition statistics of the Chang 8 oil layer group in the study area.

W [1]	Q (%) [2]	Feldspar (%) [3]			Debris (%) [4]				M (%) [5]	T (%) [6]
		OC	PC	T	MR	MPR	SR	T		
ZJ5	35.2	13.6	17.8	31.4	15.8	3.4	1.67	20.87	3.2	90.67
ZJ18	40	7.5	15	22.5	14	3.5	6	23.5	4	90
ZJ19	26.7	11.55	17.2	28.75	13.8	5.78	3.83	23.41	5.58	84.44

Table 1. *Cont.*

W [1]	Q (%) [2]	Feldspar (%) [3]			Debris (%) [4]				M (%) [5]	T (%) [6]
		OC	PC	T	MR	MPR	SR	T		
ZJ21	25	9	16.5	25.5	15.88	7.13	4.63	27.63	5.75	83.88
HH23	28	12.67	24.19	36.86	15.62	4	1.8	21.42	4	90.32
HH24	32	10	9	20	14	6	4	24	5	80
HH26	32.8	12.38	16.5	28.88	16.25	4.5	1.4	22.15	4.75	88.53
HH103	25.5	10.7	20	30.7	16.3	7.9	5.8	30	3.9	90.1
average	29	11	16	26	15	6	4	24	5	85

[1] Well (W); [2] quartz (Q); [3] orthoclase (OC); plagioclase (PC); the total of feldspar (T); [4] magmatic rock (MR); metamorphic rock (MPR); sedimentary rock (SR); the total of debris (T); [5] mica (M); [6] the total of rock composition.

3. Diagenetic Evolution Analysis of the Reservoir

Diagenesis is defined as the evolution process where sediments solidify into rocks through a series of physical, chemical, and biological reactions. This process is affected by many factors, such as burying rate, pressure, local temperature distribution, and sediment composition. Hence, it can greatly influence the physical properties of the reservoir. Therefore, diagenesis is closely related to the hydrocarbon accumulation mechanism [10–14,25].

Zhenjing area is located in the continental facies lacustrine deposit environment. There are many types of diagenesis of the Chang 8 reservoir [5,6]. In this work, both the diagenesis and the diagenetic sequence of the Chang 8 reservoir were studied systematically by using slice authentication, SEM and XRD measurements, as well as vitrinite reflectance and rock pyrolysis. The major diagenesis effects include compaction, cementation, dissolution, metasomatism, and rupturing. Among them, dissolution and rupturing have a positive impact on the improvement of the physical properties of the reservoir, while compaction and cementation facilitate the compactness of the reservoir.

3.1. Diagenetic Analysis

3.1.1. Compaction

Due to the pressure of the overlying rocks, the process that makes the reservoir structure tighter and tighter is called compaction. It is regarded as the most influential and the most common diagenesis type in the diagenetic evolution of the reservoir [26–35]. The manifestation of the compaction effect of the Chang 8 reservoir in the Zhenjiang area is obvious, especially as the bending deformation of the plastic mineral particles is concerned, due to compaction in the early diagenesis. For example, minerals such as mica developed deformation of plastic particles after experiencing strong mechanical compactness (Figure 4a). Mineral particles are also compacted and filled into spaces among primary pores, thus decreasing the physical properties of the reservoir significantly. As the compaction continues, the lattices at the particle contact points may be deformed and even be dissolved. As a result, the contact relation among the particles may change from the original point contact to the linear contact and even to the linear-concave-convex contact. Many clastic particles like mica are aligned in a direction that forms texture layers (Figure 4b). Compaction is the major cause of tightening and sharp reduction of the primary pores, and could lead to the deterioration of the physical properties of the Chang 8 reservoir in the Zhenjing area.

3.1.2. Cementation

The change during the process of minerals precipitation in the pores of fragmental deposits into authigenic minerals is called cementation, which renders sediments solidified into rocks [36,37]. The role of cementation is to fill the pores and it is considered an important cause of decreasing the porosity in the reservoir layer. However, volumes along particles may not decrease due to cementation, which is obviously different from compaction [28,30–35]. From a general point of view, the inter-granular pores are filled by authigenic minerals, which has a negative impact on changes in the physical properties of the reservoir. However, the early filling of authigenic minerals can inhibit compaction to some extent. In the retained inter-granular pores, solvent-sensitive types of cement from secondary pores are formed as a response to dissolution. Hence, cementation has some positive influence on changes in the physical properties of the reservoir to some extent. In particular, early cementation has dual contributions to the physical properties of the reservoir. As an important factor of compaction of the Chang 8 reservoir in the Zhenjing area, the cementation procedure can be divided into authigenic clay mineral cementation, siliceous cementation, and carbonate cementation according to the types of the cement. It is mainly influenced by the fluid features in pores, the sedimentation environment, and composition [32,35].

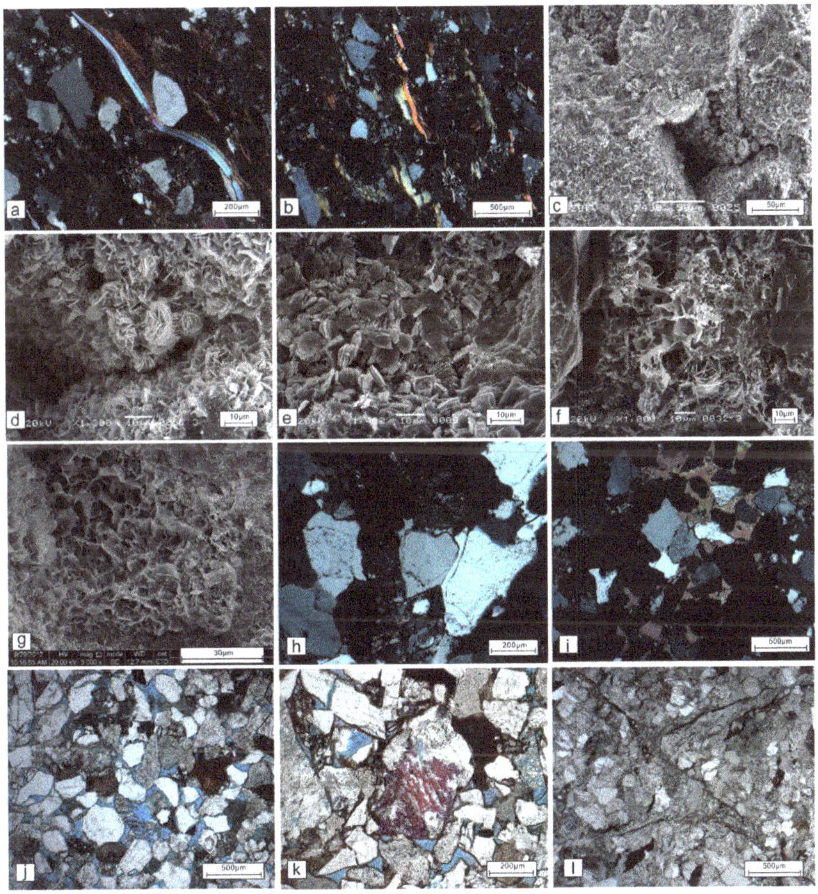

Figure 4. Thin-section microscopic analysis of the casting of the Chang 8 reservoir in Zhenjing area: (**a**) Plastic deformation of mica, Well JH36, 1376 m, orthogonal light, 100×; (**b**) directional arrangement of mica, forming laminae, Well HH157, 2037 m, Orthogonal light, 50×; (**c**) chlorite film is attached to the surface of the detrital particles and filled inside the intergranular pores, Well HH92, 2267 m, SEM, 1300×; (**d**) leaf-like chlorite film is attached to the surface of the detrital particles, Well HH92, 2267 m, SEM, 430×; (**e**) book-like kaolinite filled in intergranular pores, Well HH78, 2400 m, SEM, 1400×; (**f**) flake-like and hair-like illite filled between detrital particles, Well

HH107, 2436 m, SEM, 1000×; (**g**) honeycomb-like illite mixed layer, Well HH111, 2035 m, SEM, 3000×; (**h**) secondary enlargement of quartz, Well HH193, 2298 m, positive cross light, 100×; (**i**) Calcite pore cementation, Well HH193, 2299 m, orthogonal light, 50×; (**j**) dissolution pores formed by dissolution of feldspar, Well HH193, 2295 m, single polarized light, 50×; (**k**) calcite metasomatic feldspar, Well HH188, 2413 m, single polarized light, 100×; (**l**) intersecting microfractures, Well HH166, 2396 m, single polarized light, 50×.

① Authigenic clay mineral cementation

According to the casting slice and SEM analysis, chlorite and kaolinite are authigenic clay minerals in the Chang 8 reservoir of Zhenjing area, accompanied by some illite/smectite formation and illite. Chlorite, which was developed in the early diagenetic periods, was found in slice and foliated structures, manifested as looped lining cementation and pore-lining of cementation (Figure 4c,d). The supporting framework that was formed by the looped lining cementation has the ability to protect pores from cement filling effectively and it can also inhibit mechanical compaction to some extent. Moreover, it can be used as a separation layer between the silica-containing fluid and quartz particles to inhibit the nucleation of SiO_2 on quartz granules, as well as the secondary expansion of authigeneic quartz, and promote the storage of primary pores. This mechanism has a constructive effect on the evolution of pores (Figure 5a).

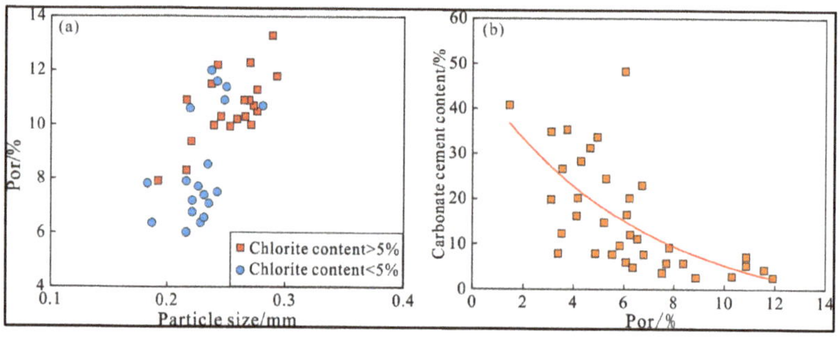

Figure 5. Analysis of clay mineral cementation in the Chang 8 reservoir [28]: (**a**) Relationship between chlorite content and porosity and particle size; (**b**) relationship between carbonate cement content and porosity.

Kaolinite cement appeared as book-like and worm-like under SEM imaging (Figure 4e). Its formation is closely related to the alteration of feldspar. Feldspar is one of the most commonly observed clastic particles in the region and it may produce kaolinite after dissolving in acid water. In other words, the occurrence of kaolinite is accompanied by the dissolution of feldspar.

The authigenic illite has a relatively low content in the Chang 8 reservoir of the Zhenjing area and it is frequently adhered onto particle surfaces as thin films, schistose, hair-like, and fibrous structures (Figure 4f). Montmorillonite illite may cause illite/smectite formation, which appeared as honeycomb shapes under SEM imaging (Figure 4g) [28].

② Siliceous cementation

Siliceous cement is the product precipitated by siliceous materials in acid fluid and it is formed by the surrounding quartz particles or regions with poor development of chlorite film. In the study area, there's a universal development of siliceous types of cement in the Chang 8 reservoir, which is mainly manifested as the secondary expansion of quartz (Figure 4h). The thickness of the expanded sides ranges between 0.02~0.2 mm. The quartz particles still exist in residual inter-granular pores after the second expansion is generated.

③ Carbonate cementation

In the study area, calcite is regarded as the major carbonate cement in the Chang 8 reservoir, with an average content of 9.56%. Carbonate types of cement produced substrate cementation, which were formed after the pore-lining chlorite in early diagenesis. Calcite is cemented among clastic particles (Figure 4i), thus forming a compacted reservoir. According to an intersecting analysis of the porosity and carbonate cement, a negative correlation was discovered (Figure 5b). Hence, it is believed that carbonate cement is the major cause of the compactness of the Chang 8 reservoir in the study area.

3.1.3. Dissolution

Secondary pores, which are produced upon dissolution of mineral components and cement in the reservoir, facilitate the large-scale expansion of the reservoir spaces. It is the most important diagenesis process to improve the physical properties of the reservoir [5,26,31,34]. After experiencing tectonic lifting and oil–gas emplacement in the Chang 8 reservoir section of the Zhenjing area, feldspar and rock debris will be dissolved and eroded by organic acids (Figure 4j), thus forming mold pores, intragranular pores, and inter-granular pores. This effect mainly occurs at the end of early diagenesis and Phase-A of middle diagenesis.

3.1.4. Metasomatism

The occurrence of metasomatism is closely related to the local temperature and pressure distributions, as well as the fluid properties in pores and it occurs in all diagenetic periods [29,32,35]. Metasomatism is a process of mutual replacement of minerals. It is accompanied by the dissolution and sedimentation effects. Hence, the influence of the metasomatism process on the physical properties of the reservoir is very small and even can be ignored. Calcite metasomatism is regarded as the most common type of metasomatism in the study area (Figure 4k), with local metasomatism of kaolinite and clay minerals.

3.1.5. Rupturing

According to the core observation and analysis of slice data under SEM imaging, microcracks that are produced by rupturing in the Chang 8 reservoir were developed greatly (Figure 4l). These microcracks connect pores in the reservoir effectively, which play an important role in the improvement of the physical properties of the reservoir (decompaction), especially the migration and settlement of oil and gas [5,30].

3.2. *Diagenetic Sequence and Diagenetic Periods*

The various diagenesis types in each diagenetic period are different and the duration of diagenesis also varies. Based on the SEM analysis and the casting slicing observation, a comprehensive analysis of the diagenetic sequence of sandstones in the Chang 8 reservoir of the Zhenjing area was carried out by combining the diagenesis theoretical knowledge [29–31].

In the early diagenetic period A, the Chang 8 reservoir section in the study area becomes more and more compact, thus resulting in the plastic deformation of minerals such as mica. The contact relation among particles changes from the original point contact into the linear contact. According to SEM observation, clastic particles such as mica align toward oriented array to form a texture layer. In this period, the porosity of the reservoir declines sharply due to compaction. In the early diagenesis period B, compaction continues to increase and chlorite film begins to produce in pores. In this period, the production of cement-like kaolinite and sparry calcite decreases the porosity of the reservoir continuously. Compared with the early diagenetic period, the influence of compaction in the middle diagenetic period A on the physical properties of the reservoir has been very small and the secondary quartz development expands. In the reservoir, production of organic acids occurs, while carbonate cement and feldspar are dissolved, thus forming multiple secondary pores. The dissolution that consumes acid water and pore water becomes increasingly alkaline, thus facilitating changes in illite/smectite formation toward illite. During this period, diagenesis plays an important role in improving the physical properties of the Chang 8 reservoir. In the middle diagenetic period B, illite/smectite formation continues to change toward illite. Cementation and dissolution occur alternatively, while the porosity tends to be stable gradually. To sum up, the diagenetic sequences of the Chang 8 reservoir in the study area are determined as follows: mechanical compaction → early sedimentation of chlorite clay mineral → early cementation of sparry calcite → authigenic kaolinite precipitation → quartz secondary expansion → dissolution of carbonate cement → dissolution of feldspar → late cementation of minerals like ferrocalcite (Figure 6).

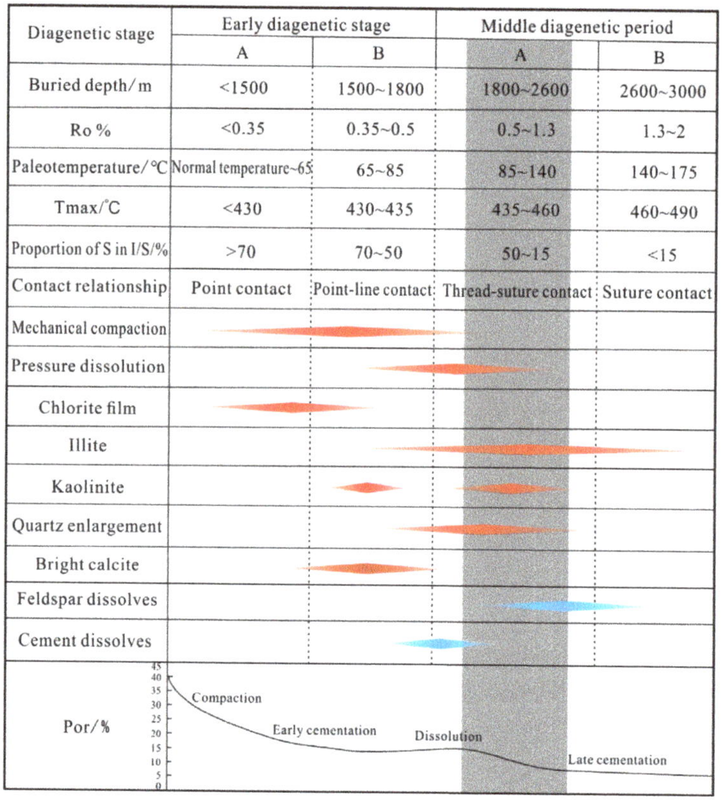

Figure 6. Diagenetic evolution of the Chang 8 reservoir in the Zhenjing area.

In this work, the J&M TIDAS PMT IV&MSP200 vitrinite reflectance test system was applied to test the vitrinite reflectance of ten rock samples (including mudstone, oil shale and coal) in the study area. The minimum and maximum reflectance values are 0.67% and 1.34%, respectively, averaging at 0.90% (Table 2). The maximum paleogeotemperature in the rock burying process was tested by acoustic emission, which ranges between 109.3~125.3 °C, averaging at 120.32 °C (Table 3). The extracted Tmax of mudstone in the Chang 8 reservoir of the study area ranges between 445~463 °C, averaging at 454 °C [38]. According to the XRD analysis, the montmorillonite content in the illite/montmorillonite formation is between 15~30%, averaging at 20% (Table 4). Moreover, according to the comprehensive analysis of the diagenetic sequence, both vitrinite reflectance and paleogeotemperature tests were performed based on the acoustic emission. With references to the relevant literature and industrial standards, it is believed that the Chang 8 reservoir in Zhenjing area is currently in the middle diagenetic period A (Figure 6).

Table 2. Table of vitrinite reflectance in Zhenjing area.

Well	Lithology	Depth (m)	Ro (%)
HH112	Gray black mudstone	2123.61	0.88
HH128-14	Dark gray mudstone	2294.66	0.91
HH128-15	Dark gray mudstone	2324.43	0.80
HH147	Gray black mudstone	2411.9	1.14
HH149	Dark gray mudstone	2354.35	0.95
HH156	Dark gray mudstone	1856.5	0.96
HH157	Grey black oil shale	2005.46	1.34
HH183	Gray black mudstone	2222.34	0.82
HH185	Dark gray mudstone	1794.83	0.88
HH198	Black coal	1819.24	1.00

Table 3. Maximum paleo-geotemperature (acoustic emission) of Chang 8 reservoir.

Well	Depth (m)	Layer	T (°C)
HH109	2335.12	C81	122
ZP1	2341.99	C82	109.3
HH106	2418.63	C82	124.7
HH107	2451.11	C82	125.3

Table 4. Statistical table of X-ray diffraction data of Chang 8 clay minerals in Zhenjing area.

Sample Number	Layer	I (%)	I/S	S (%)
HH78-2	C81	11	26	20
HH78-12	C81	10	34	20
HH78-21	C81	9	21	20
HH78-31	C81	9	24	20
HH78-41	C81	9	30	20
HH92-4	C81	6	14	20
HH92-9	C81	6	7	20
HH92-13	C81	9	11	25
HH92-24	C81	4	5	20
HH92-32	C81	7	8	25
HH92-41	C81	6	9	20
HH107-03	C81	14	36	20
HH107-11	C81	12	57	30
HH107-17	C81	13	39	20
HH107-26	C81	11	28	20
HH107-35	C81	13	41	20
HH107-39	C81	14	43	15
HH107-52	C81	16	38	20
HH107-65	C82	14	24	20
HH107-78	C82	11	26	15
HH107-91	C82	12	20	15
HH107-104	C82	13	20	15
HH109-3	C81	6	13	20
HH109-14	C81	7	6	20
HH109-25	C81	6	14	20
HH109-33	C81	7	23	20

4. Determination of Hydrocarbon Accumulation Periods

In this work, hydrocarbon accumulation periods of the Chang 8 reservoir in the study area were studied systematically by applying the indirect limiting method of the thermal evolution history-inclusion temperature measurement and the direct dating method of illite [39–44]. First, the freezing-point and homogenization temperatures of rock samples from 12 wells (including HH188 and HH156) in the Chang 8 reservoir section were tested.

The LINKM600 cold-heat table in the Thermal Chronology Laboratory of Northwestern University was used under the enforcement of 10.5 V of voltage, 26 °C of indoor temperature, and 65% of humidity. Meanwhile, the corresponding salinity was calculated. In this section, the saline inclusion, which coexists with the hydrocarbon inclusion was chosen as the test object. It is mainly reserved at the quartz expansion edges or inside of fracture and quartz. According to the analysis of the experimental results, the homogeneous temperature of inclusion has a wide range of 69~155 °C, with peaks ranging between 100~125 °C. This result and the reproduction diagram of the burying history were used together for mutual calibration, through which the accumulation period of the Chang 8 reservoir was 110~120 Ma (Figure 7).

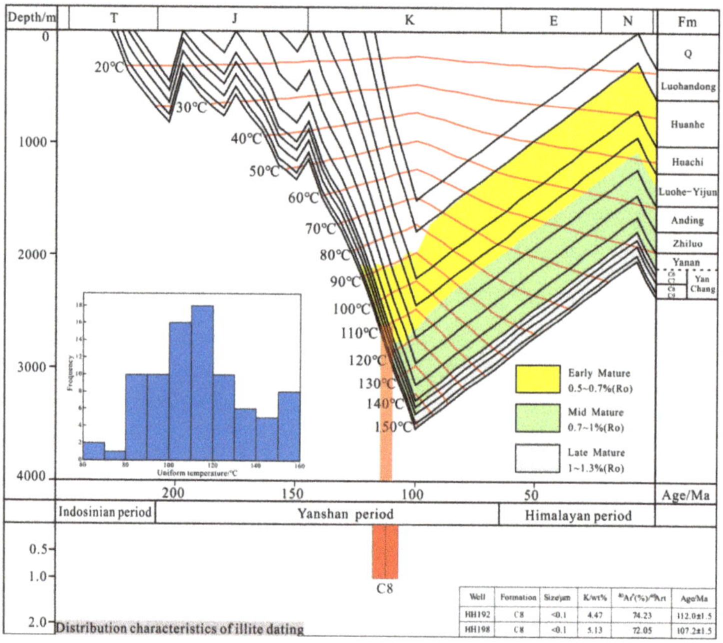

Figure 7. Analysis of thermal evolution history of Well HH155.

When both oil and gas enter the reservoir, the growth of authigenic illite stops. Hence, this period is viewed as the time for hydrocarbon accumulation [44–50]. In this work, illite test analysis was accomplished in the Geochemistry Laboratory of China University of Petroleum (Beijing). The 38AR diluent was put in accurately by using the VG3600 mass spectrometer while melting the samples under 1500 °C. Later, the isotope ratios of (38Ar/36Ar) and (40Ar/38Ar) were tested. The radioactive factor 40Ar of the samples was calculated and the corresponding age was calculated according to the K content [44,47–50]. Based on the above principle, a dating analysis was carried out on the oil-containing sandstone illite in HH192 and HH198. The acquired results showed that the accumulation period of the Chang 8 reservoir was 105~115 Ma (Figure 7). Consistent with the inclusion analysis outcomes, it is concluded that the Chang 8 reservoir in the study area is a phase-I continuous accumulation process and the accumulation period is 105~125 Ma, which is the middle Early Cretaceous Epoch (Figure 7).

5. Conclusions

(1) According to the borehole core observation, SEM and XRD analysis, the lithology of the Chang 8 reservoir section in the study area is mainly dominated by feldspar rock debris, and rock debris arkose. Quartz, feldspar, and rock debris are the major clastic constituents. Among them, quartz has the highest content (about 29%), followed by feldspar (26%). The content of the rock debris is the lowest, averaging at 24%. Among the sandstone clay minerals, chlorite content is the highest, averaging at 36.3%. The illite/smectite formation and kaolinite possess the lower content and the illite content is the lowest, averaging at 6.85%.

(2) According to the SEM, XRD, vitrinite reflectance and rock pyrolysis analysis, the major diagenesis effects of the Chang 8 reservoir section in the Zhenjing area mainly include the following procedures: compaction, cementation, dissolution, metasomatism, and rupturing. Consequently, the proposed diagenetic sequence is the following: mechanical compaction → early sedimentation of chlorite clay mineral film → early cementation of sparry calcite → authigenic kaolinite precipitation → quartz secondary expansion → dissolution of carbonate cement → dissolution of feldspar → late cementation of minerals like ferrocalcite. Now, the reservoir is in the middle diagenetic period A.

(3) Based on the thermal evolution history analysis and the illite dating method, the hydrocarbon accumulation periods of the Chang 8 reservoir are analyzed comprehensively. The Chang 8 reservoir in the study area is a phase-I continuous accumulation process and the accumulation period is 105~125 Ma, which is the middle Early Cretaceous Epoch.

Author Contributions: Supervision, Z.R. and K.Q.; Writing—review & editing, G.Y. All authors have read and agreed to the published version of the manuscript.

Funding: This research was funded by National Natural Science Foundation of China (No. 41630312), National Key R&D Projects (No. 2017YFC0603106) and National Major Projects (No. 2017ZX05005002-008).

Institutional Review Board Statement: Not applicable.

Informed Consent Statement: Informed consent was obtained from all subjects involved in the study.

Acknowledgments: We appreciate encouragement and guidance from Zhanli Ren, during the formulation and drafting of this paper.

Conflicts of Interest: The authors declare no conflict of interest.

References

1. Zhang, Y.Y.; Ren, Z.L.; He, F.Q.; Cui, J.P.; Yang, G.L.; Wang, K.; Ji, Y.Y. Meso-Cenozoic tectonic features in the structural turning area of the Craton Basin and its significance for controlling reservoirs: Taking the Yanchang Formation in Zhenjing area in the southwest of the Ordos Basin as an example. *Chin. J. Petrol.* **2020**, *36*, 3537–3549.
2. Chen, Q.H. Study on the Upper Paleozoic Sedimentary System and Oil and Gas Accumulation Law in the Ordos Basin. Ph.D. Thesis, Northwestern University, Xi'an, China, 2007.
3. Liu, Y. Fracture Characteristics of Low Permeability Reservoirs and Their Control on Oil and Gas Enrichment. Ph.D. Thesis, Chengdu University of Technology, Chengdu, China, 2013.
4. He, F.Q.; Liang, C.C.; Lu, C.; Yuan, C.Y.; Li, X.W. Identification and description of fault-fracture bodies in tight-low permeability oil reservoirs in the transition zone in the southern margin of the Ordos Basin. *Oil Gas Geol.* **2020**, *41*, 710–718.
5. Liang, C.C.; Guo, J.X. Diagenesis and reservoir characteristics of tight sandstone in Chang 81 sublayer of Yanchang Formation, Honghe Oilfield, Ordos Basin. *Oil Gas Geol. Recovery* **2017**, *24*, 57–63.
6. Guo, M.L.; Chen, Y.; Zheng, Z.H.; Li, J.; Yu, L.; Liu, L.Q. Rapid evaluation method of recoverable reserves probability in tight oil reservoirs—Taking Chang 8 reservoir in Honghe Oilfield as an example. *Pet. Exp. Geol.* **2021**, *43*, 154–160.
7. Xue, C.; Qi, G.M.; Wei, A.J.; Hu, X.Q. Research progress and application of sedimentary diagenesis. *Groundwater* **2014**, *36*, 186–189.
8. Zhu, Y.X.; Jin, Z.K.; Jin, K.; Guo, Q.H.; Wang, H.; Lv, P.; Wang, X.Y.; Shi, Y. Petrological characteristics and diagenetic evolution of fine-grained sedimentary rocks in continental lake basins in China: A case study of the Lower Jurassic Da'anzhai Member in Yuanba area, Sichuan Basin. *Pet. Nat. Gas Geol.* **2021**, *42*, 494–508.
9. Liu, D.K. Diagenetic Evolution and Response Mechanism of Hydrocarbon Charging and Microscopic Pore Throat Structure in Tight Sandstone Reservoirs. Ph.D. Thesis, Northwestern University, Xi'an, China, 2019.

10. Song, H.Y.; Ji, Y.L.; Zhou, Y. Reservoir characteristics and diagenetic evolution sequence of the third member of Shahejie Formation in Dongpu Sag. *Zhongwai Energy* **2021**, *26*, 27–34.
11. Gao, H. Diagenetic evolution characteristics of buried hill tight sandstone reservoirs in Dongpu Sag. *Fault Block Oil Gas Fields* **2021**, *28*, 295–299, 317.
12. Wang, E.Z.; Wu, Z.B.; Song, Y.C.; Shi, K.B.; Liu, H.Y.; Liu, B. Diagenetic evolution and pore structure characteristics of tight sandstone in Chang 7 Member in Qingcheng area, Ordos Basin. *J. Peking Univ. (Nat. Sci. Ed.)* **2022**, 1–13. [CrossRef]
13. Zhu, R.J.; Li, R.X.; Liu, X.S.; Yang, M.Y.; Qin, X.L.; Wu, X.L.; Zhao, B.S.; Liu, F.T. Diagenetic evolution characteristics and physical property evolution of Upper Paleozoic tight sandstone gas reservoirs in the southwestern Ordos Basin. *J. Lanzhou Univ. (Nat. Sci. Ed.)* **2021**, *57*, 637–649, 658.
14. Wang, R.Y.; Hu, Z.Q.; Bao, H.Y.; Wu, J.; Du, W.; Wang, P.W.; Peng, Z.Y.; Lu, T. Diagenetic evolution and storage control of key minerals in the Upper Ordovician Wufeng Formation-Lower Silurian Longmaxi Formation shale in the Sichuan Basin. *Pet. Exp. Geol.* **2021**, *43*, 996–1005.
15. Ren, D.Z.; Sun, W.; Wei, H. Diagenetic characteristics of Chang 81 sandstone reservoir in Huaqing Oilfield. *Geol. Sci. Technol. Inf.* **2014**, *33*, 72–79.
16. Shi, J.A.; Wang, J.P.; Mao, M.L. Study on diagenesis of reservoir sandstone in Chang 6-8 member of Triassic Yanchang Formation in Xifeng Oilfield, Ordos Basin. *Chin. J. Sediment.* **2003**, *21*, 373–379.
17. Li, Z.; Han, D.L.; Shou, J.F. Diagenesis system of sedimentary basin and its spatiotemporal properties. *Chin. J. Petrol.* **2006**, *22*, 2151–2164.
18. Bu, J.; Li, W.H.; Zeng, M. Reservoir diagenesis and its influence on pores in the Middle Jurassic Yan 9 oil formation in Longdong area, Ordos Basin. *Pet. Geol. Eng.* **2010**, *24*, 24–27.
19. Zheng, Q.H.; Liu, Y.Q. Diagenesis and diagenetic facies of Chang 4 + 5 tight oil layer in Yanchang Formation, Zhenbei area, Ordos Basin. *Chin. J. Sediment.* **2015**, *33*, 1000–1012.
20. Liu, R.C.; Ren, Z.L.; Ma, Q.; Zhang, Y.Y.; Qi, K.; Yu, C.Y.; Ren, W.B.; Yang, Y. Study on hydrocarbon accumulation periods of Yanchang Formation in the southern Ordos Basin. *Mod. Geol.* **2019**, *33*, 1263–1274.
21. Tang, J.Y.; Zhang, G.; Shi, Z.; Zhang, X.; Chen, Y.B. Characteristics of fluid inclusions and hydrocarbon accumulation stages in the Yanchang Formation in the rich Sichuan area, Ordos Basin. *Lithol. Reserv.* **2019**, *31*, 20–26.
22. Song, S.J.; Liu, S.; Liang, Y.X. Stages and periods of hydrocarbon accumulation in Chang 8 tight oil layer in southwestern Ordos Basin. *Fault Block Oil Gas Fields* **2018**, *25*, 141–145.
23. Luo, C.Y.; Luo, J.L.; Luo, X.R.; Bai, X.J.; Lei, Y.H.; Cheng, M. Characteristics of fluid inclusions and hydrocarbon accumulation period in Chang 8 sandstone in the central and western Ordos Basin. *J. Geol. Univ.* **2014**, *20*, 623–634.
24. Zhang, X.L. Analysis of Reservoir Characteristics and Controlling Factors of Oil and Gas Enrichment in Chang 8 Member of Yanchang Formation in Honghe Oilfield, Ordos Basin. Master's Thesis, Northwestern University, Xi'an, China, 2018.
25. Li, C. Characteristic Evaluation of Chang 6 Tight Sandstone Reservoirs in Huangling Block, Ordos Basin. Ph.D. Thesis, Northwestern University, Xi'an, China, 2020.
26. Jiang, H.X.; Wu, Y.S.; Luo, X.R. Formation of Chang 8 oil layer of Triassic Yanchang Formation in central and southern Ordos Basin and its control on reservoir physical properties. *Sediment. Tethys Geol.* **2007**, *27*, 107–114.
27. Liang, Y.; Ren, Z.L.; Shi, Z.; Zhao, X.Y.; Yu, Q.; Wu, X.Q. Hydrocarbon accumulation period of Yanchang Formation in Fuxian-Zhengning area, Ordos Basin. *Chin. J. Pet.* **2011**, *32*, 741–748.
28. Xiong, D.; Ding, X.Q.; Zhu, Z.L.; Le, J.P. Study on diagenesis of Chang 8 tight sandstone reservoir in Zhenjing area. *Lithol. Reserv.* **2013**, *25*, 31–36, 43.
29. Xu, M.L.; He, Z.L.; Yin, W.; Wang, R.; Liu, C.Y. Characteristics and main controlling factors of tight sandstone reservoirs in Chang 8 member of Yanchang Formation in Zhenjing area, Ordos Basin. *Oil Gas Geol.* **2015**, *36*, 240–247.
30. Wang, F.B.; Yin, W.; Chen, C.F. Genetic mechanism of "sweet spots" in tight sandstone reservoirs of Chang 8 oil formation in Honghe Oilfield, Ordos Basin. *Pet. Exp. Geol.* **2017**, *39*, 484–490.
31. Liu, C.L.; Liu, X.; Zhang, L.N.; Chen, Z.L. Clastic diagenesis and its influence on reservoirs: A case study of Zhenjing area, Ordos Basin. *Pet. Exp. Geol.* **2017**, *39*, 348–354.
32. Zhang, M.T.; Li, H.; Li, W.H.; Bai, J.L.; Tian, W.; Qi, K. Study on diagenesis and pore quantification of Chang 81 reservoir in Yanchang Formation, Jingchuan area, Ordos Basin. *Geol. Sci. Technol. Inf.* **2017**, *36*, 98–105.
33. Wang, M.P.; Xia, D.L.; Wu, Y.; Pang, W.; Zou, M. Diagenetic characteristics of Chang 8 tight sandstone reservoirs in Honghe Oilfield, Ordos Basin. *Pet. Exp. Geol.* **2018**, *40*, 786–792 + 835.
34. Xiao, H.; Wang, H.N.; Yang, Y.D.; Ke, C.Y.; Zhe, H.Q. Pore evolution characteristics of tight sandstone and its influence on reservoir quality by diagenesis—Taking Chang 8 reservoir of Malingnan Yanchang Formation in Ordos Basin as an example. *Pet. Exp. Geol.* **2019**, *41*, 800–811.
35. He, Y.C.; Zhao, J.X.; Guan, D.B.; Jia, H.C. Reservoir characteristics of Chang 8 and Chang 6 members in Zhenjing area and the transformation of pores by diagenesis. *J. Chengdu Univ. Technol. (Nat. Sci. Ed.)* **2021**, *48*, 217–225.
36. Zhao, Z.M. Study on the Heterogeneity and Development Characteristics of Chang 4+5-Chang 6 Reservoirs in Hujianshan Area. Master's Thesis, Xi'an Shiyou University, Xi'an, China, 2018.
37. Zhang, B. Reservoir Characteristics of Dujiatai Oil Layer in Shu 103 Block in Liaohe Depression and Its Controlling Effect on Oil and Gas Distribution. Master's Thesis, Northeast Petroleum University, Daqing, China, 2020.

38. Ji, Y.Y.; Gao, Y.L.; Zheng, K. Diagenetic evolution and densification of Chang 812 low-permeability clastic reservoirs in the northeast of Zhenjing area. *Unconv. Oil Gas* **2020**, *7*, 11–17.
39. Guo, F.F. Characteristics and accumulation stage of fluid inclusions in the Hetaoyuan Formation of the Paleogene Hetaoyuan Formation in Nanyang Sag, Nanxiang Basin. *Nat. Gas Geosci.* **2022**, 1–11. [CrossRef]
40. Li, B.; Cui, J.P.; Li, Y.; Li, J.S.; Zhao, J.; Chen, Y.W. Analysis of hydrocarbon accumulation period of Yanchang Formation in Wuqi area of Yishan Slope. *Lithol. Reserv.* **2021**, *33*, 21–28.
41. Zhang, Z.Q.; Liu, H.; Ma, L.C.; Liu, J.D.; Guo, Z.Y. Reservoir Stage and Process of Buried Hill Oil and Gas Reservoirs in Jiyang Depression, Bohai Bay Basin: Evidence from Reservoir Fluid Inclusions. *Pet. Exp. Geol.* **2022**, *44*, 129–138.
42. Ma, L.Y.; Qiu, G.Q.; Liu, C.Y.; Hu, C.Z.; Luo, Y. Reservoir densification and petroleum accumulation in the Yanchang Formation of Honghe Oilfield, Ordos Basin. *Chin. J. Sedimentol.* **2020**, *38*, 620–634.
43. Chen, R.Q.; Liu, G.D.; Sun, M.L.; Cao, Y.S.; Liu, X.B.; Li, Q. Study on Mesoproterozoic Fluid Inclusions and Hydrocarbon Accumulation Stages in the Northern Jibei Depression. *Geol. J. Univ.* **2022**, *28*, 64–72.
44. Zhao, Y.D.; Qi, Y.L.; Luo, A.X.; Cheng, D.X.; Li, J.H.; Huang, J.X. Reconstruction of hydrocarbon charging history of Jurassic reservoirs in the Ordos Basin using fluid inclusions and authigenic illite dating. *J. Jilin Univ. (Earth Sci. Ed.)* **2016**, *46*, 1637–1648.
45. Yu, M.D.; Wang, P.J.; Shi, C.R.; Zhang, H.; Tang, H.F.; Li, F.X. Inclusion characteristics and illite dating in Yanqi Basin as an indication of hydrocarbon accumulation stage. *J. Jilin Univ. (Earth Sci. Ed.)* **2009**, *39*, 45–52.
46. Cui, J.P.; Ren, Z.L.; Chen, Q.H.; Xiao, H. Analysis of hydrocarbon accumulation periods in Wuerxun Sag, Hailar Basin. *J. Northwest. Univ. (Nat. Sci. Ed.)* **2007**, *37*, 465–469.
47. Xiang, C.F.; Feng, Z.H.; Wang, F.D.; Zhang, S.; Peng, P.; Liang, X.D. Tectonic-controlled rapid hydrocarbon accumulation in late stage: Evidence of fluid inclusions and authigenic illite in Daqing Changyuan, Songliao Basin. *Chin. J. Geol.* **2012**, *86*, 1799–1808.
48. Li, J.J.; Wang, Y.; Li, H.L.; Zhang, W.B. Application of isotope dating in the study of hydrocarbon accumulation period. *Pet. Exp. Geol.* **2012**, *34*, 84–88.
49. Chen, G.; Xu, L.M.; Ding, C.; Zhang, H.R. Using authigenic illite dating to determine the hydrocarbon accumulation period of the Permian in the northeastern Ordos Basin. *Pet. Nat. Gas Geol.* **2012**, *33*, 713–719, 729.
50. Zhang, Y.Y.; Horst, Z.; Liu, K.Y.; Luo, X.Q. Comparison of authigenic illite K-Ar and Ar-Ar dating techniques and prospect of their application—Taking Sulige gas field as an example. *Chin. J. Pet.* **2014**, *35*, 407–416.

energies

MDPI

Article

Quantitative Analysis of Cenozoic Extension in the Qiongdongnan Basin, South China Sea: Insight on Tectonic Control for Hydrocarbon Reservoir Accumulation and Formation

Yan Zhang [1,2], Li Zhang [1,2,*], Lijun Mi [3], Xiangyang Lu [4], Shiguo Wu [4], Lishan Tang [5], Jie Zhou [5], Xiaofeng Xiong [5] and Jitian Zhu [5]

1 Guangzhou Marine Geological Survey, National Engineering Research Center for Gas Hydrate Exploration and Development, Guangzhou 511458, China; zyan1@mail.cgs.gov.cn
2 Southern Marine Science and Engineering Guangdong Laboratory (Guangzhou), Guangzhou 511458, China
3 Beijing Institute of Research, China National Offshore Oil Corporation (CNOOC) Limited, Beijing 571000, China; milijun@cnooc.com.cn
4 Institute of Deep-Sea Science and Engineering, Chinese Academy of Sciences, Sanya 572000, China; luxy@idsse.ac.cn (X.L.); swu@idsse.ac.cn (S.W.)
5 Hainan Branch of China National Offshore Oil Corporation (CNOOC) Limited, Haikou 571000, China; tanglishan@cnooc.com.cn (L.T.); zhoujie@cnooc.com.cn (J.Z.); xiongxiaof@cnooc.com.cn (X.X.); zhujitian@cnooc.com.cn (J.Z.)
* Correspondence: zhangli@mail.cgs.gov.cn

Abstract: Cenozoic extension rates were calculated based on 20 seismic profiles across the Qiongdongnan Basin, South China Sea. The results confirmed that the Cenozoic rifting in the Qiongdongnan Basin exhibited multistage extension and spatiotemporal variation. In terms of the N–S striking seismic profiles, the structural forms of the northern and southern sags of the basin were characterized by narrow half grabens, while the structure at the center sag of the basin was characterized by wide and gentle grabens. The fault strikes in the west of the basin were mainly northeast–southwest trending, whereas those in the east of the basin changed from northeast–southwest trending to nearly east–west trending. The extension rate in the east sag was higher than that in the west area. The extension rate in the middle part was lower relative to the east and west sags. This was because the rifting was controlled by the distribution of the main boundary fault along the basin. Temporally, the Cenozoic extension could be divided into three periods: Eocene, Oligocene, and Miocene. The amount of stretching in the three extension stages was unevenly distributed in the entire basin. The maximum was mainly in the Oligocene Lingshui and Yacheng Formations. The Oligocene extension occurred in the entire basin, and the Eocene extension was limited to the Ledong and Changchang sags. Significant fault activity could be observed during the deposition period of the Yacheng and Lingshui Formations and could be attributed to strong extensional activity. The rifting tectonics controlled the distribution of source rocks and oil-generating window as well as hydrocarbon generation, reservoir formation, and accumulation.

Keywords: extension; structure; fractal; hydrocarbon reservoir; Qiongdongnan Basin; South China Sea

Citation: Zhang, Y.; Zhang, L.; Mi, L.; Lu, X.; Wu, S.; Tang, L.; Zhou, J.; Xiong, X.; Zhu, J. Quantitative Analysis of Cenozoic Extension in the Qiongdongnan Basin, South China Sea: Insight on Tectonic Control for Hydrocarbon Reservoir Accumulation and Formation. *Energies* **2022**, *15*, 4011. https://doi.org/10.3390/en15114011

Academic Editor: Renato Somma

Received: 7 April 2022
Accepted: 20 May 2022
Published: 30 May 2022

1. Introduction

In recent years, mathematical methods have been applied in petroleum geology. Quantitative calculation data play an increasingly important role in petroleum systems [1,2]. Studies on petroleum geology are conducted using complete data models [3,4]. A complete data model is typically adopted to contrast between petroleum geological data. Although the volume of data reported herein does not reach the standards of mass (significant amounts of data), the data are nonetheless geological big data (focus on the whole, on efficiency, and on the correlation). The conventional method is based on causality, whereas the method of big data studies correlations [5]. Typically, reasoning involves human factors, and the result of reasoning may be incorrect when analyzing causes and results. However, petroleum geology data are objective and are not affected by human factors. Human factors are involved only when a causal relationship is explored. Big data research can aid conventional data research and thus promote the progress of science and technology [6]. The transformation from conventional data research to big data research and from theory-driven models to data-driven models represents a significant change in research methods and ideas, paving the way for new scientific innovations in the field of petroleum geology [6].

The fault system in the Qiongdongnan Basin can be divided into three groups based on the strikes: NE-trending, NE-trending, and nearly EW-trending faults (Figure 1). The NE-trending faults are dominant. The fault strikes in the western part are mainly NNE trending, while those in the eastern part are mainly EW trending. Seismic data have shown a decrease in the fault activity since the Middle Miocene. Spatially, the fault strikes are mainly east trending in the north of the west area and nearly east–west trending in the east area (Figures 2 and 3). Currently, the Qiongdongnan Basin has a wide exploration area with various datasets including massive amounts of geological, geophysical, drilling well, logging, cutting, and on-site observation data [7–11]. Such data are not only expected to open new directions for the study area, but can also promote the interdisciplinary field of mathematical geology and petroleum geology [12].

Basin extension is an important parameter in quantitative extensional basin modeling, which describes the relationship between lithospheric dynamics and basin structural characteristics, and how to determine the basin extension is an important link [13,14]. Generally, when the restoration of balance is carried out, only the extension of the master fault caused by the Earth's crust is calculated, and small faults are often ignored, particularly unrecognized faults that are difficult to identify from the seismic wave profile. Previous studies have shown that the cumulative extension of small faults that are difficult to identify at the seismic section can be as high as 30% of the basin extension [15]. If these small faults are ignored, the calculated basin extension can have considerable errors, which can affect the analysis results of the tectonic evolution of the entire basin. Therefore, it is necessary to calculate the amount of displacement of small faults.

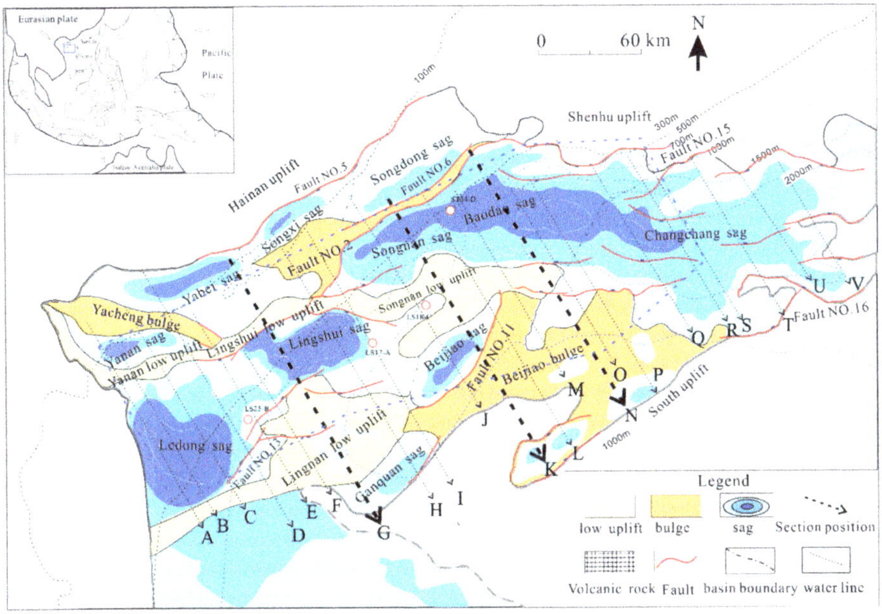

Figure 1. Tectonic unit of the Qiongdongnan Basin and 20 seismic track lines mentioned in the discussion (A to V indicate the section direction in sequence). Coarse line is given in Figure 2.

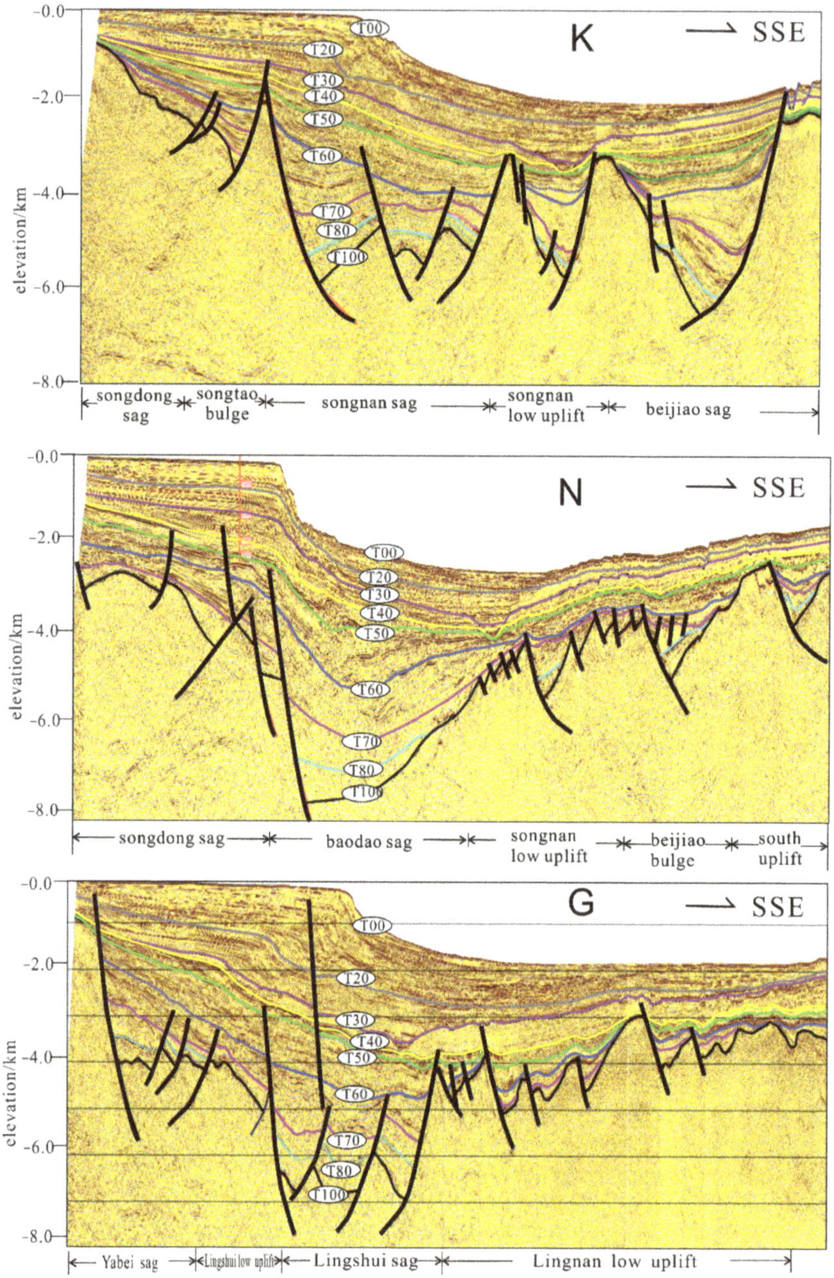

Figure 2. Seismic profiles across the Qiongdongnan Basin. Upper: seismic section across the Songtao bulge, Songnan sag, Yanan low uplift, Songnan low uplift, and Beijiao sag; down: seismic section across the Yabei sag, Lingshui low uplift, Lingshui sag, and Lingnan low uplift. See position in Figure 1. (The three sections from top to bottom correspond to the three lines K, N, and G, respectively, in Figure 1).

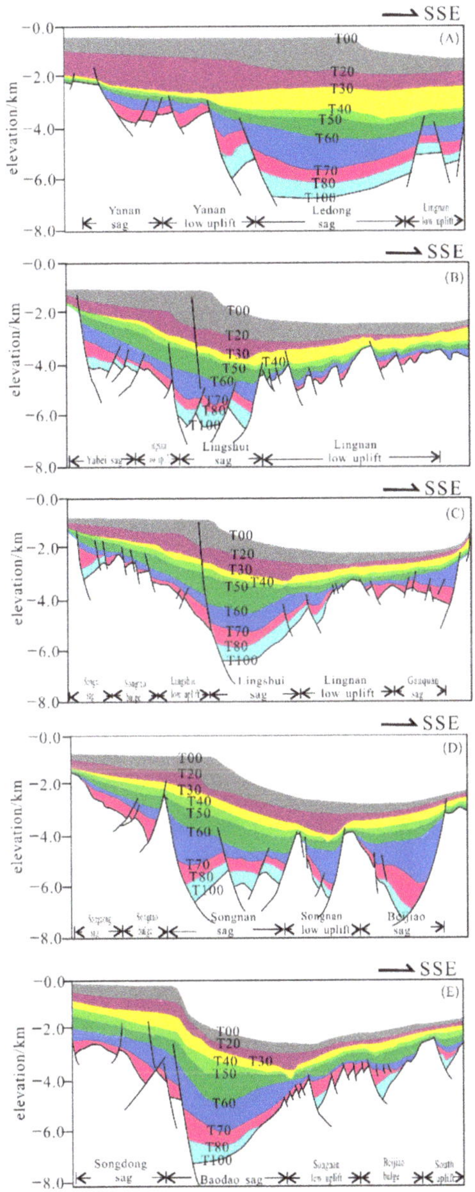

Figure 3. *Cont.*

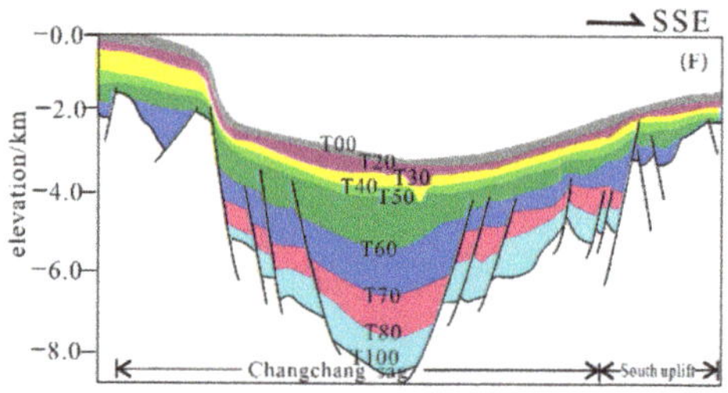

Figure 3. Interpretation of seismic profiles across the Qiongdongnan Basin. (**A**) Seismic section across the Yanan sag, Yanan low uplift, Ledong sag, and Lingnan low uplift, (**B**) seismic section across the Yabei sag, Lingshui low uplift, Lingshui sag, and Lingnan low uplift, (**C**) seismic section across the Songxi sag, Songtao bulge, Lingshui low uplift, Lingshui sag, Lingnan low uplift, and Ganquan sag, (**D**) seismic section across the Songdong sag, Songtao bulge, Songnan sag, Songnan low uplift, and Beijiao sag, and (**E**) seismic section across the Songdong sag, Baodao sag, Songnan low uplift, Beijiao bulge, and south uplift. (**F**) Seismic section across the Changchang sag, and south uplift. See position in Figure 1.

In this study, the Cenozoic extension rates were calculated based on 20 seismic profiles across the Qiongdongnan Basin, South China Sea. The fractal stretching quantity calculation method was used to estimate the cumulative extension of the small faults in this region. Additionally, we discuss the contribution of the rifting tectonics to the distribution of source rocks and oil-generating window as well as the accumulation and formation of hydrocarbon reservoirs in the basin.

2. Geological Background

The Qiongdongnan Basin is a Cenozoic rift basin that developed above the pre-Paleogene basement and experienced syn-rifting and post-rifting thermal subsidence [7,14,16–18]. Generally, it is shown as the east–west structural zone. The sags in the two sides of this basin mainly comprise half grabens, e.g., the Yabei sag and Beijiao sag. However, some sags in the central part of the basin can be characterized by graben structures (Figures 1 and 2). In the east–west direction, the western basin is dominated by half grabens, whereas the eastern part has graben structures [3,4,19]. Four first-order structural units are the northern depression belt, the central uplift zone, the central depression zone, and the southern uplift zone from north to south, respectively (Figures 1–3).

The formation and evolution of the Qiongdongnan Basin are affected by regional tectonics such as the Indo–Eurasian plate collision, south China continental margin rift, and South China Sea spreading. An initial rift developed in the northern margin since the Eocene, mainly distributed in the NNE–NE direction [20–22]. In the early Oligocene, the subsidence of the Qiongdongnan Basin was intensified by the Indo–Eurasian plate rotary extrusion and Indochina extrusion [23–25]. In the late Oligocene, the fault activity gradually diminished, mainly in the NW direction. The eastern region was affected by the expansion effect of the South China Sea, the mantle uplift was significant, and magmatic intrusion activity increased [9,26]. The Neogene entered the post-rift depression period, and the early basin underwent regional subsidence. Since the late Miocene, the concave boundary fault (No. 2 fault) in the northern part of the early fault accelerated the activities, which contributed to the development of a deepwater continental slope and the formation of a typical continental shelf-slope sedimentary system. The central depression belt and the southern uplift area became deepwater areas [27].

The Lower Miocene Sanya Formation and Middle Miocene Meishan Formation were mainly shallow sea deposits (Figure 4). However, the Upper Miocene Huangliu Formation in most of the basin formed a typical continental shelf-slope system and developed deepwater continental slope deposits. The central canyon in the middle part of the depression was in the peak developmental period; the turbidite channel sand and mass flow were widely distributed. The Pliocene–Quaternary Yinggehai Formation formed semi-deep-sea deposits, and large submarine fans were developed locally in the Yinggehai Formation [28].

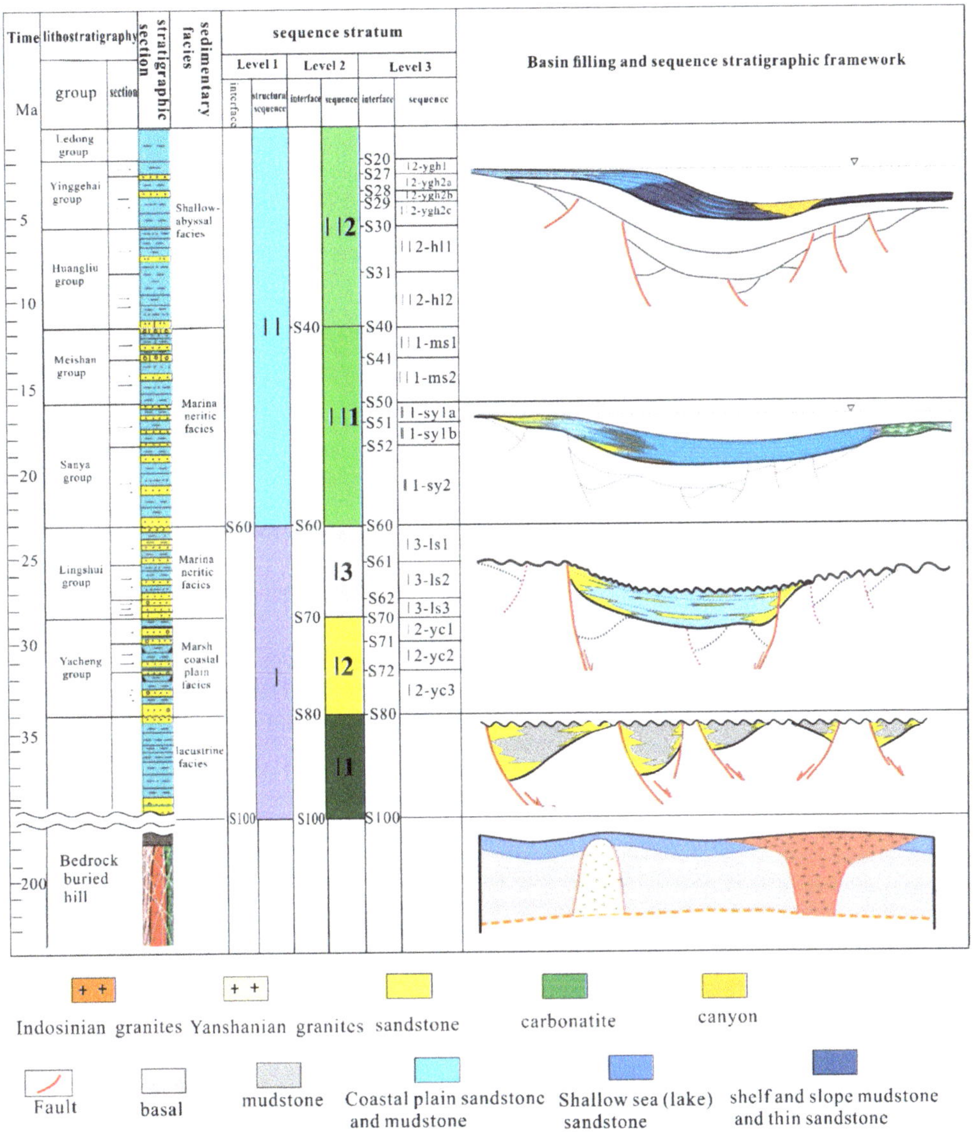

Figure 4. Cenozoic sequence framework in the Qiongdongnan Basin (blue represents mudstone, and yellow represents sandstone. Interface means the stratum interface. Sequence stratigraphy is divided into first-order, second-order, and third-order sequences. Levels 1, 2, and 3 represent the first-order, second-order, and third-order sequences, respectively.).

In the past ten years, significant progress has been made in the basic research of basin tectonic evolution, basin formation, hydrocarbon generation, and reservoir formation [29–31]. New seismic data verified that the deep detachment in the northern continental margin of the South China Sea controls the formation of large-scale basin groups, and large-scale detachment leads to strong thinning of the crust and uplift of the asthenosphere. Thereafter, the faults lead to crustal thinning and high heat flow in deepwater areas [8,18,31], and there is large-scale rapid hydrocarbon generation from source rocks [11]. Significant breakthroughs have been made in basin exploration and large- and medium-size gas fields; for example, Lingshui 17-A and 25-B gas fields have been discovered, with exploration in the Qiongdongnan Basin entering a fast developmental period. However, the drilling results over the past two years have been rather unsatisfactory. The Lingshui 18-C medium-size gas field was discovered around the central canyon and its surrounding deepwater areas. In addition, the Songtao 34-D gas-bearing structure was discovered in the Songtao uplift north of the Songnan sag in a shallow water area. Despite the achievements made, for further breakthroughs in exploration, it is necessary to re-study the basic geological conditions of the basin and summarize and reflect the conditions and scenarios of hydrocarbon accumulation. Studying the structure and regional difference is key to hydrocarbon generation, reservoir formation, and accumulation as well as being an important factor for oil and gas exploration in the basin.

3. Data and Methods

3.1. Dataset

The NW–SE striking seismic profiles across the Qiongdongnan Basin were acquired from the CNOOC Zhanjiang Branch. Two-dimensional seismic data corresponding to an area of 36,670 km^2 in combination with 3D seismic data were used in the calculation and discussion. The bin spacing of the 3D volumes was 12.5 m in the in-line direction and 25 m in the cross-line direction. Prestacked time-offset 2D seismic profiles with densities ranging from 1 km $\times$ 1 km to 3 km $\times$ 8 km were collected at the edge of the depression without 3D seismic data coverage. The parameters for the 2D line acquisition were as follows: 7.5 km long streamer, 12.5 m track distance, 2 ms sampling interval, and record length of 12 s in TWT. All the seismic data are displayed with zero phase, prestack depth migration, and SEG positive polarity, and the acoustic impedance increases downward [32].

Based on the latest drilling data from the deep waters east and west of the Qiongdongnan Basin, a comparative analysis of the oil and gas indicators, namely methane, ethane, propane, n-butane, isobutane, n-pentane, isopentane, equilibrium ratio, characteristic ratio, micromigration index, heavy hydrocarbon, light hydrocarbon to heavy hydrocarbon ratio, and drying ratio, in these waters was conducted. This was conducted to indicate the difference in hydrocarbon accumulation between the east and west.

3.2. Methods

3.2.1. Calculation of the Amount of Extension

The Move software has been widely used in structural evolution analyses. The balanced profile recovery [33] typically adopts a 2D module in the Move software, and the selected profile must be in the vertical direction toward the regional tectonic line to calculate the basin extension.

Parameters, such as the extension amount, rate of extension, and extension rate, are typically used to characterize the horizontal extension amount of the extensional basin. The profile length before deformation is denoted by L_0, the length after extensional tectonic deformation is denoted by L_1, and the profile length changes from L_0 to L_1 [34].

The formulae for the amount, rate, and rate of elongation are as follows:

$$\Delta L_i = L_{1i} - L_{0i} \quad (1)$$

$$R_i = \Delta L_i / \Delta L \quad (2)$$

$$r_i = \Delta L_i / L_{0i} \quad (3)$$

$$v_i = \Delta L_i / \Delta t_i \quad (4)$$

where ΔL_i is the extension of the layer i, L_{1i} and L_{0i} are the length of the section after the extension and the original section length, respectively, ΔL is the total extension, R_i is the proportion of the extension rate of layer i to the total extension, R_i is the extension rate of layer i, v_i is the average stretching rate of layer i, and t_i is the stretching time of layer I [12].

3.2.2. Fractal Computing

Basin extension is calculated using three methods: basin geothermal subsidence analysis, crustal thickness change measurement, and fault equilibrium restoration [35]. The first two methods provide results close to the actual extension. Because the third method involves producing the seismic profile, it is simple and therefore often used. The lack of consideration of the cumulative extension of small faults that cannot be resolved using seismic data is the main reason for the errors [36]. Previous studies have shown that the cumulative extension of small faults that cannot be determined by the reflected seismic data can reach 30–40% of the actual extension of the basin [15]. Therefore, the extension of these small faults cannot be ignored. As such, the method proposed by [36] was used, and the fractal theory was applied to calculate the extension amount with one survey line in the Qiongdongnan Basin as an example, yielding good results.

The number of faults and the distribution of fault moments obey the power exponential law, and they have a scale invariance; in other words, they have statistical fractal characteristics [12,37,38]:

$$N(D) \propto D^{-C} \tag{5}$$

In Equation (5), the cumulative number of faults is represented by $N(D)$, and C is the fractal dimension.

The horizontal direction is sorted from small to large, the horizontal direction is displayed on the horizontal coordinate, and the ordinate is the corresponding serial number. A logarithmic coordinate system was established, the least squares method was used for fitting, and the negative value of the slope is the C value [3,4,39].

The fractal dimension was calculated based on the scale invariance of the fault moment distribution.

Based on the identified large faults, a fractal analysis was conducted to determine the amount of extension due to the unrecognized small faults in the reflection seismic section.

$$\Delta L = h_1 + h_2 + h_3 + \ldots\ \ldots + h_N = \varepsilon h_i \tag{6}$$

$$h_e = h_N[C/(1 - C)](N + 1)[N/N + 1)]^{1/C} \tag{7}$$

$$D_{\text{total}} = \varepsilon h_{\text{i}} + h_e \tag{8}$$

where ΔL is the cumulative slip, its unit is m, h_N is the horizontal offset of the Nth major fault in the section, its unit is m, and N is the fault label value, dimensionless.

4. Results

4.1. Cenozoic Nonuniform Extension in the Qiongdongnan Basin

Based on the results of the seismic section, main fractures such as fault No. 2, 5, and 11 were mainly developed in the Qiongdongnan Basin. Fault No. 2 could be divided into three segments: Ledong member, Lingshui member, and Baodao-Changchang member. As the main fault running through the Qiongdongnan Basin, from the Ledong sag in the west to the Changchang sag in the east, it is one of the most important faults in the basin. The fault began to move in the Eocene and reached its maximum rate in the Oligocene. The maximum activity rate of the fault in the Ledong sag and the Baodao-Changchang sag reached 300 m/Ma. The calculation results showed that, during 28.4–23 Ma, the extension rate was 3.02%, and the extension rate was 0.82 mm/a; during 36–28.4 Ma, the extension rate was 12.27%, and the extension rate was 2.106 mm/a, reaching the maximum; and during 45–36 Ma, the extension rate was 4.35%, and the extension rate was 0.604 mm/a (Figure 5).

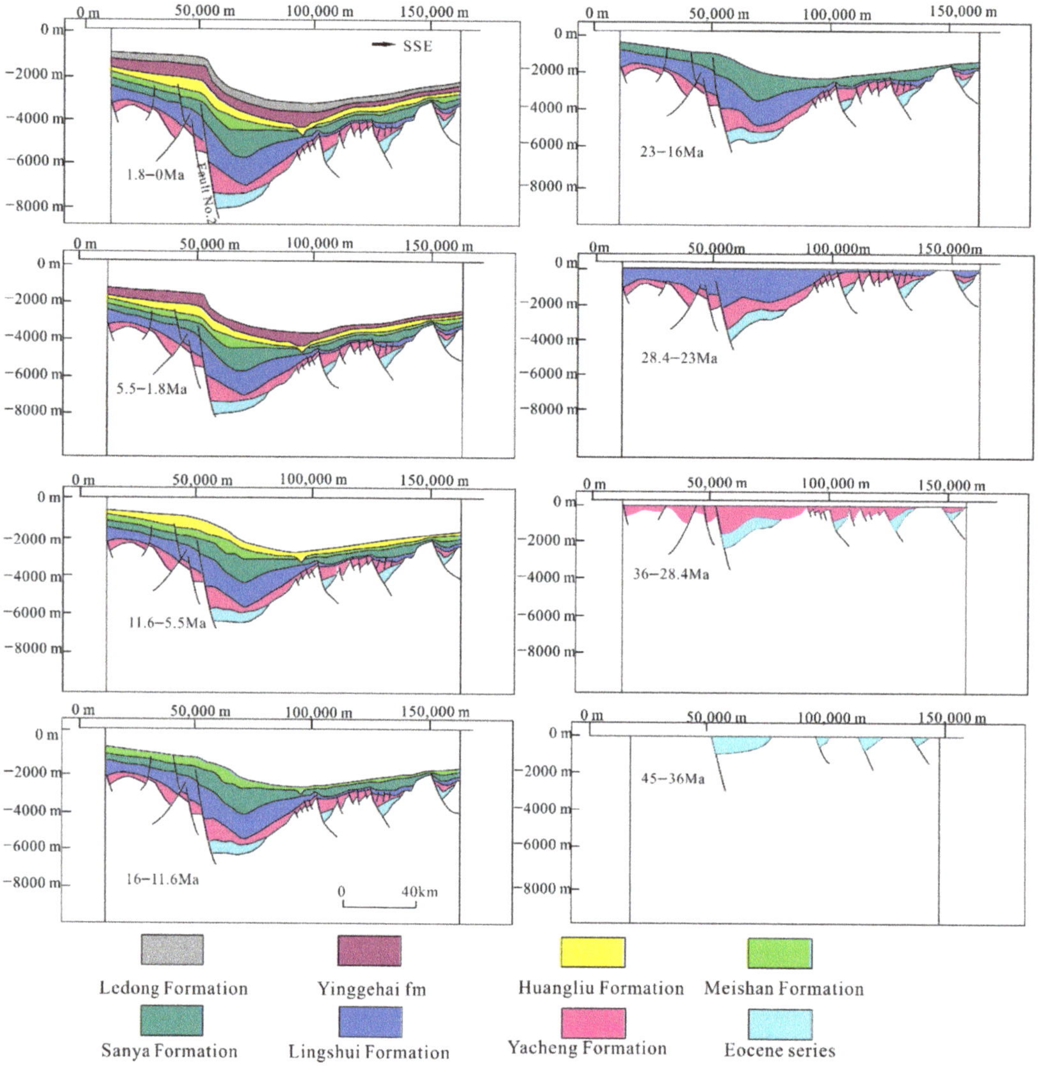

Figure 5. Cenozoic tectonic evolution of the Yabei–Lingshui sag–Lingnan low uplift in Qiongdongnan Basin.

The seismic sections through the Songdong sag–Baodao–Songnan low uplift–Changchang sag D–southern uplift revealed the geometry of the fractures and basin type. Vertically, the strata of the Qiongdongnan Basin could be divided into four layers from bottom to top: Eocene fault sag structure, Oligocene depression structure, Lower Miocene fault sag structure, and Middle Miocene quaternary depression structure. In the Eocene, there was a half graben structure with a fault in the north and overrunning in the south. In the Oligocene, the sag presented a recalcitration fault structure. In the late Miocene, it presented a butterfly depression structure (Figures 4 and 6).

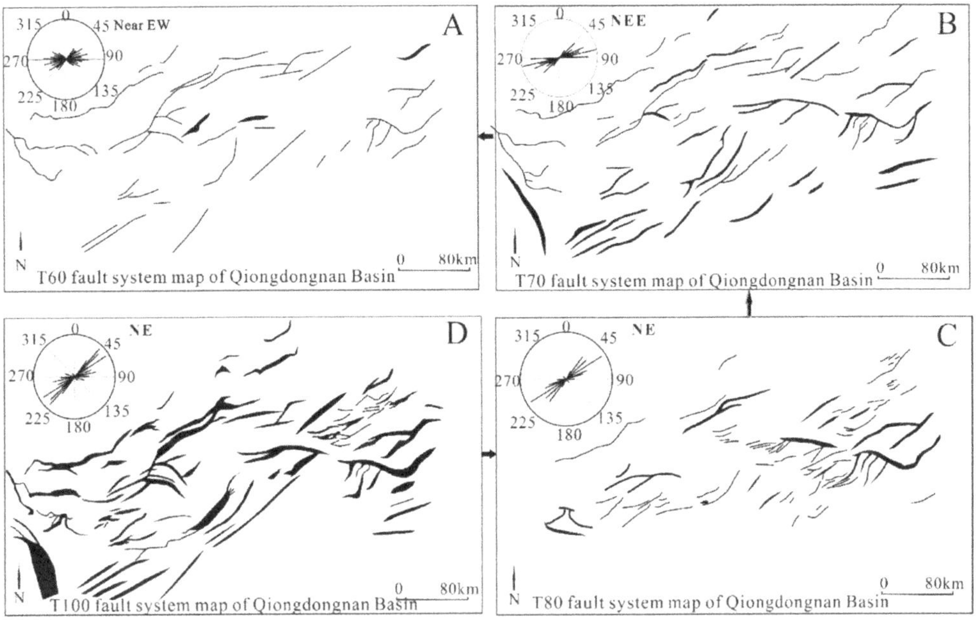

Figure 6. Fracture system diagram of each tectonic layer in Qiongdongnan Basin. The faults could be divided into three groups: NE-trending, NE-trending, and nearly EW-trending faults. (**A**) T60 fault strikes were mainly east–west; (**B**) T70 strikes were nearly NEE; (**C**) fault T80 strikes were nearly NE; and (**D**) fault T100 strikes were nearly NE.

The fractal dimension of one of the survey lines in the Baodao sag was 1.3 (Figure 7). The extension was calculated to be 51.19 km (Table 1). Compared with the result (46.58 km) calculated using the Move software (Figure 6), it was found that the cumulative extension of small faults that cannot be determined by the seismic data was 9% of the actual extension; the error was within the controllable range [40]. This showed that the Move software is generally capable of modeling the extensional tectonics of the Qiongdongnan Basin [37].

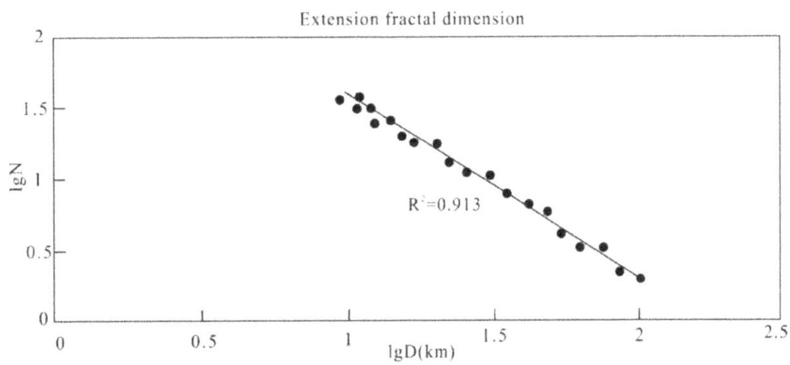

Figure 7. Logarithm plot of the horizontal fault distance D of Line 07e31033 fault vs. cumulative fault number N.

Table 1. Basin extension comparison based on the fractal method.

Survey Line	Current Length of Line (km)	Horizontal Graph$_N$ (km)	εh_i (km)	h_e (km)	Calculation of Total Elongation by Fractal Method (km)	Calculation of Total Extension by Equilibrium Recovery Method (km)
07e31033	198.53	1.7	47.28	3.91	51.19	46.58

4.2. Multistage Extension of Qiongdongnan Basin

An analysis of the Cenozoic cumulative extension amount curve of the Qiongdongnan basin (Figure 8) showed that the evolution of the Cenozoic basin occurred in four stages: collapse period, depression period, fault-depression period, and passive continental margin [18]. From the Paleocene to the Eocene, the subduction direction of the Pacific plate shifted from northwest to west, which reduced the convergence rate to the Eurasian plate and formed the south China continental margin in the NW–SE extensional stress field. Under this stress field, the Qiongdongnan Basin began to crack in the Eocene to form a series of NE-trending faults, and basement faults, such as No. 2 and 5, were activated in the basin. A series of fault lacustrine basins with NE-trending grabens and half grabens developed. The lacustrine deposits formed during this period are important source rocks in the basin [27].

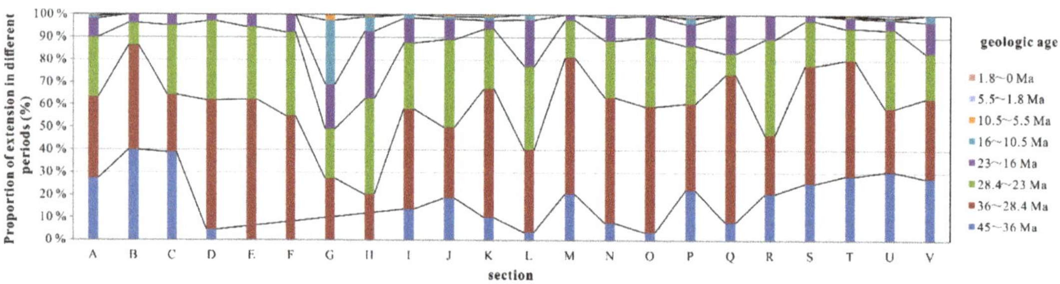

Figure 8. Extension amounts from different times and different sections in the Qiongdongnan Basin.

In the Oligocene, under the control of the regional stress field, the basin rifting scope was further expanded, and the basin turned into a depression fault structure. At the end of the Middle Eocene (approximately 42.5 Ma), the Indian plate collided completely and gradually wedged into the Eurasian plate [41]. The Indosinian block rotated and extruded to form a large-scale escape structure. As the western fault of the Qiongdongnan Basin, the red river fault exhibited a strike-slip, forming multiple extension and compression zones along both the sides of the strike-slip fault. In addition to the extensional stress field, the west area of the Qiongdongnan Basin was subjected to the tensile effect induced by the left-lateral strike-slip of the red river fault. The stress action in these two regions was mainly along the NW–SE direction, and the overall structure of the basin moved along the NE–SW direction; therefore, the extension amount of the west area along the No. 1 fault was much greater than those of the central and eastern regions. The expansion of the north–west sub-basin mainly affected the eastern part of the Qiongdongnan Basin. Several nearly EW-trending faults were developed in the eastern part of the Lingshui Formation.

In the early Miocene, the spread center of the South China Sea transited from the northwest sub-basin to the southwest sub-basin, resulting in sea movement and causing regional regression [42].

Based on the proportion diagram of the extension in different periods corresponding to each section (Figure 8), the maximum extension periods were 36–28.4 Ma and 28.4–23 Ma. Studies have shown that the period from 45 to 23 Ma was the main period of extension and deformation in the study area, and there were significant differences in the extension and deformation of the different blocks in the basin (Figure 8). Based on the proportion of extension in each period, the cumulative extension in 36–23 Ma was the highest, and it was mainly located in the Baodao-Changchang sag.

4.3. Migration of Subsidence Center in Qiongdongnan Basin

The depositional depressions in the Qiongdongnan Basin migrated regularly and spatially over time [43], and the lateral migration in each period shifted from east to west, controlling the migration of the subsidence center.

The Qiongdongnan Basin was under the control of the NE-trending active faults and NW-trending prior faults in the Changchang sag during the Eocene. The tectonic activity in the basin during the depression period was still extremely active, and a rapid subsidence occurred, particularly in the early Oligocene. Taking the subsidence analysis of the Y19 well area as an example (Figure 9), under the control of the regional stress field, the range of basin rifting was further expanded, and some areas became depression structures.

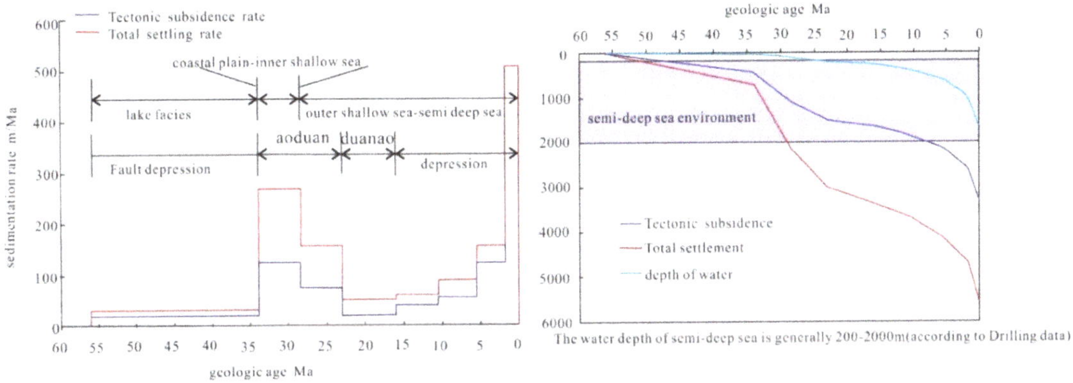

Figure 9. Subsidence history in Well Y19.

Because the spread center of the South China Sea migrated from the northwest sub-basin to the southwest sub-basin in the early Miocene, the basin entered a transition period from syn-rift to thermal subsidence. The tectonic activity in the basin weakened, the thermal subsidence strengthened, the number of faults decreased, the development of the depression was uncontrolled, and the water depth rapidly increased.

In the early Oligocene, influenced by the expansion of the South China Sea, seawater entered the basin and formed a marine sedimentary environment. At this time, the basin entered the depression stage, and the fracture still played a major role in controlling the basin development. The sag fracture distribution was controlled, and each sag was relatively independent in the Oligocene. The sedimentary center was mainly located near the main fault of each sag. The basin was surrounded by Hainan uplift, Zhongjian uplift, Xisha uplift, and Shenhu uplift. The four major provenance supplies were sufficient, and thick marine–continental transitional facies and marine strata were formed in the basin. After the Miocene, the basin entered the stage of fault depression–depression, where the depression was further enhanced, the sea level increased further, and the basin was integrated. Since the Xisha uplift and Shenhu uplift gradually failed to enter the water, the supply of provenance in the south and east weakened, whereas the Hainan uplift in the north and Kunsong uplift in the west continued to provide sufficient provenance, resulting in the continuous advancement of the shelf-slope break to the southeast and the deposition of thick Miocene strata in the west of the basin. In comparison, the Miocene strata in the east gradually weakened, i.e., the sedimentary center moved to the west of the basin after the Miocene.

This migration of the sedimentary center led to a difference in the stratigraphic thickness between the eastern and western regions. In the plane, there was no significant difference in the thickness of the strata in the Paleogene at the center of each sag in the eastern and western regions of the central depression, and the strata in the Neogene were thick in the west and thin in the east. Vertically, the strata in the eastern area were thick in the Paleogene and thin in the Neogene. The thickness of the strata in the western area was not significantly different from that observed in the Paleogene.

5. Discussion

5.1. Mechanism of Uneven Distribution of the Extensional Capacity within the Qiongdongnan Basin

The rifting in the Qiongdongnan Basin was reflected in the uneven temporal and spatial distributions of the extensional amount. We inferred that the reason for the anomalous development of rifting was the varying crust thickness, and the fact that the Moho surface in the west was low, while it was high in the east [33,44,45]. The pre-existing faults in the basin also controlled the extension of the fractures.

The regional tectonic stress field deviated from the basin structure during the Cenozoic (Figures 2 and 6).

An anomalous crustal structure existed between the eastern and western parts of the Qiongdongnan Basin. The upper crust layer was thin in the east and thick in the west. The low Moho surface in the west and high Moho surface in the east resulted in a difference in the basement properties between the east and west of the basin, and the difference in the basement properties led to an evident difference in the fracture structure morphologies between the east and west sags of the basin in the Paleogene. At the section, the eastern depression was V-shaped with a large number of faults, small fault spacing, and narrow depressions, while the western depression was W-shaped with a small number of faults, large scale of faults and related folds, and broad depressions. The basic reason for this difference lies in the nature of the basement and the difference in the pre-existing faults. The crust was thick in the west and thin in the east, the Moho plane was low in the west and high in the east, and the pre-existing fault was NE trending in the west and EW trending in the east. As a result, the basement lithology was rigid in the west, ductile in the east, W-shaped in the west, V-shaped in the east, and more faults were present in the east. The magmatic activity was weak in the west and strong in the east, and the heat flow field was cold in the west and hot in the east. Carbon dioxide risk was low in the west and high in the east.

No. 2 fault is a basin-controlling first-order fault running through the Qiongdongnan Basin. The activity rate map (Figure 10) shows evident differences in the activities of the faults. No. 2 fault could be divided into Ledong, Lingshui, Songnan, and Baodao-Changchang sections. The maximum activity rate of the fault in the west of the Ledong sag and the middle of the Baodao-Changchang sag was 300 m/Ma, which is conducive to the deposition of marine source rocks. The high activity rate of No. 2 fault may be related to the strike-slip of the red river fault in the west and high extension strength in the east.

The east–west block pattern in the southern part of the basin was closely related to the differences between the pre-existing faults and basement properties. In addition to the differences on the plane, there were differences longitudinally. Affected by the Shenhu movement, the Pearl–Qiong movement, the South China Sea movement, and the neotectonic movement, the basin experienced four evolution stages: collapse period, depression period, fault-depression period, and passive continental margin. This led to the formation of four different structural strata in the basin. The tectonic evolution process and regional tectonic events in the basin had a profound impact on hydrocarbon generation, reservoir formation, and accumulation.

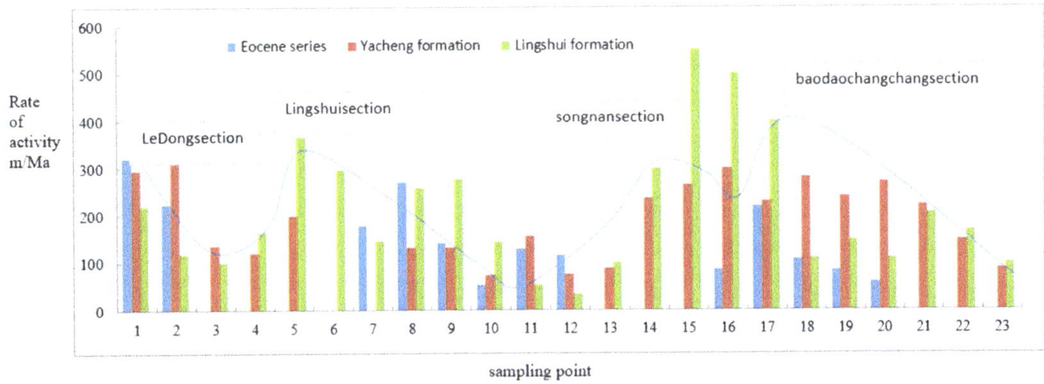

Figure 10. Activity analysis of No. 2 fault (No. 2 fault: as the main fault running through the Qiongdongnan Basin, No. 2 fault is one of the most important faults, running from Ledong sag in the west to Songnan-Baodao sag in the east. The fault started to be active in the Eocene and reached its maximum activity rate in the Oligocene).

5.2. Tectonic Control on Hydrocarbon Source, Reservoir Accumulation, and Formation in Qiongdongnan Basin

The uneven distribution of the Cenozoic extension in the Qiongdongnan Basin has a certain influence on hydrocarbon reservoir, which includes the oil source, reservoir, migration, and accumulation.

First, notably, the tectonics controlling the hydrocarbon source rocks reflected temporal and spatial characteristics (Figure 6). Temporally, the tectonic evolution of the basin controlled the development sequence of three source rocks in the basin. Affected by the Pacific plate subduction and retreat and the Indo–Eurasian plate collision, a NW–SE-trending regional tensile stress field was formed along the continental margin of south China since the Paleogene. Under the action of this stress field, the basin began to collapse in the Mesozoic basement, a series of NE–SW faults were formed in the continental margin, and the Eocene lacustrine source rocks were developed in the basin controlled by the rifting faults. Although there has been no drilling to reveal the formation, the seismic and oil and gas data indicate its existence. Seismic data showed a set of low-frequency, medium-continuous, and strong reflection strata on the basement of the basin, consistent with the seismic characteristics of the middle-deep lacustrine facies of the Eocene in the Qiongdongnan Basin. Crude oil rich in C30-4 methyl steranes and large amounts of oil-type gases with an ethane carbon isotope distribution between −28‰ and −33‰ were found in several wells in the northern and southern basins, confirming the existence of source rocks (Figure 2).

Spatially, the tectonic effect manifested in the control of the structure on the distribution and types of source rocks. The development of two groups of pre-existing faults could be seen in the basin, and they struck nearly EW in the east area and NE–SW in the west area. During the fault depression period, the basin was affected by the NW–SE-trending tensile stress field, and the pre-existing fault strike in the western region was perpendicular to the main stress field, thus forming a NE–SW-trending fault system. This system controlled the NE–SW distribution of the Eocene lacustrine source rocks in the western region, and the maximum deposit center of the source rocks was located near the fault. The strike of the pre-existing fault in the eastern region obliquely intersected with the main stress field; hence, the distribution of the lacustrine source rocks in the Eocene did not show a NE–SW pattern; however, the thickness center was mostly near the fault. On the other hand, because of the weak basement and earlier and stronger extension in the eastern region, the Eocene source rocks were more widely distributed in the eastern region. Therefore, in the fault-depression period, this control effect was mainly reflected in the distribution of the lacustrine source rocks by fault differential activities. In the depression period, this controlling effect was mainly reflected the control of the structure on the distribution and type of source rocks. In the early Oligocene, influenced by the expansion of the South China Sea, seawater entered the Qiongdongnan Basin from the east of the Xisha trough, which was bound to cause the development of marine source rocks in the eastern basin earlier than that in the western basin. In addition to this regional tectonic event, the difference in the fault activity in the basin had a profound impact on the development of hydrocarbon sources. For example, fault No. 2 and 11 in the early Oligocene continued to exhibit strong activity, resulting in a significant uplift and erosion of the central low uplift area. Here, fault No. 2 was active in the southeast part of Songnan, fault No. 11 was more active in the middle part than at both sides, and fault No. 11 had a stronger activity than fault No. 2 in Songnan. This led to a greater uplift of the west side of the Songnan low uplift than the east side, and the uplift of the south side was higher than that of the north side, i.e., the southwest was high, and the northeast was low. The southwest was uplifted and eroded more strongly. Several land sources were transported to the northeast along the slope, and seven (fan) deltas and coastal clastic sedimentary bodies were formed in the northeast. Terrigenous organic matter was also transported to the Songnan–Baodao sag by the river delta, forming a high-quality terrigenous marine hydrocarbon source.

Second, the four-stage tectonic forms had a controlled hydrocarbon reservoir model (Figure 11). The four types of reservoirs are as follows: (i) buried hill reservoirs above the basement high, such as Yongle 8 area of the Songnan low uplift and Yacheng uplift; (ii) fan delta reservoir at the center of the fault basin, such as the southern slope belt of the Songxi sag; (iii) turbidite channel-submarine fan reservoir; and (iv) carbonate reservoir at the basement uplift or seamounts in the south uplift. Tectonic uplifts controlled the occurrence of the buried hill and fracture system and carbonate platform development. The tectonic subsidence in the sag could control the submarine fan delta reservoir. The Neogene turbidite channel system had been limited by the west–east fracture structures [9].

Third, oil and gas migration has been considered a key factor for reservoir oil and gas migration in the Qiongdongnan Basin (Figure 11). The fluid potential field is the energy field controlling underground fluid migration, and the structure is an important factor controlling the fluid potential. The Cenozoic multistage tectonic movement controlled the formation and re-regulation of the pore fluid pressure inside the system, thereby changing the elastic energy and fluid potential difference. The center of the depression was a high-elastic-energy area. The low-potential area was the slope-low bulge–bulge area. The opening and closing of the faults and the continuity in the sand body distribution manifested in the varying interface energy due to the change in the pore throat, and finally, the fluid potential in the system varied. The Songnan low uplift was a relatively weak area of diagenesis in the region, and it was also a low potential area for the interfacial energy. Multistage tectonic movement controls the formation and transformation of the topography in deepwater areas, resulting in an elevation difference between the source and the circle, and the fluid between them has a gravitational potential energy. Under the control of the Cenozoic tectonic movement, the basin system witnessed the formation of a fault-sand body-tectonic ridge, other transport channels, and various transport system frameworks, indirectly controlling the spatial form of the fluid potential.

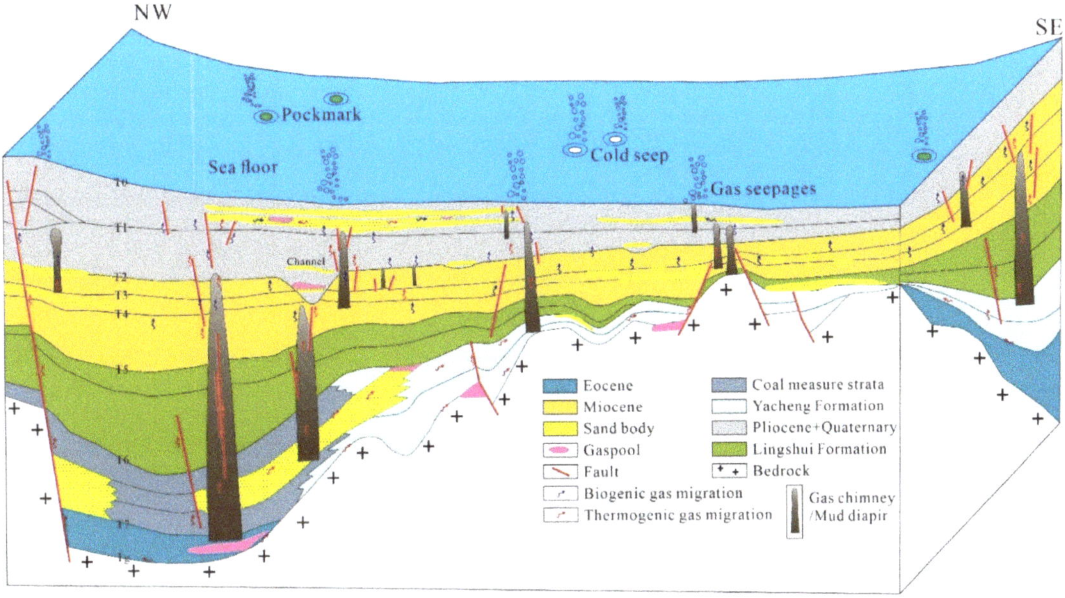

(a)

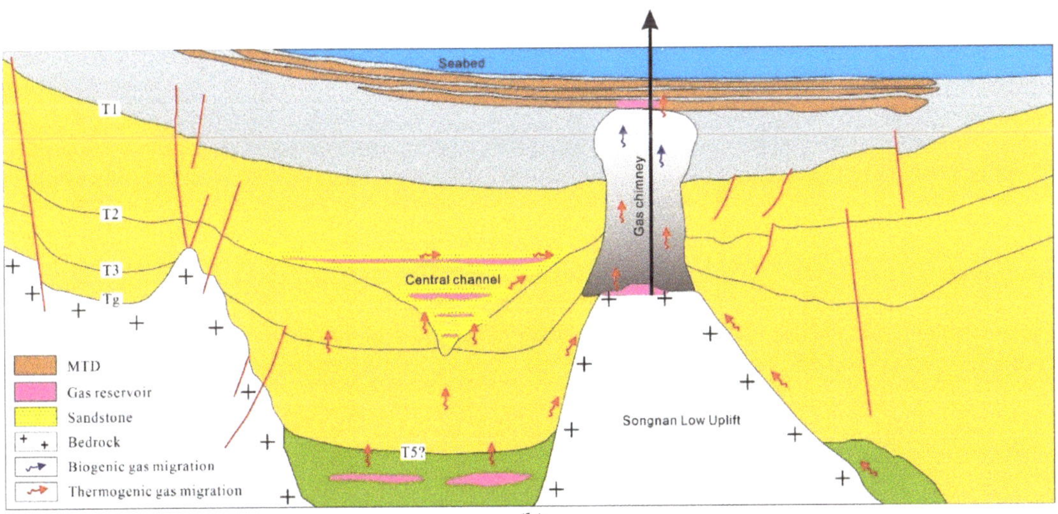

(b)

Figure 11. Hydrocarbon reservoir model and fluid migration in the Qiongdongnan Basin. (**a**) Source rock, fluid flow and hydrocarbon reservoir in the Qiongdongnan basin. (**b**) Reservoir model of the QDN central channel and buried Mesozoic granite hill.

6. Conclusions

The rifting of the Qiongdongnan Basin had been in a continuous extension process during the Cenozoic, with different extension rates. The period of 45–23 Ma was the main period of extension deformation in the study area, and there was an extension peak during 36–23 Ma, after which the extension and extension rate gradually decreased. Up to 23 Ma, the fault activity was weak, the strength of the extension deformation was low, the extension and extension rate were low, and there was little difference between the different regions. Since the Miocene (10.5 Ma), the evolution of the entire basin had entered a new stage, where the slow thermal subsidence after cracking was different and the basin sedimentation rate suddenly accelerated; in particular, the shelf-slope broke into deep waters, and the basin settlement evidently accelerated, accommodating more space; however, because of the influence of the source factors, the sedimentary strata in the southern part of the basin were relatively thin.

The spatiotemporal distribution of the Cenozoic extension in the Qiongdongnan Basin was uneven. Spatially, the extension rate in the west depression was higher than that in the east depression, mainly due to the fault displacement of the main boundary of the basin, and the extension corresponding to the larger displacement of the main boundary fault was high. The horizontal stretching movement could be divided into three periods: Eocene, Oligocene, and Miocene.

The displacement distribution of the faults had a self-similar structure, called the fractal feature. This feature provides the fractal dimension method for calculating the extensional amount of extensional basins, outlining the total contribution of small faults, and compensating for the difference in the results of the equilibrium restoration method. Some structural factors affected the source rock, accumulation, and reservoir model in the Qiongdongnan basin.

Author Contributions: Conceptualization, L.Z. and J.Z. (Jie Zhou); methodology, Y.Z.; software, Y.Z.; validation, Y.Z., S.W. and L.M.; formal analysis, X.L.; investigation, J.Z. (Jitian Zhu) and X.X.; resources, L.T.; data curation, J.Z. (Jie Zhou); writing—original draft preparation, Y.Z.; writing—review and editing, Y.Z.; visualization, Y.Z.; supervision, L.Z. and L.M.; project administration, L.Z.; funding acquisition, L.Z. All authors have read and agreed to the published version of the manuscript.

Funding: This study was financially supported by the Natural Science Foundation of China (U1701245), the China Geological Survey Project South China Sea Oil and Gas Resources Survey (DD20190213), the China Geological Survey Projects (DD20221700, DD20221705, DD20221708, and DD20220224), Key Special Project for Introduced Talents Team of Southern Marine Science and Engineering Guangdong Laboratory (Guangzhou) (grant number GML2019ZD0102), Key-Area Research and Development Program of Guangdong Province (2020B1111030003), and the National Science and Technology Project (2016ZX05026-002). The first author gives her thanks to the Hainan branch of CNOOC for permission to release the seismic data.

Institutional Review Board Statement: Not applicable.

Informed Consent Statement: Not applicable.

Data Availability Statement: Not applicable.

Conflicts of Interest: The authors declared that they have no conflict of interest to this work. We declare that we do not have any commercial or associative interest that represents a conflict of interest in connection with the work submitted.

References

1. Jiao, S.T.; Zhou, Y.Z.; Zhang, Q.; Jin, W.J.; Liu, Y.P.; Wang, J. Study on intelligent discrimination of tectonic settings based on global gabbro data from GEOROC. *Acta Petrol. Sin.* **2018**, *34*, 3189–3194. (In Chinese)
2. Zuo, R.G.; Xia, Q.L.; Wang, H.C. Compositional data analysis in the study of integrated geochemical anomalies associated with mineralization. *Appl. Geochem.* **2013**, *28*, 202–211. [CrossRef]
3. Zhou, Y.Z.; Chen, S.; Zhang, Q.; Xiao, F.; Wang, S.G.; Liu, Y.P.; Jiao, S.T. Advances and prospects of big data and mathematical geoscience. *Acta Petrol. Sin.* **2018**, *34*, 255–263. (In Chinese)
4. Zhou, Y.Z.; Wang, J.; Zuo, R.G.; Xiao, F.; Shen, W.J.; Wang, S.G. Machine learning, deep learning and Python language in field of geology. *Acta Petrol. Sin.* **2018**, *34*, 3173–3178. (In Chinese)
5. Luo, Z.J.; Xiong, Y.H.; Zuo, R.G. Recognition of geochemical anomalies using a deep variational autoencoder network. *Appl. Geochem.* **2020**, *122*, 104710. [CrossRef]
6. Xiong, Y.H.; Zuo, R.G. Recognizing multivariate geochemical anomalies for mineral exploration by combining deep learning and one-class support vector machine. *Comput. Geosci.* **2020**, *140*, 1044484. [CrossRef]

7. He, C.; Long, G.Y.; Wu, S.M. Esimation of the stretching factors of the Qiongdongnan Basin and domino-style fault model. *Geotecton. Et Metallog.* **2012**, *36*, 204–208. (In Chinese)
8. Shi, X.; Jiang, H.; Yang, J.; Yang, X.; Xu, H. Models of the rapid post-rift subsidence in the eastern Qiongdongnan Basin. South China Sea: Implications for the development of the deep thermal anomaly. *Basin Res.* **2017**, *29*, 340–362. [CrossRef]
9. Sun, Z.; Wang, Z.; Sun, Z.P.; Wang, Z.W.; Zhang, W.; He, L.J. Structure and kinematic analysis of the deepwater area of the Qiongdongnan Basin through a seismic interpretation and analogue modeling experiments. *Acta Oceanol. Sin.* **2015**, *34*, 32–40. [CrossRef]
10. Zhu, W.L.; Zhong, K.; Li, Y.C. Characteristics of hydrocarbon accumulation and exploration potential of the northern South China Sea deepwater basins. *Chin. Sci. Bull* **2012**, *57*, 3121–3129. [CrossRef]
11. Zhu, W.L.; Xie, X.N.; Wang, Z.F.; Zhang, D.J.; Zhang, C.L.; Cao, L.C.; Shao, L. New insights on the origin of the basement of the Xisha Uplift, South China Sea. *China Earth Sci.* **2017**, *60*, 2214–2222. [CrossRef]
12. Zhang, Y.; Pei, J.X.; Zhou, Y.Z.; Zhu, J.T.; Chen, Y. The application of nonlinear technique in provenance analysis and reservoir parameter prediction at the Lingshui17-2 gas field of Qiongdongnan Basin. *Acta Geol. Sin.* **2016**, *90*, 559–568. (In Chinese)
13. Xie, X.N.; Müller, R.D.; Li, S.T.; Gong, Z.; Steinberger, B. Origin of anomalous subsidence along the northern South China Sea margin and its relationship to dynamic topography. *Mar. Pet. Geol.* **2006**, *23*, 745–765. [CrossRef]
14. Zhao, Z.; Sun, Z.; Liu, J.; Pérez-Gussinyé, M.; Zhuo, H. The continental extension discrepancy and anomalous subsidence pattern in the western Qiongdongnan Basin, South China Sea. *Earth Planet. Sci. Lett.* **2018**, *501*, 180–191. [CrossRef]
15. Walsh, J.; Watterson, J.; Yielding, G. The importance of smallscale faulting in regional extension. *Nature* **1991**, *351*, 391–393. [CrossRef]
16. Ding, W.W.; Schnabel, M.; Franke, D.; Aiguo, R.; Wu, Z. Crustal structure across the northwest margin of South China Sea: Evidence for magma-poor rifting from wide-angle seismic profile. *Acta Geol. Sin.* **2012**, *86*, 854–866.
17. Morley, C.K. Major unconformities/termination of extension events and associated surfaces in the South China Seas: Review and implications for tectonic development. *J. Asian Earth Sci.* **2016**, *120*, 62–86. [CrossRef]
18. Zhang, Y.Z.; Xu, X.D.; Gan, J.; Zhu, J.T.; Guo, X.X.; He, X.H. Study on the geological characteristics, accumulation model and exploration direction of the giant deepwater gas field in the Qiongdongnan Basin. *Acta Geol. Sin.* **2017**, *91*, 1620–1633. (In Chinese)
19. Hu, B.; Wang, L.; Yan, W.; Liu, S.; Cai, D.; Zhang, G.; Zhong, K.; Pei, J.; Sun, B. The tectonic evolution of the Qiongdongnan Basin in the northern margin of the South China Sea. *J. Asian Earth Sci.* **2013**, *77*, 163–182. [CrossRef]
20. Briais, A.; Patriat, P.; Tapponier, P. Updated interpretation of margnetic anomalies and seafloor spreading stages inthe South China Sea: Implications for the tertiary tectonics of Southeast Asia. *J. Geophys. Res.* **1993**, *98*, 6299–6328. [CrossRef]
21. Taylor, B.; Hayes, D.E. The tectonic evolution of the South China Basin. In *The Tectonic and Geologic Evolution of Southeast Asian Seas and Islands: Part 2*; Hayes, D.E., Ed.; Geophysical Monograph Series 27; American Geophysical Union: Washington, DC, USA, 1983; pp. 23–56.
22. Yang, J.; Cheng, Q.M. A comparative study of independent component analysis with principal component analysis in geological objects identification. Part II: A case study of Pinghe District, Fujian, China. *J. Geochem. Explor.* **2015**, *149*, 136–146. [CrossRef]
23. Pellegrino, A.G.; Zhang, B.; Speranza, F.; Maniscalco, R.; Yin, C.; Moreno, H.C.; Winkler, A. Tectonics and Paleomagnetic Rotation Pattern of Yunnan (24° N–25° N, China): Gaoligong Fault Shear Versus Megablock Drift. *Tectonics* **2018**, *37*, 1524–1551. [CrossRef]
24. Speranza, F.; Pellegrino, A.G.; Zhang, B.; Maniscalco, R.; Chen, S.; Moreno, C.H. Paleomagnetic Evidence for 25–15 Ma Crust Fragmentation of North Indochina (23–26° N): Consequence of Collision With Greater India NE Corner? *Geochem. Geophys. Geosyst.* **2019**, *20*, 5425–5448. [CrossRef]
25. Tapponnier, P.; Peltzer, G.; Dain, A.Y.L.; Armijo, R. Propagating extrusion tectonics in Asia: New insights from simple experiments with plasticine. *Geology* **1982**, *10*, 611–616. [CrossRef]
26. Xia, S.; Zhao, F.; Zhao, D.; Fan, C.; Wu, S.; Mi, L.; Sun, J.; Cao, J.; Wan, K. Crustal pluming system of post-rift magmatism in the northern margin of the South China Sea: New insight from intergrated seismology. *Tectonopysics* **2018**, *744*, 227–238. [CrossRef]
27. Wu, S.G.; Yuang, S.Q.; Zhang, G.C.; Ma, Y.B.; Mi, L.J.; Xu, N. Seismic characteristics of a reef carbonate reservoir and implications for hydrocarbon exploration in deepwater of the Qiongdongnan Basin, northern South China Sea. *Mar. Pet. Geol.* **2009**, *26*, 817–823. [CrossRef]
28. Zhao, Z.; Sun, Z.; Wang, Z.; Sun, Z.; Liu, J.; Zhang, C. The high resolution sedimentary filling in Qiongdongnan Basin, Northern South China Sea. *Mar. Geol.* **2015**, *361*, 11–24. [CrossRef]
29. Franke, D.; Savva, D.; Pubellier, M.; Steuer, S.; Mouly, B.; Auxietre, J.L.; Meresse, F.; Chamot-Rooke, N. The final rifting evolution in the South China Sea. *Mar. Pet. Geol.* **2013**, *58*, 704–720. [CrossRef]
30. Savva, D.; Pubellier, M.; Franke, D.; Chamot-Rooke, N.; Meresse, F.; Steuer, S.; Auxietre, J.L. Different expression of rifting the South China sea margins. *Mar. Pet. Geol.* **2014**, *58*, 579–598. [CrossRef]
31. Zhao, Z. The deep mantle upwelling beneath the northwestern South China Sea: Insights from the time-varying residual subsidence in the Qiongdongnan Basin. *Geosci. Front.* **2021**, *12*, 101246. [CrossRef]
32. Wang, L.J.; Sun, Z.; Yang, J.H.; Sun, Z.P.; Zhu, J.T.; Zhuo, H.T.; Stock, J. Seismic characteristics and evolution of post-rift igneous complexes and hydrothermal vents in the Lingshui sag (Qiongdongnan basin), northwestern South China Sea. *Mar. Geol.* **2019**, *418*, 106043. [CrossRef]
33. Schonborn, G. Balancing cross sections with kinematic constrains: The dolomites (northern Italy). *Tectonics* **1999**, *18*, 527–545. [CrossRef]

34. Qi, J.F.; Lu, K.Z.; Zhang, Y.W.; Yang, Q.; Liu, G.C.; Li, J.Z. The formation and evolution of kongdian high in huaghua Basin. *Acta Pet. Sin.* **1994**, *15*, 27–33. (In Chinese)
35. Allen, P.A.; Allen, J.R. *Basin Analysis: Principles & Application*; Blackwell Scientific Publications: Oxford, UK, 1992; pp. 90–92.
36. Marrett, R. Amount of extension on "small" faults: An example from the Viking graben. *Geology* **1992**, *20*, 47–50. [CrossRef]
37. Li, T.; Zuo, R.G.; Chen, G.X. Investigating fluid-rock interaction at the hand-specimen scale via ITRAX. *J. Geochem. Explor.* **2019**, *204*, 57–65. [CrossRef]
38. Ma, J.Y.; Xie, S.Y.; Tang, H.Y.; Ma, M.; Carranza, E.J.M.; Han, J.; He, Z.L.; Jiao, C.L.; Zhang, H. Evolution Mechanisms of Carbonate Reservoirs Based on Dissolution Rates and Multifractal Analysis of Microscopic Morphology. *Nat. Resour. Res.* **2020**, *29*, 2843–2865. [CrossRef]
39. Cheng, Q.M. Fractal Calculus and Analysis for Characterizing Geoanomalies caused by singular Geological processes. *J. Earth Sci.* **2021**, *32*, 276–278. [CrossRef]
40. Liu, S.W.; Wang, L.S.; Liu, B. Fractal Analysis of Extensional Quantity of the Extensional Basin. *Geol. Rev.* **2001**, *47*, 229–233.
41. Yao, B.C.; Wan, L.; Liu, Z.H. Tectonic dynamics of Cenozoic sedimentary basins and hydrocarbon resources in the South China Sea. *Earth Sci. J. China Univ. Geosci.* **2004**, *29*, 543–549.
42. Li, C.F.; Shi, X.; Zhou, Z.; Li, J.; Geng, J.; Chen, B. Depths to the magnetic layer bottom in the South China Sea area and their tectonic implications. *Geophys. J. Int.* **2010**, *182*, 1229–1247. [CrossRef]
43. Kang, B. Cenozoic Subsidence and Thermal History Modelling of Qiongdongnan Basin. Ph.D. Thesis, China University of Geosciences, Wuhan, China, 2014; pp. 1–87. (In Chinese)
44. Clift, P.; Lin, J.; Barckhausen, U. Evidence of low flexural rigidity and low viscosity lower continental crust during continental break-up in the South China Sea. *Mar. Pet. Geol.* **2002**, *19*, 951–970. [CrossRef]
45. Qiu, X.L.; Ye, S.Y.; Wu, S.M.; Shi, X.B.; Zhou, D.; Xia, K.Y.; Flueh, E.R. Crustal structure across the Xisha trough, northwest South China Sea. *Tectonophysics* **2001**, *341*, 179–193. [CrossRef]

energies

Article

Study of the Appropriate Well Types and Parameters for the Safe and Efficient Production of Marine Gas Hydrates in Unconsolidated Reservoirs

Yuan Chen [1,3], Shiguo Wu [1,3,4,*], Ting Sun [2] and Shu Jia [2]

1 Institute of Deep-Sea Science and Engineering, Chinese Academy of Sciences, Sanya 572000, China; chenyuan@idsse.ac.cn
2 College of Safety and Ocean Engineering, China University of Petroleum, Beijing 102200, China; ting.sun@cup.edu.cn (T.S.); jiashu3110136566@163.com (S.J.)
3 University of Chinese Academy of Sciences, Beijing 100049, China
4 South Marine Science and Engineering Guangdong Laboratory, Zhuhai 519000, China
* Correspondence: swu@idsse.ac.cn

Abstract: The majority of marine hydrates are buried in unconsolidated or poorly consolidated marine sediments with limited cementation and strength. As a result, hydrate decomposition during production may cause significant subsidence of the formation, necessitating a halt in production. The numerical model of unconsolidated hydrate formation, based on geomechanics, was established in order to elucidate the depressurization production process. The sensitive factors of unconsolidated hydrate production were determined by analyzing the influence of formation parameters and production parameters on gas production. Then, a safety formation subsidence was proposed in this paper, and the appropriate well type and parameters for the safe and efficient production of hydrates in unconsolidated formations of various saturations were determined. The sensitivity of gas production to the formation parameters was in the order of formation porosity, hydrate saturation, and buried depth, while the effects of the production parameters were BHP (bottom hole pressure), horizontal length, and heat injection, in descending order. For hydrate reservoirs in the South China Sea, when hydrate saturation is 20%, a horizontal well is necessary and the appropriate horizontal length should be less than 80 m. However, when hydrate saturation is more than 30%, a vertical well should be selected, and the appropriate bottom hole pressure should be no less than 3800 kPa and 4800 kPa for 30% and 40% saturation, respectively. Based on the simulation results, hydrate saturation was the key factor by which to select an appropriate production technique in advance and adjust the production parameters. The study has elucidated the depressurization production of marine unconsolidated hydrate formations at depth, which has numerous implications for field production.

Keywords: depressurization; formation subsidence; numerical simulation; sensitivity analysis; safe and efficient production

Citation: Chen, Y.; Wu, S.; Sun, T.; Jia, S. Study of the Appropriate Well Types and Parameters for the Safe and Efficient Production of Marine Gas Hydrates in Unconsolidated Reservoirs. *Energies* **2022**, *15*, 4796. https://doi.org/10.3390/en15134796

Academic Editors: Jalel Azaiez and Pål Østebø Andersen

Received: 17 May 2022
Accepted: 22 June 2022
Published: 30 June 2022

1. Introduction

Gas hydrate represents a new type of clean energy, with vast reserves and wide distribution, which is primarily buried in permafrost and deep-sea environments. It is a crystalline compound that is formed by natural gas and water at high pressure and low temperature [1]. In recent years, investigations into gas hydrates continue to rise. On the one hand, hydrates are of high energy density and heat value and will release methane gas after decomposition, which has lower carbon content compared with other fossil fuels, and, thus, can decrease the carbon footprint. On the other hand, a hydrate reservoir is an effective place to store carbon dioxide, which plays an important role in alleviating the effects of global climate change. The basic idea behind utilizing gas hydrate is to allow hydrates to decompose into free gas and water before extracting gas from the formation using traditional methods. Gas hydrates can be exploited in seven ways: depressurization, heat injection, CO_2 replacement, N_2 injection, chemical reagent injection, solid fluidization, and combined methods [2–5].

Marine hydrates are often buried in unconsolidated sediments that are less than 400 m beneath the seabed of the continental slope. Unlike conventional oil and gas reservoirs, hydrates serve as the formation's cementation, which means that hydrate decomposition will decrease the physical and mechanical properties of the sediments, such as the shear strength and the bearing capacity. Besides this, the hydrates will release a substantial amount of gas after decomposition, causing a sudden increase in pore pressure and a decrease in the effective stress of the sediments [6]. Under the action of pore pressure and overburden pressure, the sediments will experience significant deformation during production, which may result in a series of problems, such as subsea landslides, formation subsidence, and wellbore instability [2,7]. Furthermore, secondary hydrates and the ice that is generated as a result of the temperature drop induced by the endothermic decomposition process would clog the fluid seepage channel, reducing the production efficiency [8]. More dangerously, the two-phase flow and the change in effective stress may cause the sediment particles to move, resulting in substantial sand production and damage to the production equipment [9].

At present, many production tests have been conducted by five countries around the world. The most successful example was the Messoyakha gas field production project in the former Soviet Union, in 1969 [10]. Pure gas hydrate production tests occurred in the Mackenzie Delta, Canada in 2002, 2007, and 2008 [11,12]. The Ingnik Sikumi gas hydrate field trial in the US realized the joint development of conventional gas and gas hydrate in 2012 [13]. Marine NGH production tests were initiated in the eastern Nankai Trough in Japan in 2013 and 2017 [14,15]. Two more successful ocean production tests were conducted in the Shenhu area in the South China Sea, in 2017 and 2020 [16–18].

However, only the Messoyakha gas field has achieved commercial development. Due to various factors, such as sand production and formation subsidence, the duration of the other production tests was short, with the longest trial production time being only 60 days [18]. As a result, it is crucial to figure out how to reduce risk while increasing gas production, in order to achieve the safe and efficient development of hydrate reservoirs. This is related to the production behavior and mechanical properties of the formation. Many hydrate experiments and numerical simulation investigations into these two issues have been carried out over the years.

Although many problems exist in marine gas hydrate production, numerical simulation is a useful approach to unraveling the key factors in the production of marine unconsolidated hydrate reservoirs. In terms of hydrate production behavior, Li utilized TOUGH + HYDRATE to simulate the effects of different heat-injection temperatures and rates on gas recovery performance, under the combined exploitation methods of heat injection and depressurization [19]. Merey et al., conducted depressurization simulations using HydrateResSim (the United States) at different depressurization pressures, and the gas production and distribution of pressure, temperature, and saturation were compared [20]. Chen et al., used TOUGH + HYDRATE to simulate hydrate production in the Shenhu area in China and found that fracturing could increase the permeability of the sediments and gas production rate; cumulative gas production was also improved [21]. The production behaviors of hydrate reservoirs adopting a horizontal well [22] and a multi-branch well were studied [23]. Jin et al., numerically simulated the production behavior of depressurization, using horizontal wells combined with thermal stimulation, and the results showed that horizontal wells could achieve greater gas production [24]. Feng et al., studied hydrate decomposition using a horizontal well system known as LRHW (left well for injection and right well for production). The simulation results showed that the gas production using this well placement was higher than that of ULHW (upper well for production and lower well for injection) [25]. Almost all the research ignored the influence of formation subsidence on gas production. However, most marine hydrate sediments are unconsolidated and formation subsidence will occur during production, which will decrease the porosity and permeability of the formation and further influence gas production.

As for the mechanical response of the formation during production, Gong et al., used FLAC3D (Itasca International Inc., Minneapolis, MI, USA) to simulate the seafloor subsidence caused by hydrate production in the Shenhu area of the South China Sea [26]. Wan et al., established a coupled mathematical model of thermo-fluid-solid-chemical fields, based on hydrate drilling data in the Shenhu area, and analyzed reservoir subsidence, stress distribution, and seabed stability [27]. Jin et al., simulated the depressurization production of hydrates using a horizontal well in the Shenhu area via TOUGH + HYDRATE (Lawrence Berkeley National Laboratory, Los Angeles, CA, USA) and found that formation subsidence in the early stages of production accounted for more than half of the overall subsidence [28]. Jin et al., reported that the elastic–plastic relationship of hydrate-bearing sediments was important when studying their geomechanical behavior during production [28]. Li investigated the influence of excess pore pressure caused by hydrate decomposition on the slope stability of submarine soil, using the finite element strength reduction method [29]. Lee et al., carried out a field-scale numerical simulation study using the cyclic depressurization method in the Ulleung Basin of the Korea East Sea; the vertical displacement during production was calculated via the Geomechanics module of CMG (Computational Modeling Group, Calgary, AB, Canada) [30]. Sun et al., established a coupling model of seepage field, temperature field, and deformation field in the process of gas hydrate depressurization exploitation and analyzed the law of seabed subsidence [31]. These studies focused on the change and distribution of stress–strain and formation subsidence caused by hydrate decomposition during production.

Although the previous research analyzed the production performance and mechanical response of hydrate reservoirs in detail by numerical simulation, there is no research that considers the interaction according to these two aspects and that tries to find how to maximize gas production and minimum formation subsidence at the same time, to achieve safe and efficient production.

In this paper, a production model for unconsolidated marine hydrate sediments was developed. The formation subsidence and the sensitivity of gas production to formation and production parameters were analyzed, and the optimal production methods and parameters aiming at the safe and efficient production of hydrate sediments with different saturations in the South China Sea were determined.

2. Methodology and Procedures

The numerical simulation approach is adopted to elucidate the depressurization production process. The reservoir simulator CMG STARS (v2015, Computational Modeling Group, Calgary, AB, Canada) was utilized to build the reservoir model and to perform all the scenarios discussed in this paper. The STARS (Steam Thermal and Advanced Process Reservoir Simulator) is a reservoir simulator developed by the Computational Modeling Group Ltd., primarily for modeling the flow of three phases in the form of multi-component fluids. It can be used to model compositional, seam, geomechanical, and dispersed components, and the in situ combustion process. SRATS uses a discretized wellbore model that improves the modeling by discretizing the wellbore and solving the resulting coupled wellbore and reservoir flow problem simultaneously. It can also handle the formation and decomposition of hydrates [32,33]. Therefore, CMG STARS is adopted in this paper to build a model of unconsolidated hydrate reservoirs in the South China Sea.

The model is simplified to be a cuboid, assuming that the overburdening and underlying rock layers are impermeable and that saturation, permeability, and porosity are considered homogeneous, which are different from the real and complicated reservoir. The simulation parameters were determined according to the production test conducted in the South China Sea [29] and are simplified to improve the calculation speed: the water depth is 1500 m, the buried depth is 120 m, and the underlying rock strata is 20 m. This model was 2280 m in length, 1600 m in width, and 200 m in thickness. There is a free gas layer under the hydrate-bearing layer. The non-consolidation is achieved through the Geomechanics module in STARS (Figure 1). Although hydrate decomposition temporarily increases the formation porosity, the overburdening rock layers will compact the sediments and decrease the porosity. Moreover, the formation strength will also decrease with hydrate decomposition by changing the cohesion value, which will lead to substantial formation subsidence. The Carmen–Kozeny formula was used to calculate permeability, according to porosity. The model is shown in Figure 2, with a vertical well in the center. The detailed parameters are listed in Table 1, below.

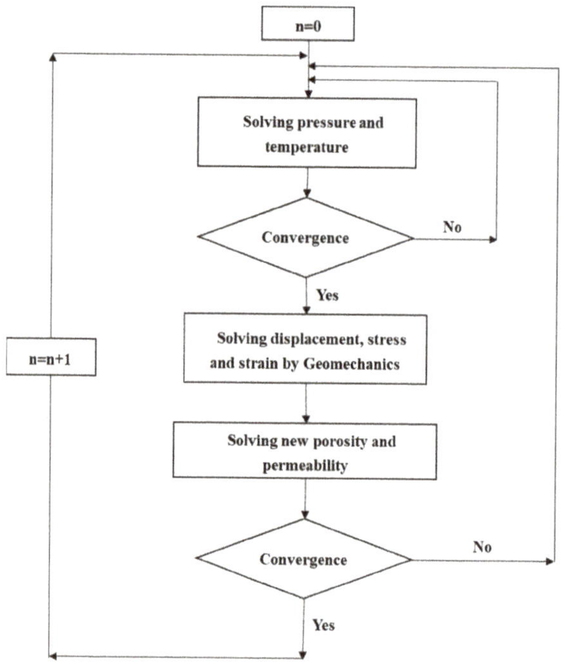

Figure 1. Flow chart of the iterative calculation of the coupled geomechanics module.

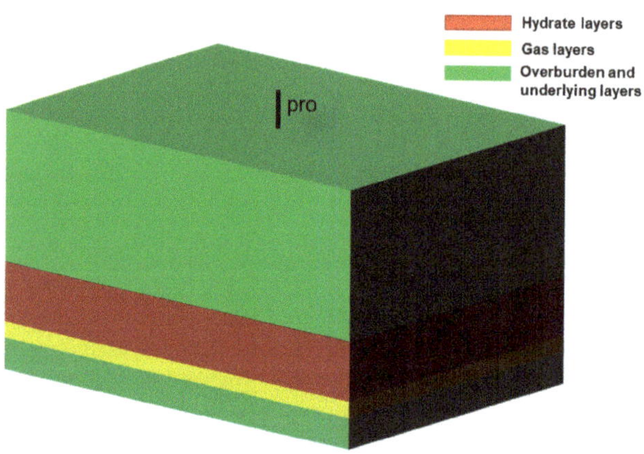

Figure 2. Diagram of the production model of unconsolidated hydrate-bearing strata in the deep-water basin.

Table 1. Simulation parameters.

Parameters	Value
Initial temperature, °C	20
Initial pressure, kPa	19,000
Thickness of hydrate, m	50
Thickness of underlying gas, m	10
Initial hydrate saturation	0.4
Initial water saturation	Hydrate layer = 0.6 Underlying gas layer = 0.5
Initial gas saturation	Hydrate layer = 0 Underlying gas layer = 0.1
Porosity	0.4
Permeability, Md	$k_i = k_j = k_k = 0.1$
Gas composition	100% CH_4
Thermal conductivity of the sediments, J/(m·day·C)	1.728×10^5 (constant)
Yield criterion	Mohr-Coulomb
Initial cohesion, kPa	850
Elasticity modulus, kPa	3.6×10^6
Poisson's ration	0.35

The model was verified by comparing it with the known data of trial production tests, and the distribution of formation subsidence after production was analyzed briefly.

Then, the numerical simulation approach can be divided into two steps, to determine the safe and efficient production methods for unconsolidated hydrate formations with various parameters.

Firstly, a series of scenarios were run in an attempt to analyze the sensitivity of gas production to different parameters, including formation parameters and production parameters. A vertical well was adopted for sensitivity analysis, except when considering the sensitivity regarding the horizontal length. All the scenarios in this paper were simulated for a production duration of 1100 days.

The formation parameters are defined once the target reservoir has been identified, and consist of buried depth, initial hydrate saturation, and formation porosity. The production parameters refer to the well type and the associated parameters that can be intentionally changed. There are two types of wells: vertical and horizontal. Bottom hole pressure is the most important parameter for vertical wells, while horizontal wells should be concerned with horizontal length. Moreover, heat injection can be used with depressurization to boost gas production to some extent in both well types.

In the sensitivity analysis, it was only the parameter to be analyzed that varied, while all the other parameters remained constant. Several groups of simulations were run for each parameter. The relationship curve between cumulative gas production and the parameter to be investigated was linearly fitted and the slopes of these lines were compared, to find out which parameters had the greatest influence on gas production.

Secondly, the appropriate well types and parameters were determined. According to exploration data in the South China Sea, the average porosity of hydrate reservoirs is approximately 40%, with hydrate saturation ranging from 10% to 40% [18,19]. However, hydrate sediments with 10% saturation are not considered promising targets for production [34] and are ignored in the subsequent simulations. A horizontal well and vertical well were adopted. Under various production parameters, cumulative gas production and formation subsidence were obtained. Finally, the appropriate methods and parameters for maximum cumulative gas production were selected, while not exceeding the safety subsidence for various hydrate reservoirs in the South China Sea, based on the various economic and safety concerns.

3. Results and Discussion

3.1. Formation Subsidence of Unconsolidated Hydrate Reservoirs during Production

Unconsolidated hydrate sediments will experience great subsidence during production. Until now, only Japan and China have conducted marine hydrate production tests. According to the trial production test of marine hydrates around the world, the first production test in Japan in 2013 lasted 6 days and was halted due to serious sand production. The monitored seafloor subsidence was 0.1 m to 0.3 m after depressurization, which was combined with heat-injection production [35]. There were no reference data for the formation subsidence of the production test in the South China Sea. Figure 3 shows the subsidence distribution of the unconsolidated hydrate sediments after depressurization production. The largest subsidence of 0.123 m (the blue box in Figure 3) occurred around the well-bore, which was close to the trial test value for Japan, meaning that the model and CMG simulator were feasible for predicting the geomechanical response. The distribution of subsidence was axisymmetric with the production well because hydrates near the well-bore decomposed first and the decomposition area extended outward gradually.

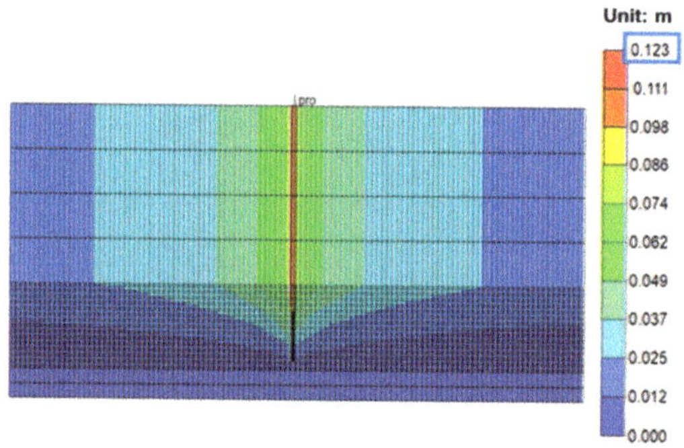

Figure 3. The subsidence distribution of unconsolidated hydrate formation during depressurization.

3.2. Sensitivity to Formation Parameters

3.2.1. Effect of Buried Depth

As illustrated in Figure 4, at the beginning of production, the gas rate increased with the buried depth. The reason is that as the buried depth increases, the initial formation pressure rises, resulting in a larger production pressure differential. Thus, the hydrate decomposition rate would be higher. However, in the case of a deeply buried hydrate reservoir, the decomposition rate would decline dramatically over time, due to the rapid reduction of the residual hydrate, as shown by the yellow curve in Figure 4.

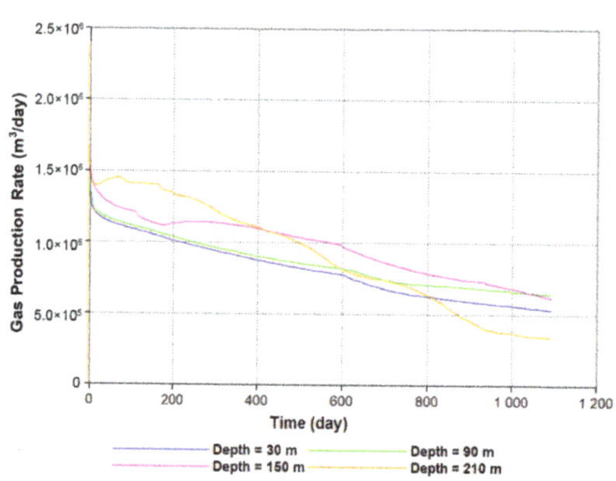

Figure 4. Gas production rates of various buried depths.

Figure 5 depicts the relationship between cumulative gas production and buried depth, which is roughly positive. The curve was linearly fitted, with a slope of 0.00794, indicating that the buried depth had a minor impact on gas production.

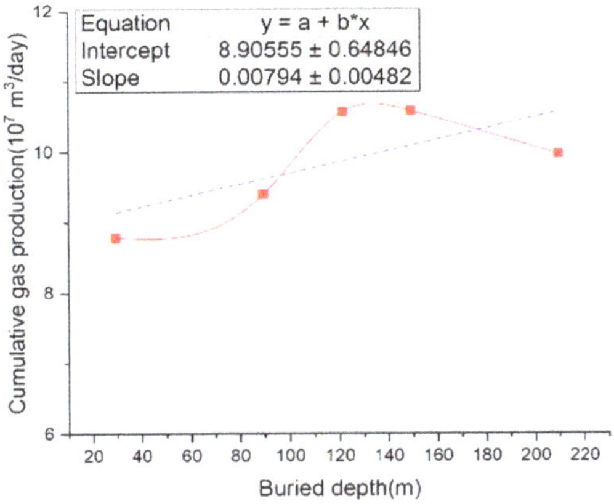

Figure 5. Relationship between cumulative gas production and buried depth.

3.2.2. Effect of Hydrate Saturation

As shown in Figure 6, the gas production rate increases with the decrease in hydrate saturation (denoted by S_h below). As production progressed, the gas rate of the formation with a low S_h began to decline rapidly.

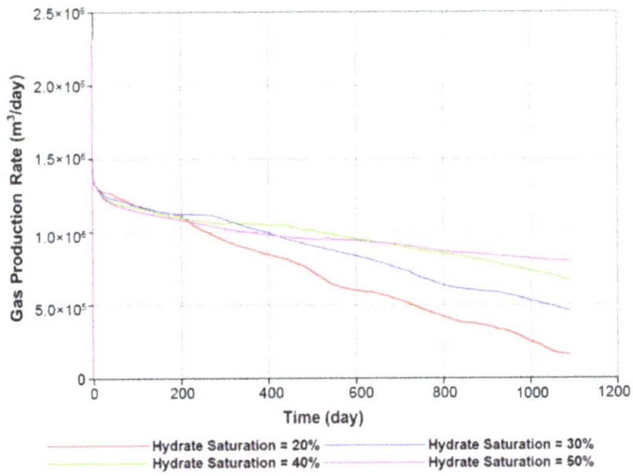

Figure 6. Gas production rates of various hydrate saturations.

This is analogous to ice melting. The distribution of hydrate in the formation is dispersed in reservoirs with a low S_h. The hydrates will rapidly decompose and release methane once the formation pressure declines. Because there is not much hydrate remaining in the formation during the middle and late phases, the gas production rate will drop drastically. In the case of a reservoir with a high S_h, the distribution of hydrate in the formation is rather concentrated, the majority of it being massive. Only the outermost hydrate decomposes at a slow pace at the start of production. As time goes on, the hydrate progressively becomes more dispersed and gas production continues to rise, exceeding the rate of reservoirs with a low S_h.

Therefore, the influence of hydrate saturation on gas production varied, depending on the stages. Reservoirs with a low S_h could generate a large quantity of gas quickly, but the production duration was quite limited. While reservoirs with a high S_h had a long production duration, it required time to achieve significant gas production.

Figure 7 shows the relationship between cumulative gas production and S_h. The slope of the linearly fitted curve was 0.0989, indicating that S_h had a greater effect on gas production than buried depth.

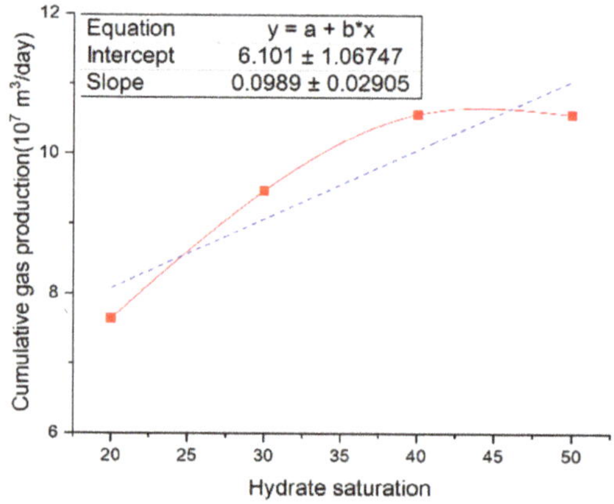

Figure 7. Relationship between cumulative gas production and hydrate saturations.

3.2.3. Effect of Formation Porosity

The larger the porosity (denoted by φ below), the faster the fluid flowed in the formation, and so the pressure drop spread quickly. As a result, the decomposition rate of hydrate would be faster and more gas would be produced. That is, the rate of gas production was proportional to φ, as shown in Figure 8.

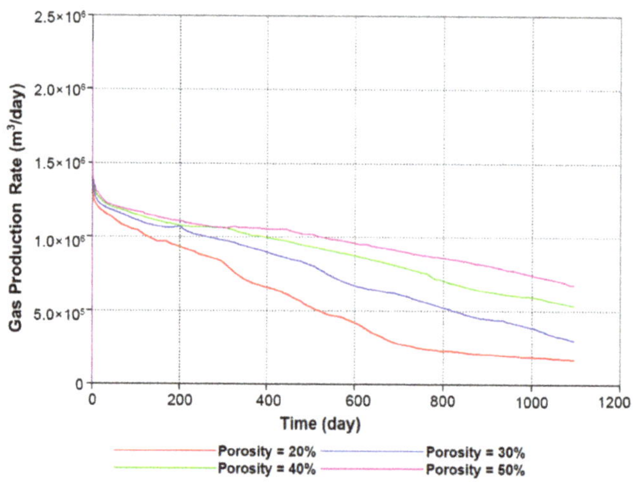

Figure 8. Gas production rates of various formation porosities.

The relationship between gas production and φ is seen in Figure 9. The slope of the curve was 0.1561, suggesting that φ had a greater impact on gas production than S_h.

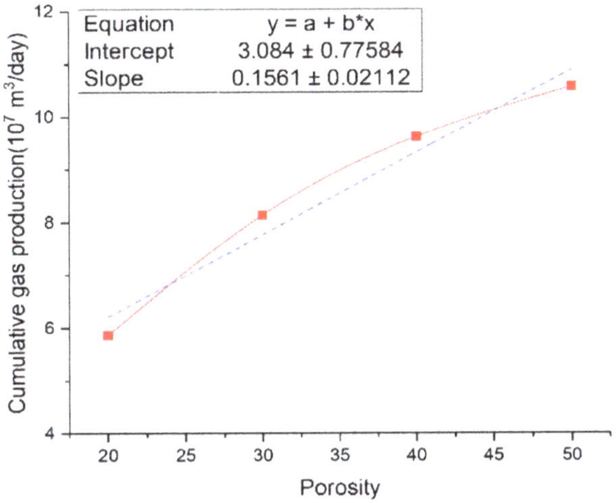

Figure 9. Relationship between cumulative gas production and the formation porosities.

3.3. Sensitivity to Production Parameters

3.3.1. Effect of Bottom Hole Pressure

As shown in Figure 10, the gas rate is inversely proportional to BHP, which implies that if the safety aspect of production is ignored, gas production will increase with the decrease in BHP.

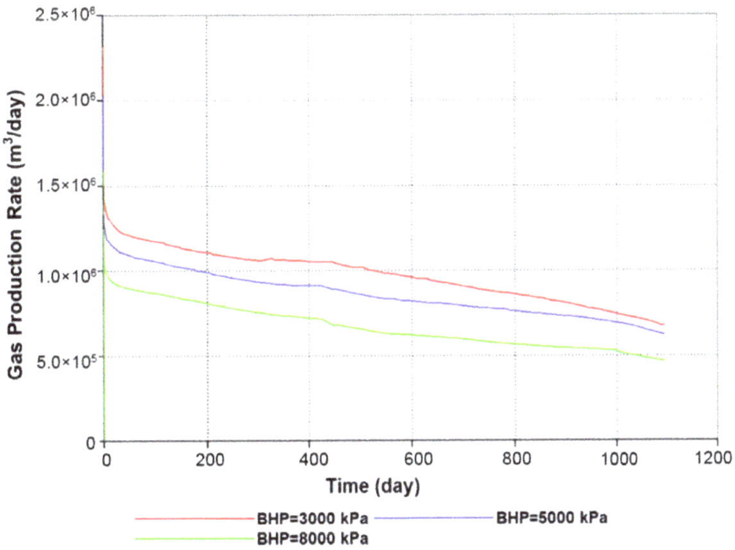

Figure 10. Gas production rates at different bottom-hole pressures.

Figure 11 depicts the connection between gas production and BHP. The curve was linearly fitted, and the slope was −0.6605, indicating that BHP had a significant impact on gas production.

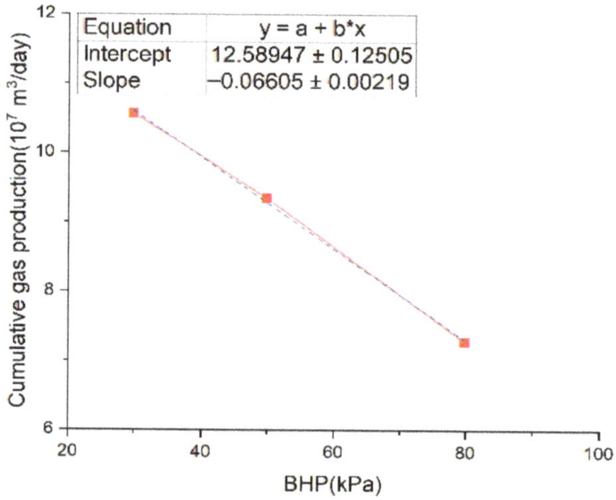

Figure 11. Relationship between cumulative gas production and bottom hole pressures.

3.3.2. Effect of Horizontal Length

As shown in Figure 12, gas production increases with the horizontal length (denoted by L, below) due to the faster pressure drop propagation induced by the increase in the exposed surface.

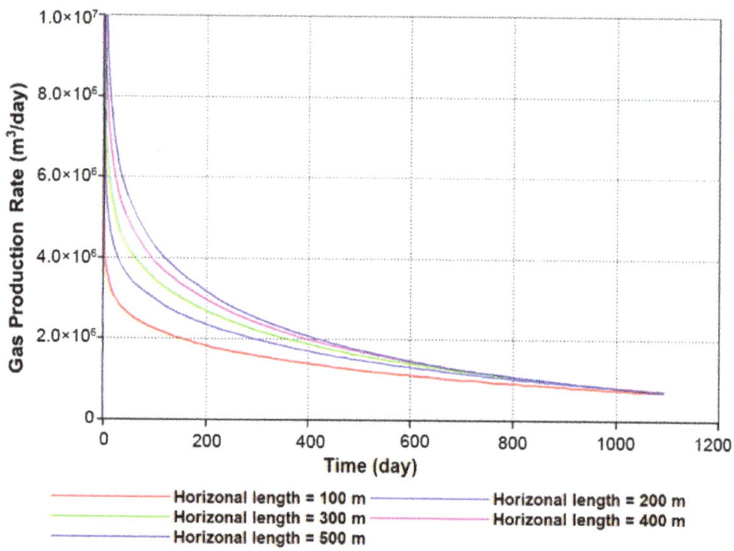

Figure 12. Gas production rates at different horizontal lengths.

Figure 13 displays the correlation between gas production and L. The slope of the curve was 0.02196, demonstrating that L had a great influence on gas production, although not as much as BHP.

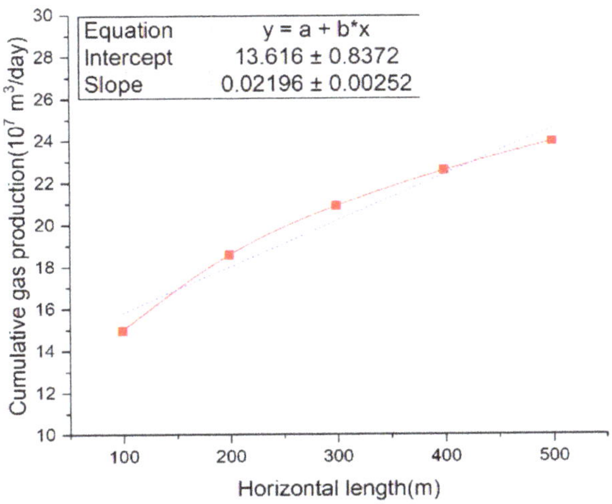

Figure 13. The relationship between cumulative gas production and horizontal length.

3.3.3. Effect of Injection Heat

As shown in Figure 14, injecting heat into the formation can boost the gas rate, albeit only slightly. When the injected heat surpassed a specific threshold (1×10^9 J/day), the gas rate would decrease compared with the rate when there is no heat injection. In other words, within a specific range, the gas rate was positively correlated with heat injection.

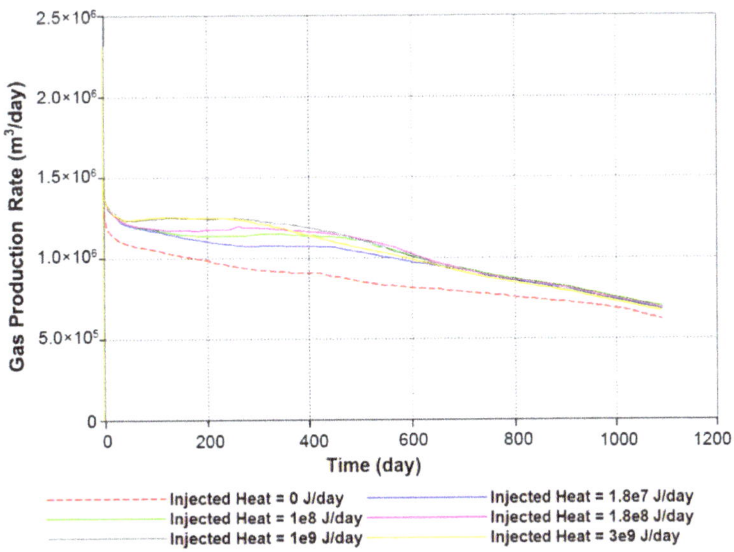

Figure 14. Gas production rates under different heat injection specifications.

Figure 15 shows the relationship between gas production and Q. The curve was linearly fitted with a slope of 0.0021, indicating that Q had a negligible impact on gas production.

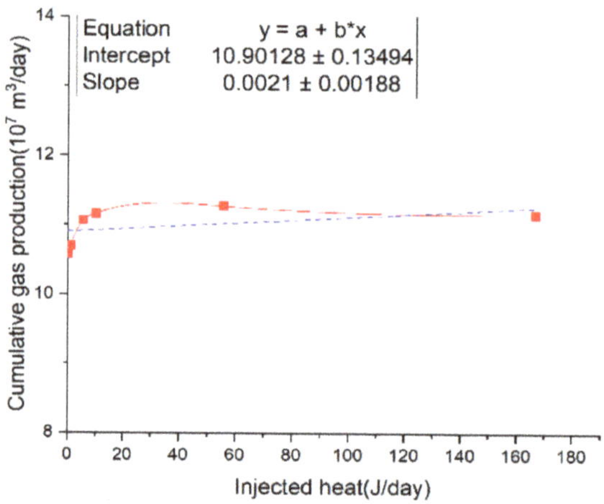

Figure 15. The relationship between cumulative gas production and injected heat.

Overall, when all other parameters were assumed to be constant, the sensitivity of gas production to formation parameters was as follows: formation porosity > hydrate saturation > buried depth. The impact of the production parameters was: BHP > horizontal length > injected heat, as shown in Table 2.

Table 2. Summary of sensitivity analysis.

Parameters		The Slope	Gas Production Sensitivity to the Parameters
Formation parameters	Formation porosity	0.1561	Formation porosity > hydrate saturation > buried depth
	Hydrate saturation	0.0989	
	Buried depth	0.00794	
Production parameters	BHP	−0.6605	BHP > horizontal length > injected heat
	Horizontal length	0.02196	
	Injected heat	0.0021	

3.4. Appropriate Well Types and Parameters

Figures 16 and 17 illustrate the simulation results of hydrate reservoirs of different saturations in the South China Sea. The solid line denotes cumulative gas production, whereas the dotted line represents formation subsidence. The different colors represent different hydrate saturations. Among the simulations in this paper, gas production and formation subsidence using a horizontal well were all higher than that of a vertical well. However, what we expect is maximum gas production with minimum formation subsidence. In order to achieve this goal, it is necessary to choose an appropriate well type. In addition, gas production and formation subsidence both increased with the decrease in bottom hole pressure when a vertical well was adopted, and both increased with the increase in horizontal length when a horizontal well was adopted. Therefore, a balance between gas production and formation subsidence must be determined. That is, the appropriate parameters for the selected well type must be determined to achieve high gas production and low formation subsidence. A safety formation subsidence figure is proposed in this paper to achieve the above objectives. The safety formation subsidence was set at 0.1725 m (the average of the maximum and minimum subsidence figures in all the simulations), as shown by the dotted green line in Figures 16 and 17.

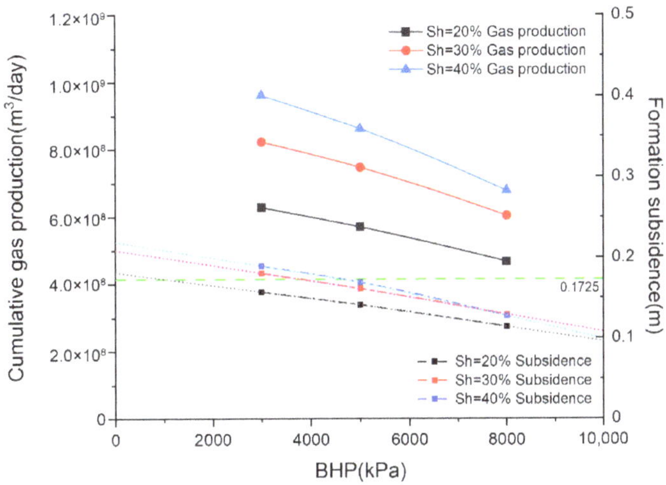

Figure 16. Cumulative gas production and formation subsidence, using vertical wells.

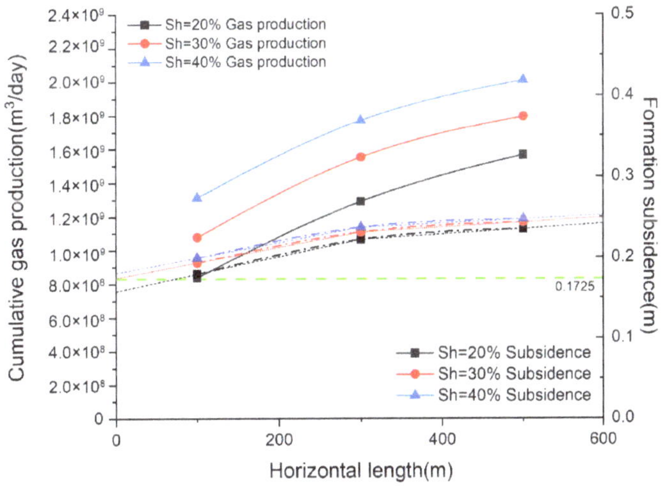

Figure 17. Cumulative gas production and formation subsidence, using horizontal wells.

At this point, the appropriate parameters for different reservoirs could be determined. Figure 16 depicts the appropriate BHP for hydrate reservoirs with various saturation values when a vertical well was used. The subsidence of hydrate reservoirs with different saturations was lower than the safety formation subsidence under a range of BHPs. When the hydrate saturation was 20%, the BHP should be at least 1060 kPa; when hydrate saturation was 30%, the BHP should be at least 3800 kPa; and when hydrate saturation is 40%, the BHP should be at least 4800 kPa.

Figure 17 shows the critical horizontal length found when a horizontal well was adopted. When the hydrate saturation was 20%, the horizontal length should not exceed 80 m. The subsidence of hydrate reservoirs with 30% and 40% saturation exceeded the safety value, whatever the horizontal length of the well, which means that a horizontal well was unsuitable for these two reservoir schemes.

Table 3 summarizes the well type and the parameters suitable for hydrate reservoirs with different saturations. For a hydrate reservoir with a saturation of 20%, gas production using a horizontal well was higher than when using a vertical well. For a hydrate reservoir with a saturation of 30% and 40%, a vertical well was suitable because the subsidence when using a horizontal well exceeded the safety value.

Table 3. Methods and parameters that are suitable for various hydrate reservoirs.

Porosity	Hydrate Saturation	Production Method	Production Parameters	Critical Gas Production
40%	20%	Horizontal well	$L \leq 80$ m	7.7×10^8 m^3/day
	30%	Vertical well	BHP $\geq$ 3800 kPa	7.92×10^8 m^3/day
	40%	Vertical well	BHP $\geq$ 4800 kPa	8.7×10^8 m^3/day

The method offered in this paper to determine the suitable production parameters for gas production is worth using as a reference, but the values still need to be adjusted, based on the consideration of drilling and operation costs in trial production projects.

4. Conclusions

A series of simulations were conducted utilizing CMG software to elucidate the depressurization production process of marine gas hydrates in unconsolidated sediments. The sensitivity analysis showed that porosity and hydrate saturation were the formation parameters that had the greatest impact on gas production and could be used to determine the type of reservoir, while bottom hole pressure and horizontal length were the production parameters that should be optimized for safe and efficient production. Based on economic and safety concerns, the appropriate production methods with the highest cumulative gas production and lowest subsidence for different hydrate reservoirs were determined. When hydrate saturation is 20%, a horizontal well is suitable, and the appropriate horizontal length is less than 80 m. When hydrate saturation is 30% or 40%, a vertical well is suitable, and the appropriate bottom hole pressure is no less than 3800 kPa and 4800 kPa for 30% and 40% saturation, respectively. However, the safety formation subsidence rate is determined merely according to the simulation results in this paper, which need to be adjusted considering other factors in the future, such as the tension on production equipment, the properties of seafloor soils, the operation costs, and so on.

Author Contributions: Conceptualization, Y.C. and T.S.; methodology, Y.C.; validation, S.J.; writing—original draft preparation, Y.C.; writing—review and editing, S.W.; supervision, S.W., T.S.; funding acquisition, S.W. All authors have read and agreed to the published version of the manuscript.

Funding: This research was funded by the Specific Research Fund of the Innovation Platform for Academicians of Hainan Province, grant number No. YSPTZX202204, the Sanya Yazhou Bay Science and Technology City Program, grant number NO. SKJC-2020-01-009, Hainan Provincial Major Science and Technology Program of China, grant number No. ZDKJ202019, and National Key Research and Development Program of China, grant number No. 2019YFC0312301.

Institutional Review Board Statement: Not applicable.

Informed Consent Statement: Not applicable.

Data Availability Statement: All data generated or analyzed during this study are included in this published article.

Acknowledgments: This study was funded by the Specific Research Fund of the Innovation Platform for Academicians of Hainan Province, grant number (No. YSPTZX202204), the Sanya Yazhou Bay Science and Technology City Program (NO. SKJC-2020-01-009), the Hainan Provincial Major Science and Technology Program of China (No. ZDKJ202019), and the National Key Research and Development Program of China (No. 2019YFC0312301).

Conflicts of Interest: The authors declare no conflict of interest.

References

1. Sloan, E.D.; Koh, C.A. *lathrate Hydrates of Natural Gases*, 3rd ed.; CRC Press: Boca Raton, FL, USA, 2008; pp. 1–5.
2. Moridis, G.J.; Collett, T.S.; Pooladi-Darvish, M.; Hancock, S.H.; Santamarina, C.; Boswell, R.; Kneafsey, T.J.; Rutqvist, J.; Kowalsky, M.B.; Reagan, M.T.; et al. Challenges, Uncertainties, and Issues Facing Gas Production From Gas-Hydrate Deposits. *SPE Reserv. Evaluation Eng.* **2011**, *14*, 76–112. [CrossRef]
3. Yuan, Y.L. Numerical Simulation of Exploitation Potential and Mechanical Stability of Marine Natural Gas Hydrate Depressurization. Ph.D. Thesis, Jilin University, Changchun, China, 2019.
4. Hassanpouryouzband, A.; Joonaki, E.; Farahani, M.V.; Takeya, S.; Ruppel, C.; Yang, J.; English, N.J.; Schicks, J.M.; Edlmann, K.; Mehrabian, H.; et al. Gas hydrates in sustainable chemistry. *Chem. Soc. Rev.* **2020**, *49*, 5225–5309. [CrossRef]
5. Ruan, X.; Li, X.-S.; Xu, C.-G. A review of numerical research on gas production from natural gas hydrates in China. *J. Nat. Gas Sci. Eng.* **2020**, *85*, 103713. [CrossRef]
6. Zander, T.; Choi, J.C.; Vanneste, M.; Berndt, C. Potential Impacts of Gas Hydrate Exploitation on Slope Stability—A Study from the Danube Fan, Black SEA. In Proceedings of the Near Surface Geoscience 2016-Second Applied Shallow Marine Geophysics Conference, Barcelona, Spain, 4–8 September 2016. [CrossRef]
7. Matsuda, H.; Yamakawa, T.; Sugai, Y.; Sasaki, K. Gas Production from Offshore Methane Hydrate Layer and Seabed Subsidence by Depressurization Method. *Engineering* **2016**, *8*, 353–364. [CrossRef]
8. Hyodo, M.; Li, Y.; Yoneda, J.; Nakata, Y.; Yoshimoto, N.; Nishimura, A. Effects of dissociation on the shear strength and deformation behavior of methane hydrate-bearing sediments. *Mar. Pet. Geol.* **2014**, *51*, 52–62. [CrossRef]
9. Hyodo, M.; Yoneda, J.; Yoshimoto, N.; Nakata, Y. Mechanical and dissociation properties of methane hydrate-bearing sand in deep seabed. *Soils Found.* **2013**, *53*, 299–314. [CrossRef]
10. Grover, T.; Holditch, S.A.; Moridis, G. Analysis of Reservoir Performance of Messoyakha Gas Hydrate Field. In Proceedings of the Eighteenth International Offshore and Polar Engineering Conference, Vancouver, BC, Canada, 6 July 2008. [CrossRef]
11. Moridis, G.J.; Collett, T.S.; Dallimore, S.R.; Satoh, T.; Hancock, S.; Weatherill, B. Numerical studies of gas production from several CH4 hydrate zones at the Mallik site, Mackenzie Delta, Canada. *J. Pet. Sci. Eng.* **2004**, *43*, 219–238. [CrossRef]
12. Yamamoto, K.; Dallimore, S. Aurora-JOGMEC-NRCan Mallik 2006–2008 gas hydrate research project progress. *Nat. Gas Oil* **2008**, *304*, 285–4541.
13. Collett, T.S.; Boswell, R.; Lee, M.W.; Anderson, B.J.; Rose, K.; Lewis, K.A. Evaluation of Long-Term Gas Hydrate Production Testing Locations on the Alaska North Slope. *SPE Reserv. Eval. Eng.* **2012**, *15*, 243–264. [CrossRef]
14. Boswell, R. Japan completes first offshore methane hydrate production test—methane successfully produced from deepwater hydrate layers. *Cent. Nat. Gas Oil* **2013**, *412*, 386–7614.
15. Konno, Y.; Fujii, T.; Sato, A.; Akamine, K.; Naiki, M.; Masuda, Y.; Yamamoto, K.; Nagao, J. Key Findings of the World's First Offshore Methane Hydrate Production Test off the Coast of Japan: Toward Future Commercial Production. *Energy Fuels* **2017**, *31*, 2607–2616. [CrossRef]
16. Zhou, S.; Zhao, J.; Li, Q.; Chen, W.; Zhou, J.; Wei, N.; Guo, P.; Sun, W. Optimal design of the engineering parameters for the first global trial production of marine natural gas hydrates through solid fluidization. *Nat. Gas Ind. B* **2018**, *5*, 118–131. [CrossRef]
17. Li, J.-F.; Ye, J.-L.; Qin, X.-W.; Qiu, H.-J.; Wu, N.-Y.; Lu, H.-L.; Xie, W.-W.; Lu, J.-A.; Peng, F.; Xu, Z.-Q.; et al. The first offshore natural gas hydrate production test in South China Sea. *China Geol.* **2018**, *1*, 5–16. [CrossRef]
18. Ye, J.L.; Qin, X.W.; Xie, W.W.; Lu, H.L.; Ma, B.J.; Qiu, H.J.; Bian, H. The second natural gas hydrate production test in the South China Sea. *China Geol.* **2020**, *3*, 197–209. [CrossRef]
19. Li, S. Numerical simulation of gas hydrate depressurization and thermal injection in mining process. Master's Thesis, Harbin Engineering University, Harbin, China, 2017.
20. Merey, S.; Sinayuc, C. Numerical simulations for short-term depressurization production test of two gas hydrate sections in the Black Sea. *J. Nat. Gas Sci. Eng.* **2017**, *44*, 77–95. [CrossRef]
21. Chen, C.; Yang, L.; Jia, R.; Sun, Y.; Guo, W.; Chen, Y.; Li, X. Simulation Study on the Effect of Fracturing Technology on the Production Efficiency of Natural Gas Hydrate. *Energies* **2017**, *10*, 1241. [CrossRef]
22. Zheng, R.C.; Yin, Z.; Khoo, B.C.; Linga, P. E Enhanced Gas Recovery from Water Saturated Hydrate Bearing Sediments Using Horizontal Wellbore. In Proceedings of the Offshore Technology Conference Asia, Kuala Lumpur, Malaysia, 20 March 2018. [CrossRef]
23. Zhang, P.; Tian, S.; Zhang, Y.; Li, G.; Zhang, W.; Khan, W.A.; Ma, L. Numerical simulation of gas recovery from natural gas hydrate using multi-branch wells: A three-dimensional model. *Energy* **2020**, *220*, 119549. [CrossRef]
24. Jin, G.; Xu, T.; Xin, X.; Wei, M.; Liu, C. Numerical evaluation of the methane production from unconfined gas hydrate-bearing sediment by thermal stimulation and depressurization in Shenhu area, South China Sea. *J. Nat. Gas Sci. Eng.* **2016**, *33*, 497–508. [CrossRef]
25. Feng, J.-C.; Li, X.; Li, G.; Li, B.; Chen, Z.-Y.; Wang, Y. Numerical Investigation of Hydrate Dissociation Performance in the South China Sea with Different Horizontal Well Configurations. *Energies* **2014**, *7*, 4813–4834. [CrossRef]
26. Gong, B.; Jiang, Y.J.; Wang, G.; Huang, N. Prediction of seabed subsidence from gas hydrate exploitation in the South China Sea. *J. Shandong Univ. Sci. Technol. Nat. Sci.* **2015**, *34*, 61–68. [CrossRef]
27. Wan, Y.Z.; Wu, N.Y.; Hu, G.W.; Xin, X.; Jin, G.R.; Liu, C.L.; Chen, Q. Reservoir stability during gas hydrate depressurization exploitation in Shenhu area, South China Sea. *Nat. Gas Ind.* **2018**, *38*, 117–128. (In Chinese) [CrossRef]

28. Jin, G.; Lei, H.; Xu, T.; Xin, X.; Yuan, Y.; Xia, Y.; Juo, J. Simulated geomechanical responses to marine methane hydrate recovery using horizontal wells in the Shenhu area, South China Sea. *Mar. Pet. Geol.* **2018**, *92*, 424–436. [CrossRef]
29. Li, L.L.; Yang, J.; Lu, B.P.; Ke, K.; Wang, L.; Chen, K.J. Research on stratum settlement and wellhead stability in deep water during hydrate production testing. *Pet. Drill. Tech.* **2020**, *48*, 61–68. (In Chinese) [CrossRef]
30. Lee, T.; Lee, J.Y.; Ahn, T.; Son, H.A. Numerical Simulation of Gas Hydrate Production Using the Cyclic Depressurization Method in the Ulleung Basin of the Korea East Sea. *Appl. Sci.* **2021**, *11*, 9748. [CrossRef]
31. Sun, J.; Wu, S.G.; Zhu, L.Q.; Liu, Y.R.; Sun, Z.Y. Characteristics and influencing factors of seabed subsidence in gas hydrate depressions mining. *J. Cent. South Univ. Sci. Technol.* **2022**, *53*, 1033–1046.
32. Chaves, G. Simulation of CO_2 sequestration in deep saline aquifers. Master's Thesis, New Mexico Institute of Mining and Technology, Socorro, NM, USA, 2011.
33. Janicki, G.; Schlüter, S.; Hennig, T.; Lyko, H.; Deerberg, G. Simulation of Methane Recovery from Gas Hydrates Combined with Storing Carbon Dioxide as Hydrates. *J. Geol. Res.* **2011**, *2011*, 1–15. [CrossRef]
34. Moridis, G.J.; Sloan, E.D. Gas production potential of disperse low-saturation hydrate accumulations in oceanic sediments. *Energy Convers. Manag.* **2007**, *48*, 1834–1849. [CrossRef]
35. Yokoyama, T.; Shimoyama, M.; Matsuda, S.; Tago, K.; Takeshima, J.; Nakatsuka, Y. Monitoring system of seafloor subsidence for methane hydrate production test. In Proceedings of the SPWLA 18th Formation Evaluation Symposium of Japan, Chiba, Japan, 27 September 2012.

energies

MDPI

Article

Numerical Simulation on Erosion Wear Law of Pressure-Controlled Injection Tool in Solid Fluidization Exploitation of the Deep-Water Natural Gas Hydrate

Yang Tang [1,2,3,*], Peng Zhao [1,3], Xiaoyu Fang [2], Guorong Wang [1,2,3], Lin Zhong [1,2,3] and Xushen Li [2]

1 School of Mechatronic Engineering, Southwest Petroleum University, Chengdu 610500, China; zp15181079658@163.com (P.Z.); 200331010023@swup.edu.cn (G.W.); zhonglin858296@163.com (L.Z.)
2 Guangdong Provincial Laboratory of Southern Marine Science and Engineering (Zhanjiang), Zhanjiang 524000, China; fangxy@zjblab.com (X.F.); lixsh@zjblab.com (X.L.)
3 Energy Equipment Institute, Southwest Petroleum University, Chengdu 610500, China
* Correspondence: tangyanggreat@126.com; Tel.: +86-134-3839-2610

Abstract: The pressure-controlled injection tool (PCIT) is the key equipment in the process of high-pressure water jet fragmentation in the solid fluidization exploitation of deep-sea natural gas hydrate (NGH). The internal flow field erosion wear numerical simulation model of PCIT is established through computational fluid dynamics software to study the influence law and main factors of the drilling fluid erosion wear of PCIT. The influence laws of different drilling fluid physical parameters and different structural parameters on PCIT erosion wear were analyzed based on the Euler–Lagrangian algorithm bidirectional coupled discrete phase model (DPM) and the solid–liquid two-phase flow model. The results show that the easily eroded areas are the cone of the sliding core, the plug transition section, the plug surface, and the axial flow passage. The sliding core inlet angle and solid particle size are the main factors affecting the PCIT erosion rate. When the inlet angle of the sliding core is 30°, the diameter of solid-phase particles in drilling fluid is less than 0.3 mm, and the erosion degree of the PCIT could be effectively reduced. The research results can provide guidance for the design and application of the PCIT and advance the early realization of the commercial exploitation of hydrate.

Keywords: natural gas hydrate; solid-state fluidization exploitation; pressure-controlled injection tool; erosion wear; numerical simulation

Citation: Tang, Y.; Zhao, P.; Fang, X.; Wang, G.; Zhong, L.; Li, X. Numerical Simulation on Erosion Wear Law of Pressure-Controlled Injection Tool in Solid Fluidization Exploitation of the Deep-Water Natural Gas Hydrate. *Energies* **2022**, *15*, 5314. https://doi.org/10.3390/en15155314

Academic Editor: Ingo Pecher

Received: 11 May 2022
Accepted: 14 June 2022
Published: 22 July 2022

1. Introduction

Natural gas hydrate (NGH) has the characteristics of high energy density and huge reserves. It is widely distributed in the world and is regarded as the alternative energy source with the most potential in the 21st century [1–3]. Deep-sea NGH test production has been carried out in many locations around the world, and many laboratories and scholars have conducted research on NGH production technology and professional production tools. At present, the main NGH exploitation methods include the chemical inhibitor injection method [4], decompression method [5], thermal stimulation method [6], and carbon dioxide (CO_2) replacement method [7]. However, those above methods have different degrees of disadvantages for the exploitation of non-diagenetic NGH with the following characteristics: buried shallow, poor cementation, and low permeability. The chemical inhibitor injection method is an effective auxiliary method for NGH exploitation, which makes use of chemical inhibitors with different properties to accelerate the decomposition of gas hydrate and the formation of CO_2 hydrate or reduce the speed of secondary formation of gas hydrate. However, the main disadvantage of the chemical inhibitor injection method is expensive chemical reagents, high production costs, and the use of large quantities of chemical reagents in gas hydrate reservoirs, which will cause a series of environmental problems. With reference to the decompression method, which harvests methane from hydrates by reducing hydrate reservoir pressure below the hydrate decomposition pressure to separate the hydrate [8], serious wellbore sand production and secondary formation of NGH in the pipeline will block the transportation channel and cause the interruption of the production process. Furthermore, the formation collapse of the goaf is easily caused by the disorderly decomposition of numerous NGHs. With regard to the thermal stimulation method, which extracts methane by heating hydrate reservoirs to decompose solid hydrates, the obvious disadvantage of this method is the inefficient utilization of thermal energy in subsea hydrate reservoirs [8]. Finally, in the CO_2 replacement method, carbon dioxide is injected into the hydrate reservoir, and the methane molecules in the hydrate are replaced by carbon dioxide molecules and further released under specific conditions [6]. However, the production cycle is too long due to the extremely low extraction efficiency, thus the method is not suitable for the large-scale commercial exploitation of NGHs [9,10]. Based on the above reasons, an innovative mining method, that is, natural gas hydrate solid-state fluidization exploitation was proposed by Zhou [11]. Figure 1 illustrates the schematic diagram of this method [11–14]. During this process, the uncontrollable decomposition of NGH in the traditional mining method is transformed into controlled water jet fragmentation and decomposition. The application of bottom-hole sediment coarse separation technology and sediment backfill technology in solid-state fluidized exploitation solves the problems of pipeline blockage and geological collapse inherent in traditional mining technology.

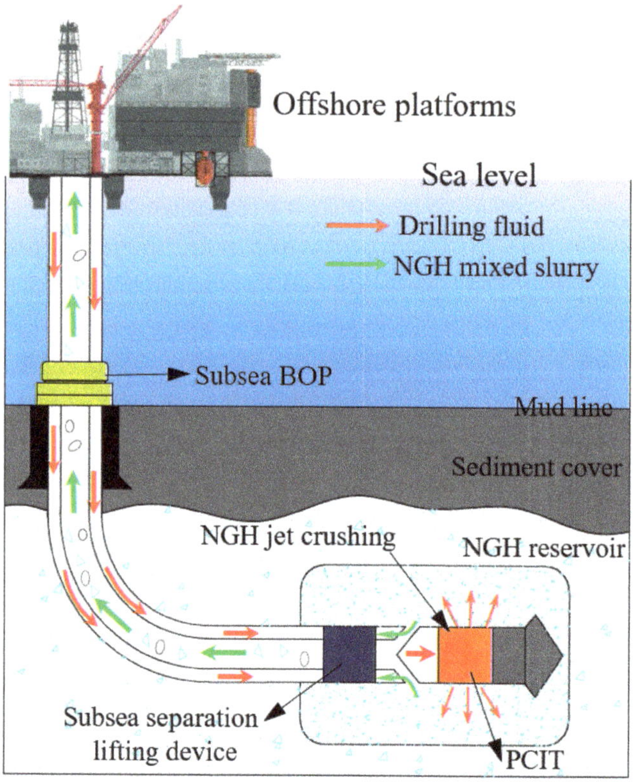

Figure 1. Solid fluidization exploitation method.

Based on the requirements of crushing and cavity making of the hydrate layer in NGH solid-state fluidized exploitation, Tang Yang et al. [15,16] innovatively designed a pressure-controlled injection tool (PCIT) that combines the principle of throttling pressure control and the inclined surface guide mechanism. The PCIT integrates the drilling process of horizontal wells and the jet breaking and cavity making process in the solid-state fluidized exploitation and realizes controllable and rapid switching between the two processes by changing the drilling fluid flow rate inside the tool. In the process of horizontal well drilling and cavity making, high-pressure injection of drilling fluid containing solid particles can effectively fragment and drill hydrate formations with soft geology and large porosity. Under the erosion wear of high-pressure drilling fluid with solid particles over time, the opening and closing process of PCIT will be affected, which will further cause the entire production process to be interrupted. Therefore, in order to ensure the long-term stable use of the PCIT, it is necessary to conduct erosion wear analysis.

With the development of computational fluid dynamics simulation software, computational fluid dynamics (CFD) has become a key tool for predicting erosion in complex flows and is a cost-effective way to mitigate erosion wear when designing drilling tools. Many scholars have done some research on fluid erosion behavior. Yi et al. [17] carried out research on the erosion of fracturing elbows in hydraulic fracturing operations and obtained the influence of fracturing fluid parameters such as flow rate, particle size, and density on the erosion rate of fracturing elbows. Huang et al. [18] established the relationship between the erosion of drill pipe and the gas injection rate and mechanical speed in the process of gas drilling through theoretical derivation and laboratory experiments. It was found that erosion will be reduced under the conditions of small gas injection rate and large mechanical speed. Huang et al. [19] carried out numerical simulation on the erosion of the faucet elbow in reverse circulation drilling, predicted the location of the puncture point of the faucet elbow, and obtained the main factors affecting the collision zone. Wang et al. [20] used the discrete phase model (DPM) and the semi-empirical material removal model to predict the mass loss and erosion distribution of drilling choke valves and greatly reduced the erosion wear of choke valves through optimized design. Jafari et al. [21] analyzed and compared the erosion properties of four wear-resistant steel plates and obtained the relationship between material hardness, erosion resistance, and wear resistance by conducting laboratory experiments. Habib et al. [22] analyzed the liquid–solid two-phase erosion problem in the reducer pipe through numerical simulation and found that the inlet liquid-phase velocity, particle size, and surface hardness of the material are the main factors affecting the erosion law of the reducer. These experiments and studies have made a great contribution to the design and application of the oil and gas exploitation tools. Meanwhile, the PCIT is the key equipment in solid fluidization exploitation. However, to the best of our knowledge, no specific study has addressed the erosion wear analysis of this kind of PCIT, which seriously limits the design and application of such tools. In order to meet the requirements of long-term and stable jet breaking and horizontal well drilling in solid fluidization exploitation, it is necessary to study the erosion wear law of this kind of tool. It is hoped that the commercial application of the natural gas hydrate solid-state fluidization process could be promoted with our proposed PCIT.

2. Working Mechanism of PCIT

The design idea of PCIT is to combine the principle of throttling pressure control and the inclined surface guide mechanism. The throttling pressure control principle is that energy loss occurs when the fluid passes through the variable diameter section, and then a pressure drop occurs at the variable section. This pressure drop in turn generates an axial force at the variable cross-section to control the axial movement of the sliding core [15]. The PCIT is mainly composed of an outer hull, a sliding core, a spring, a thrust bearing, an inclined surface guide mechanism, an axial extrusion seal, and 24 jet nozzles, as shown in Figure 2. Its work process can be divided into the following four stages.

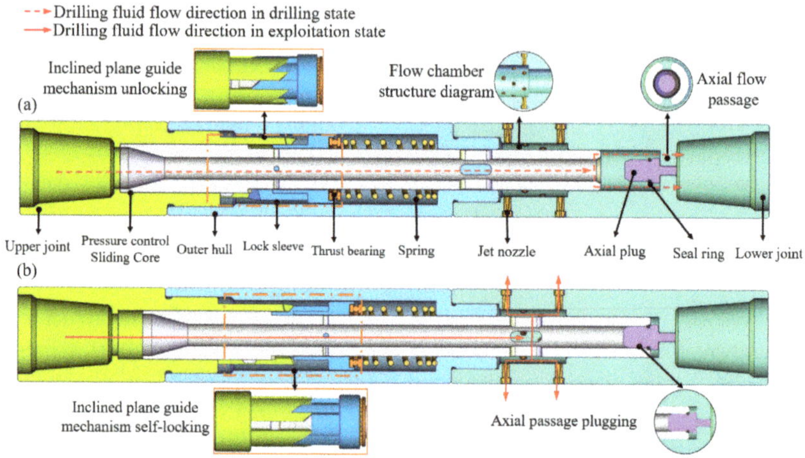

Figure 2. Structure diagram of PCIT: (**a**) PJCT in closed state, (**b**) PJCT in opend state.

2.1. *Stage 1: Horizontal Drilling Process*

The PCIT is the part of the downhole tool string in NGH solid fluidization exploitation. In the horizontal drilling process, the jet nozzle must be closed. Under low drilling fluid flow, a smaller pressure drop and small axial force is created in the slide core of the SPIT, so that the slide core does not move. Therefore, the drilling fluid provides power for lower turbine power drilling tools through the sliding core and the axial flow passage.

2.2. *Stage 2: NGH Crushing and Cavity Making*

When the horizontal well pilot hole drilling is completed, drilling fluid flow is increased by pumping from the offshore drilling platform, so greater pressure drop and axial forces occur inside the sliding core. The sliding core moves axially to compress the spring, and the inclined plane guide mechanism begins to move with the sliding core. When the sliding core moves to the predetermined position, the jet nozzle on the tool is fully opened, and the drilling fluid is ejected from the jet nozzle at a high speed for the NGH crushing and cavity-making operations. In the meantime, the axial channel is completely blocked under the action of the axial extrusion seal, and the inclined plane guide mechanism also enters the locked state, which ensures the stable process of NGH crushing and cavity making. The PCIT was slowly dragged back with the downhole tool string for NGH crushing and cavity making.

2.3. *Stage 3: Multi-Angle Directional Mining*

When the exploitation of the target reservoir in this horizontal direction is completed, the downhole tool string does not need to be taken out of the well bore. The drilling fluid flow should be reduced to 0 L/min, and then the drilling fluid flow is gradually increased to release the locked state of the inclined plane guide mechanism. The jet nozzle on the PCIT is closed, and the axial channel is reopened. After that, the whole downhole tool will change in another direction at the bottom of the wellbore and repeat Stage 1 and Stage 2 to achieve multi-angle large-scale mining.

2.4. *Stage 4: Multi-Level Mining*

When the solid-state fluidized exploitation of the target-depth hydrate reservoir has been completed, the whole downhole tool will change to another depth, and repeat Stage 1, Stage 2, and Stage 3 to realize multi-level large-scale mining.

3. Governing Equations

Numerical simulation is a cost-effective way to mitigate erosion when designing downhole tools, and computational fluid dynamics (CFD) is a highly specialized simulation tool for predicting tool erosion wear under complex flow conditions. In this study, the Eulerian–Lagrangian method was used to describe the flow behavior of solid–liquid two-phase in which liquid is regarded as the continuous phase governed by the Reynolds-averaged Navier–Stokes (RANS) equation, and solid particles were considered as the dispersed phase governed by the second law of Newton. Finally, the finite volume method (FVM) was used to solve the governing equations.

3.1. Liquid-Phase Governing Equations

Considering the solid–liquid-phase interaction, the continuous phase flow equation and the momentum equation are [23,24]

$$\frac{\partial \rho}{\partial t} + \frac{\partial \rho \overline{u_i}}{\partial x_i} = 0 \tag{1}$$

$$\rho \frac{\partial \overline{u_i}}{\partial t} + \rho \frac{\partial \overline{u_i u_j}}{\partial x_j} = -\frac{\partial\ p}{\partial x_i} + \mu \nabla^2 \overline{u_i} - \rho \frac{\partial \overline{u_i' u_j'}}{\partial x_j} + \rho g_i \tag{2}$$

where $\overline{u_i}$ is the (u, v, w) scalar components of the mean-velocity vector, $\overline{x_i}$ is the (x, y, z) scalar components of the spatial-coordinates vector, t is the time coordinate, $\overline{p}$ is the mean pressure, u_i' is the fluctuation velocity component, τ_{ij} is the scalar components of the Reynolds-stress tensor, and v is the fluid kinematic viscosity.

During the numerical simulation in this study, the liquid phase is in a turbulent state. In order to accurately simulate the liquid flow state, it is very important to choose an appropriate turbulence calculation model when performing turbulence calculation. The $k-\varepsilon$ turbulence model is suitable for a high Reynolds number, incompressible turbulent flow, which is widely used in engineering applications. Therefore, the $k-\varepsilon$ turbulence model was applied to describe the turbulent properties; the transport equations of k and ε are, respectively [25,26],

$$\frac{\partial(\rho k)}{\partial t} + \frac{\partial(\rho k u_i)}{\partial x_i} = \frac{\partial}{\partial x_i}\left[\left(\mu + \frac{\mu_t}{\sigma_k}\right)\frac{\partial k}{\partial x_j}\right] + G_b + G_k - \rho\varepsilon \tag{3}$$

$$\begin{aligned} \frac{\partial(\rho\varepsilon)}{\partial t} + \frac{\partial(\rho\varepsilon u_i)}{\partial x_i} = & \frac{\partial}{\partial x_j}\left[\left(\mu + \frac{\mu_t}{\sigma_\varepsilon}\right)\frac{\partial \varepsilon}{\partial x_j}\right] + \rho C_1 S\varepsilon \\ & + \frac{(1-C_{3\varepsilon})C_{1\varepsilon}\varepsilon}{k} G_b - C_{2\varepsilon}\rho\frac{\varepsilon^2}{k+\sqrt{v\varepsilon}} \end{aligned} \tag{4}$$

where

$$\begin{gathered} C_1 = \max\left(0.43, \frac{\eta}{\eta+5}\right),\ \eta = S\frac{k}{\varepsilon} \\ G_k = -\rho\overline{u_i' u_j'}\frac{\partial u_j}{\partial x_i}, G_b = -g_i \frac{u_t}{Pr_t}\frac{\partial \rho}{\rho \partial x_i}, u_t = \rho C_\mu \frac{k^2}{\varepsilon} \end{gathered} \tag{5}$$

where ε is the turbulent dissipation rate, m^2/s^3; k is the turbulent kinetic energy, J; μ is the dynamic viscosity, Pa · s; μ_i is the turbulent viscosity, Pa · s; S is the strain rate, G_k is the production term of the turbulent kinetic energy, G_b is the production term of the turbulent kinetic energy due to lift, P_{rt} is the Prandtl number taken as 0.85, $C_{3\varepsilon}$, $C_{1\varepsilon}$, C_2, C_μ are constants taken as 0.9, 1.44, 1.9, and 0.09, respectively, and finally, σ_k and σ_ε are turbulent Prandtl numbers taken as 1.0 and 1.3, respectively.

The turbulent intensity of the discrete phase is described by the equation of motion of the solid-phase particles. Through this equation, parameters such as the motion trajectory and velocity of the solid-phase particles can be obtained. This particle motion equation is called the DPM model [27]:

$$\frac{du_p}{dt} = \frac{C_D R_e}{24t}(u - u_p) + \frac{g_x(\rho_p - \rho)}{\rho_p} + F_{\Delta p} + F_F + F_v + F_L \tag{6}$$

$$\begin{gathered} R_e = \frac{\rho D_p |u_p - u|}{\mu} \\ C_D = \frac{24}{R_e}\left(1 + b_1 {R_e}^{b_2}\right) + \frac{b_3 R_e}{b_4 + R_e} \end{gathered} \tag{7}$$

where u_p is the velocity of the solid particle, C_d is the drag coefficient, ρ_p is the density of the solid particle, ρ is the liquid density, C_D is the drag coefficient, u is the liquid-phase velocity, $F_{\Delta P}$ is the additional force due to the pressure gradient in the fluid, F_F is the lifting force generated by the fluid, F_V is the virtual mass force required to accelerate the fluid surrounding the particle, F_L is the buoyancy force that includes gravitational and relative density effects, b_1, b_2, b_3 , b_4 are constants related to the solid-phase particles taken as 0.186, 0.653, 0.437, and 7178.741, respectively.

3.2. Erosion Model

The main material of the PCIT is 42CrMo steel. Therefore, the erosion prediction model suitable for the impact of quartz sand on the surface of carbon steel was selected [28].

$$R_{erosion} = \sum_{p=1}^{N_{particle}} \frac{m_p C(d_p) f(\alpha) v^{b(v)}}{A_{face}} \tag{8}$$

where $R_{erosion}$ is the erosion rate, $N_{particle}$ is the particle impact number, m_p is the solid-phase particle quality, $C(d_p)$ is the solid-phase particle diameter function, v is the velocity of solid particles relative to the wall, $b(v)$ is the solid-phase particle relative velocity function, α is the impact angle of the solid-phase particle path and the wall, $f(\alpha)$ is the impact angle function, when the impact angle is 0°, 20°, 30°, 45°, and 90°, $f(\alpha)$ is taken as 0, 0.8, 1, 0.5, and 0.4, respectively, A_{face} is the area of the material surface impacted by particles.

When the solid-phase particles hit the surface of the material at a certain speed, the solid-phase particles will lose part of their energy due to the impact collision, so the solid-phase particles will bounce back at a lower speed than the impact speed. The rebound coefficient is used to determine the rebound speed and rebound angle. Determining solid-phase particle trajectories requires particle bounce properties and coefficients of restitution, the tangential and normal restitution coefficients of solid particles are [29,30]

$$e_T = 0.988 - 0.029\alpha_1 + 6.43 \times 10^{-4}\alpha_2 - 3.56 \times 10^{-6}\alpha_3 \tag{9}$$

$$e_N = 0.993 - 0.0307\alpha_1 + 4.75 \times 10^{-4}\alpha_2 - 2.61 \times 10^{-6}\alpha_3 \tag{10}$$

where e_T is the tangential coefficient of restitution, e_N is the normal recovery factor.

3.3. Erosion Wear Flow Field Geometric Model and Meshing

According to the structure of the PCIT, the internal flow field calculation area of the PCIT was established, and the unnecessary structures in the flow field calculation domain were simplified. The simplified flow field geometric diagram is shown in Figure 3a, and the result of meshing the flow field is shown in Figure 3b.

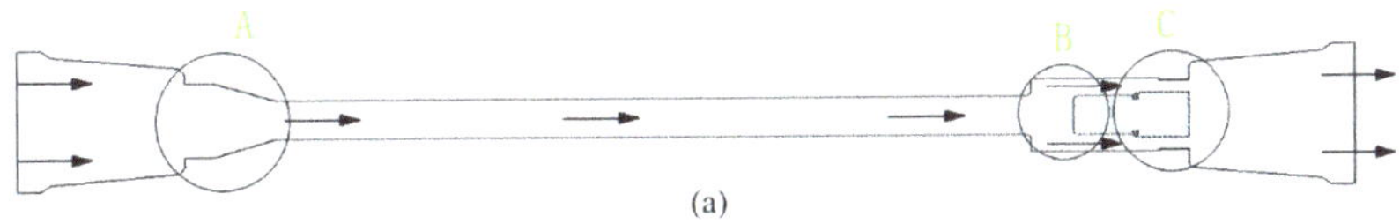

(a)

(b)

Figure 3. Flow field meshing: (**a**) the simplified flow field geometric diagram, (**b**) the result of meshing the flow field.

In Figure 3a, area A is the cone of the sliding core, area B is the transition section of the plug, area C is the surface of the plug and the axial flow passage. The above three area meshes are refined.

3.4. Simulation Parameters and Boundary Conditions

(1) In the simulation process, the velocity inlet boundary condition was used, which is determined by the flow rate and the inlet size of the PCIT. At the same time, the inlet velocity of the solid-phase particles was set to be the same as the fluid velocity, and the outlet adopted the pressure outlet boundary condition.
(2) In this paper, the standard wall function method was chosen to correct the results. The specific definition value in the reflection wall model was determined by the wall collision recovery equation. The impact angle function was defined as a linear value, as shown in Table 1.

Table 1. Impact angle function parameters.

Impact Angle (°)	$f(a)$
0	0
20	0.8
30	1
45	0.5
90	0.4

4. Analysis of Numerical Simulation Results

4.1. Prediction of Erosion Location

The specific erosion location of the PCIT must be determined to pave the way for the study of the factors that affect the erosion law. For this, the diameter of solid particles in the drilling fluid was set to d = 0.2 mm, the density was 1500 kg/m^3, the liquid inlet velocity was 10 m/s, and the viscosity was 20 mpa · s for simulation analysis.

Figure 4 shows the distribution of the PCIT's easily eroded areas. There are four main erosion areas in the PCIT as shown in Figure 4: the sliding core cone surface, the transition section of the plug, the surface of the plug, and the axial flow passage. The sliding core cone surface is located in the PCIT inlet section; if the drilling fluid flow rate increases, the cone surface will be seriously eroded by the vertical impact of solid-phase particles in the drilling fluid. The plug is located at the fluid outlet of the PCIT. The surface of the plug is directly impacted by the drilling fluid, and the average erosion rate and the maximum erosion rate are both the largest. After the fluid particles impact the plane of the plug, the particles will move to the sides of the plug. At the same time, due to the sudden shrinkage of the fluid channel and the increase of the drilling fluid flow rate, the erosion of the inner wall of the axial flow channel is also obvious.

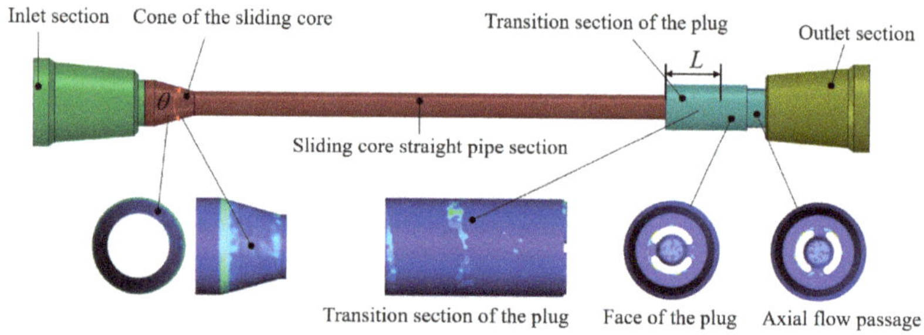

Figure 4. Erosion location distribution of PCIT.

4.2. Influence of Drilling Fluid Physical Parameters on Erosion Rate

4.2.1. Influence of Drilling Fluid Flow on Erosion Rate

The drilling fluid flow rate was set to 200, 300, 400, 500, 600, 700, and 800 L/min, respectively, and the particle diameter, plug position, and mass flow were kept in the same conditions. The maximum erosion rate and the erosion area of the PCIT under different flow rates were obtained as shown in Figures 5 and 6, respectively.

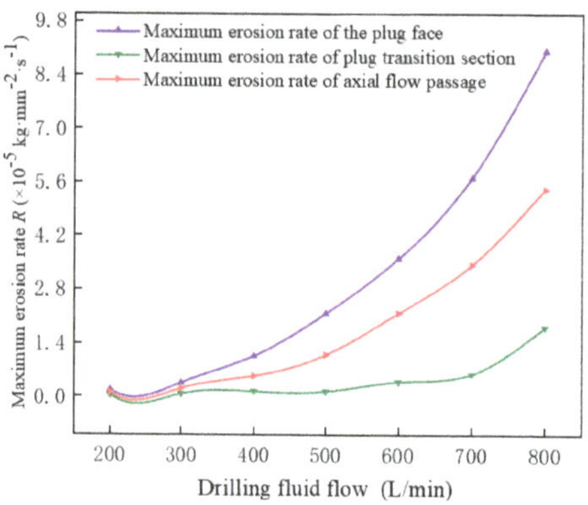

Figure 5. The maximum erosion rate changes with drilling fluid flow.

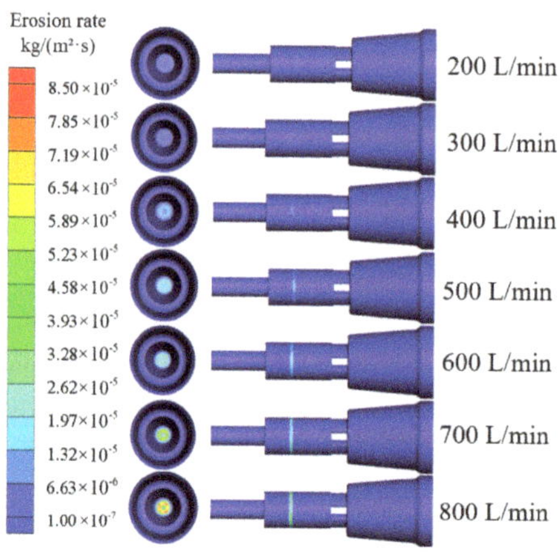

Figure 6. Erosion area changes with drilling fluid flow.

As shown in Figure 5, with the increase of the inlet drilling fluid flow, the maximum erosion rate of the plug surface, the plug transition section, and the axial flow passage all increase. In particular, the maximum erosion rate on the plug surface increases exponentially. This occurs because, as the velocity of drilling fluid flow in PCIT increases with the increase of flow rate, the impact kinetic energy of the solid particles gradually increases, and the frequency and the solid particles' impact the surface of the PCIT per unit time increases. Therefore, the erosion rate of the three erosion-prone areas increases. When the drilling fluid flow increases from 200 to 800 L/min, the average maximum erosion rate of the three erosion-prone areas increased by 65.4 times. As shown in Figure 6, the erosion area of the three erosion-prone areas did not change significantly, which indicates that the inlet flow should be controlled as much as possible during the use of the PCIT.

4.2.2. Influence of Solid-Phase Particle Mass Flow on Erosion Rate

The mass flow is the mass of fluid flowing through PCIT per unit time. Under a constant drilling fluid flow velocity and solid particles, the influence of different mass concentrations on the erosion law was analyzed. The mass flow was set as 0.001, 0.002, 0.003, 0.004, 0.005, 0.006, and 0.007 kg/s, respectively. The variation law of the maximum erosion rate and the erosion area in the erosion-prone areas in the PCIT are shown in Figures 7 and 8.

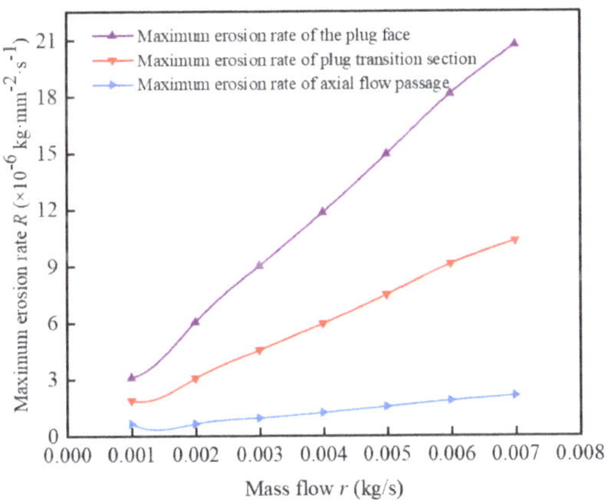

Figure 7. The maximum erosion rate changes with mass flow.

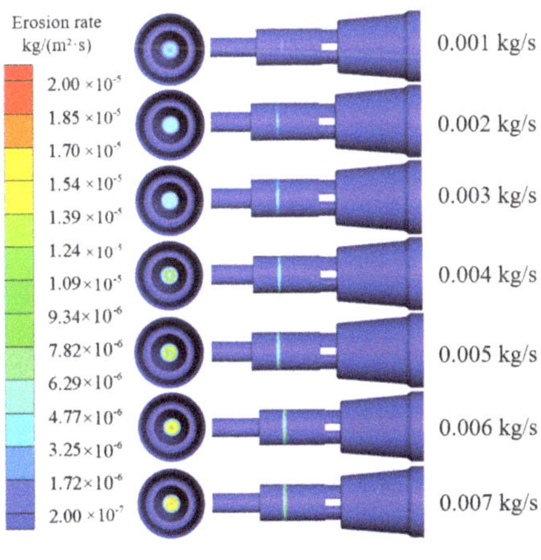

Figure 8. Erosion area changes with mass flow.

As shown in Figure 7, the maximum erosion wear rates of the three erosion-prone regions increase linearly with the increase of the fluid mass flow. The largest erosion rate at the plug surface increased by 6.05 times when the fluid mass flow increased from 0.001 to 0.007 kg/s. Due to the increase in the mass flow rate of the drilling fluid, the number of solid particles entering the PCIT per unit time increased, and the impact frequency of the PCIT by the solid-phase particles increased per unit time. The plug face was directly impacted by the solid-phase particles in the drilling fluid, and the impact angle was around 90°. Therefore, the maximum erosion rate on the plug face was always the maximum value of the three erosion-prone areas. Meanwhile, it is greatly affected by the change of mass flow. Then the solid-phase particles hitting the plug face bounce back to the plug transition section while losing kinetic energy. Thus, the erosion and wear in the plug transition section is smaller than that in the plug face. The axial flow passage has the same flow direction as the drilling fluid, and only a small amount of solid-phase particles will hit the axial flow passage. Therefore, the erosion and wear in the axial flow passage is the smallest, and it is the least affected by the increase of mass flow. As shown in Figure 8, the erosion area of the three erosion-prone areas of PCIT did not change significantly with the increase in mass flow. Therefore, it should be considered that the mass concentration of solid-phase particles in the drilling fluid can be reduced when the drilling fluid is configured, which can effectively reduce the erosion in the PCIT.

4.2.3. Influence of Solid-Phase Particle Diameter on Erosion Rate

According to the actual working conditions, the diameters of the solid-phase particles were set as 0.1, 0.2, 0.3, 0.4, 0.5, 0.6, 0.7, 0.8, 0.9, and 1.1 mm, respectively. The inlet velocity and the mass flow of the drilling fluid were kept constant. The relationship between the different diameter of solid particles and the maximum erosion rate is shown in Figures 9 and 10.

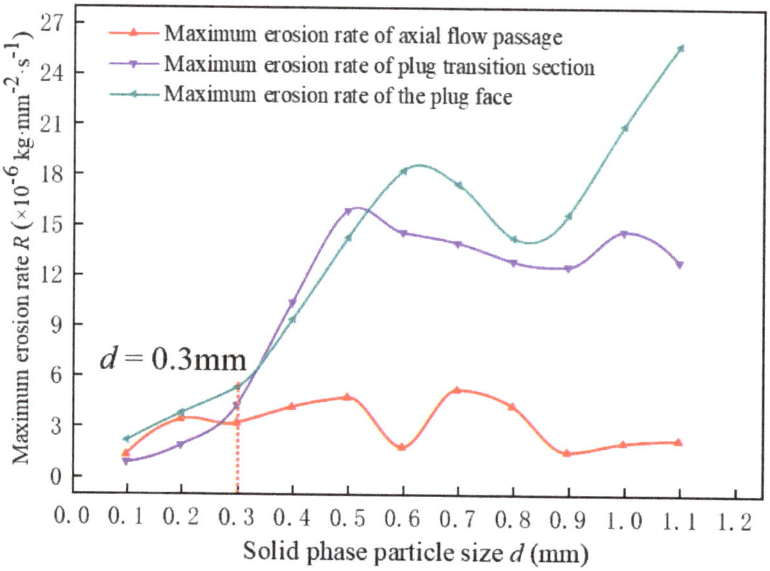

Figure 9. The maximum erosion rate changes with solid particle diameter.

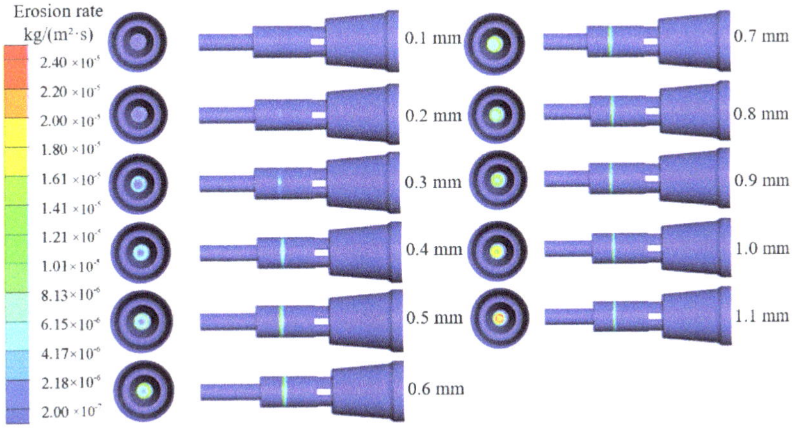

Figure 10. Erosion area changes with solid particle diameter.

As shown in Figure 9, the maximum erosion rates of the erosion-prone regions in the PCIT all increased with the increase in particle diameter. When the particle size diameter of the solid phase increased from 0.1 to 1.1 mm, the average maximum erosion rate of the three erosion-prone regions increased by 63.4 times, but the growth rate of the maximum erosion rate was different in different particle diameter ranges: When the solid-phase particle diameter was between 0 and 0.3 mm, the growth of the erosion rate of the three erosion-prone regions was relatively low. When the solid-phase particle diameter was between 0.3 and 0.7 mm, the erosion rate of the three easily eroded areas increased rapidly, especially the erosion rate on the surface of the plug increased rapidly. When diameter was greater than 0.7 mm, the maximum erosion rate of the axial flow passage decreased with the increasing particle diameter, while the maximum erosion rates of the plug transition and plug surface continued to increase linearly. At the same time, according to the erosion area diagram in Figure 10, the erosion area of the plug surface and the plug transition section gradually increased with the increase of particle diameter. The analysis results show that the diameter of the solid-phase particles in the drilling fluid should preferably be less than 0.3 mm in the actual production process and that a solid-phase particle diameter exceeding 0.7 mm accelerates the erosion damage to the PCIT.

4.3. Influence of PCIT Structural Parameters on Erosion

4.3.1. Influence of the Sliding Core Inlet Angle on Erosion Rate

The inlet angle of the sliding core is closely related to the pressure drop inside the sliding core under the same flow rate, but the variation of the erosion law at the sliding core cone caused by the change of the inlet angle cannot be ignored. The inlet velocity and mass flow rate were kept consistent to carry out numerical simulations. The relationship between the different sliding core inlet angle and the maximum erosion rate is shown in Figures 11 and 12.

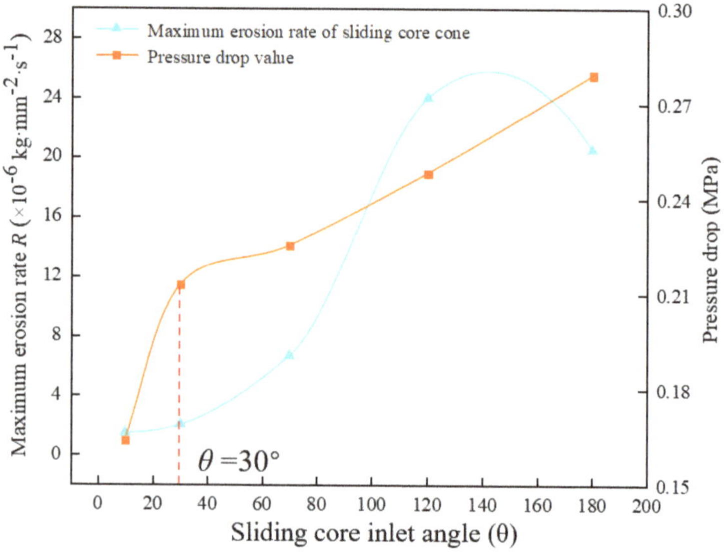

Figure 11. Maximum erosion rate and pressure drop of the sliding core cone change with θ value.

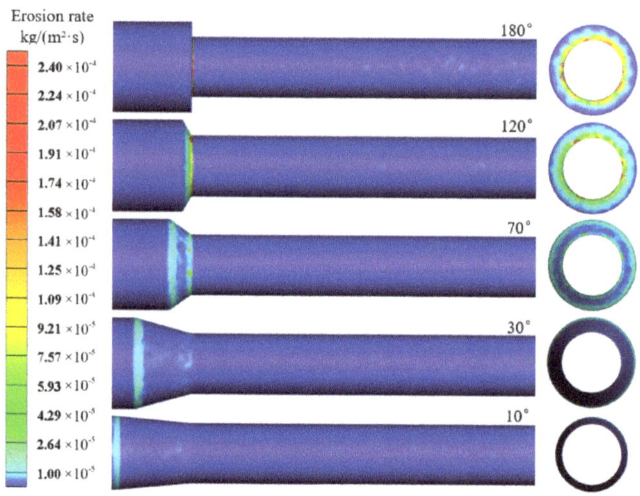

Figure 12. Erosion area changes with θ value.

As shown in Figure 12, the most serious erosion is at the sliding core cone, and the erosion area increases with the inlet angle. As shown in Figure 11, with the increase of the inlet angle, the pressure drop value inside the sliding core and the maximum erosion rate of the sliding core cone gradually increases. When the inlet angle was 10°, the maximum erosion rate of the sliding core cone was 1.45×10^{-5} kg/(m^2·s). The maximum erosion rate can reach 2.05×10^{-4} kg/(m^2·s), when the inlet angle is 120°, increasing the erosion rate by 17 times. There is a certain reduction in the maximum erosion rate when the inlet angle is greater than 120°, but the erosion severity does not improve. Therefore, the relationship between the erosion rate, the internal pressure drop, and the inlet angle of the sliding core should be taken into account. Therefore, the axial force is as large as possible under the same conditions and the erosion rate at the sliding core cone is controlled. Preferably, the inlet angle of the sliding core should be set as 30°, and the erosion rate at the variable section should be kept between 5.98 × 10^{-6} and 1.2 × 10^{-5} kg/(m^2·s), and the maximum erosion rate of the entire inlet section will be 1.99 × 10^{-5} kg/(m^2·s).

4.3.2. Influence of Plug Angle on Erosion Rate

The variation law of the erosion rate of the erosion-prone area in the PCIT was further analyzed focusing on changes in the top angle of the plug. The simulation analysis results are shown in Figures 13 and 14.

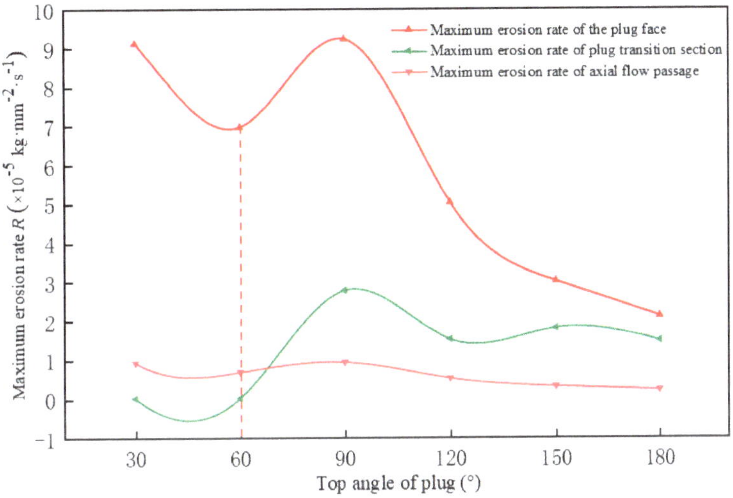

Figure 13. Maximum erosion rate changes with plug angle.

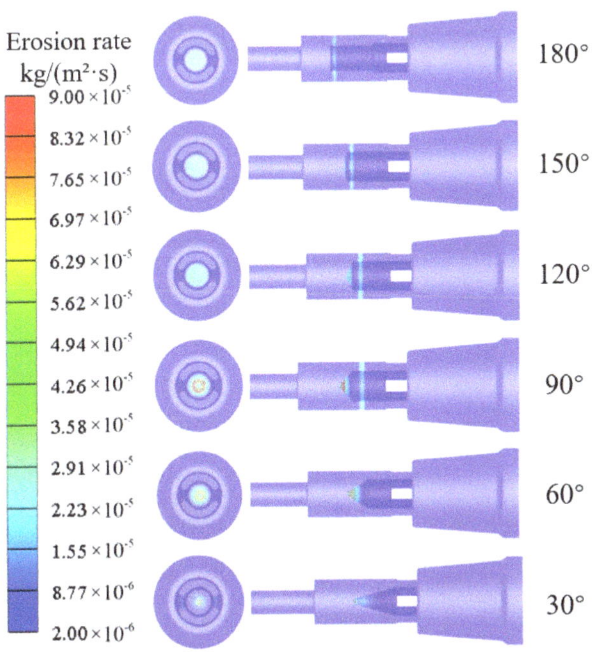

Figure 14. Erosion area changes with plug angle.

As shown in Figure 13, different top angles of the plug had obvious effects on the erosion law of the plug surface, the transition section of the plug, and the axial flow passage. The erosion rate of the plug transition section gradually increased with the gradual increase of the plug top angle, the maximum erosion rate of the plug surface gradually decreased and the maximum erosion rate of the axial flow passage was almost unchanged. When the top angle of the plug is small, the plug has a stronger diversion effect on the drilling fluid. Under the diversion effect of the plug, most of the solid-phase particles in the drilling fluid will flow directly to the axial flow passage. Therefore, it will not cause serious erosion on the plug transition section. When the top angle of the plug is large, the diversion effect on the drilling fluid is weakened, and a considerable amount of solid particles will bounce back to the plug transitional section after hitting the plug surface, thereby causing the plug erosion of excessive segments. As shown in Figure 12, when the top angle of the plug was less than 60°, almost no erosion occurred in the plug transition section. However, when the top angle of the plug was greater than 60°, the erosion area of the transition section of the plug gradually increased. Considering that it is easy to strengthen the erosion area of the plug during processing and combining the analysis results of Figures 13 and 14, the plug top angle was set to 60°. The erosion rate of the plug surface was then relatively low, and the erosion of the plug transition section was significantly improved.

4.3.3. Influence of Plug Transition Distance on Erosion Rate

The erosion of the PCIT with the transition distance was analyzed. Figures 15 and 16 show the results of the maximum erosion rate and erosion area variation of the three erosion-prone areas in PCIT.

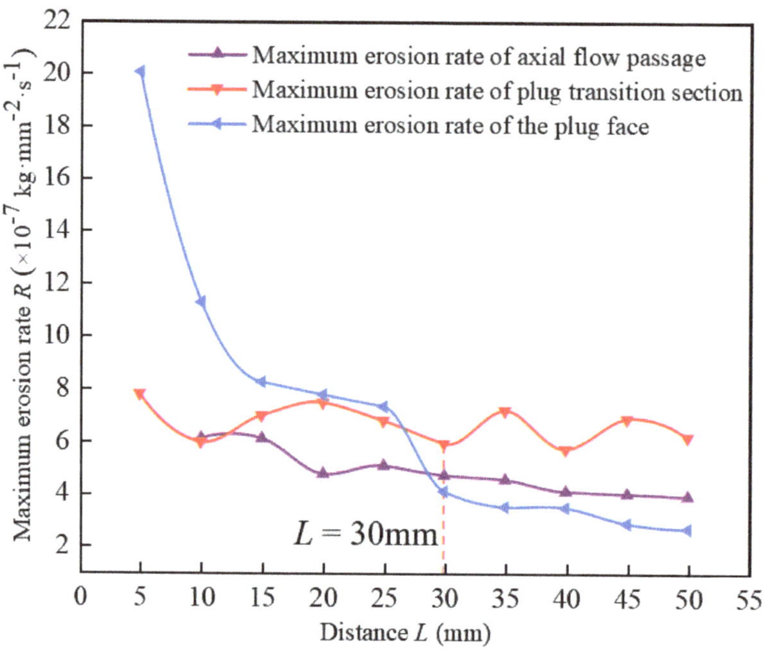

Figure 15. Maximum erosion rate changes with plug transition distance.

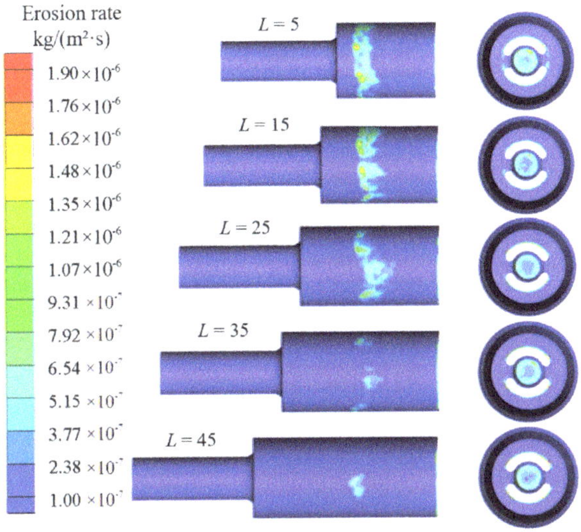

Figure 16. Erosion area changes with plug transition distance.

As shown in Figure 15, when the distance of the plug transition section was reduced from 55 to 5 mm, the maximum erosion rate of the erosion-prone areas increased by 3.8 times. The maximum erosion rate at the axial flow passage decreased slightly with the increase of the plug transition section distance. The erosion changes on the plug surface and the plug transitional section can be divided into three stages. Stage 1: when the distance of the plug transition section is between 5 and 15 mm, the maximum erosion rate on the plug surface drops sharply. Stage 2: when the distance is between 15 and 30 mm, the maximum erosion rate of the plug transition section continues to decrease. Stage 3: the erosion rate of the three easily eroded areas tends to be stable when the distance of the plug transition section is between 30 and 50 mm, the erosion rate is less affected by the distance of the plug transition section. As the distance of the plug transition section increased, the erosion area of the plug transition section decreased significantly as shown in Figure 16, while the erosion area of the plug surface and the axial flow passage did not change significantly. Therefore, when the PCIT is designed, the distance of the plug transition section should be greater than 30 mm.

5. Conclusions

In order to meet the demand for hydrate reservoir crushing and cavity making in solid fluidized production of natural gas hydrate, the novel pressure-controlled injection tool in this paper was designed. Variations in the law of erosion under different drilling fluid physical parameters and PCIT structure parameters were analyzed by computational fluid software. According to the simulation results and actual working conditions, the optimal design parameters of the PCIT were optimized. From the numerical simulation, the following conclusions were obtained:

(1) There are four erosion-prone areas in the PCIT, including the sliding core cone, the plug transition section, the plug surface, and the axial flow passage. These four areas should be considered and strengthened in the design and processing.

(2) The maximum erosion rate and erosion area in the easy erosion area increase with the increase of solid particle diameter. The average maximum erosion rate increased by 63.4 times when the particle size changed from 0.1 to 1.1 mm, which exceeded other influencing factors such as mass flow rate and drilling fluid flow rate. Therefore, it was considered that the solid particle diameter was the main factor affecting the growth of the maximum erosion rate. When the solid particle diameter was less than

0.3 mm, the erosion of each part of the PCIT was better. The particle size of solid particles should be strictly controlled when configuring the drilling fluid, which can significantly reduce erosion.

(3) With the increase of the sliding core inlet angle, the erosion of the sliding core cone is intensified. Therefore, considering the relationship between pressure drop, erosion rate, and inlet angle, the sliding core inlet angle should preferably be set to 30°. Under the same flow conditions, an inlet angle of 30° can cause the largest possible pressure drop while greatly improving the erosion of the sliding core cone. At the same time, the top angle of the plug should be set to 60°, and the distance of the plug transition section should be greater than 30 mm, so that the erosion rate of the plug surface and the plug transition section can be effectively reduced.

Author Contributions: Y.T. and P.Z. designed the simulation model and carried out the simulation of the proposed model; G.W. analyzed the data; Y.T., P.Z. and L.Z. wrote the initial draft of the manuscript; and X.L. and X.F. reviewed and contributed to the final manuscript. All authors have read and agreed to the published version of the manuscript.

Funding: This work is supported by National Key Research and Development Program (grant numbers 2021YFC2800903); the Fund of Southern Marine Science and Engineering Guangdong Laboratory (Zhanjiang) (grant numbers ZJW-2019-03); the National Natural Science Foundation of China (grant numbers 52004235); International Science and Technology Cooperation Project Funding (grant numbers 2020-GH02-00041-HZ); China Postdoctoral Science Foundation (grant numbers 2020M683359); National Science and Technology Major Project (grant numbers 2016ZX05028-001-006); China Postdoctoral Innovative Talents Support Program (grant numbers BX20190292); Sichuan Province Applied Basic Research Project (grant numbers 2020YJ0152); and the Miaozi Engineering Project of Sichuan Science and Technology Department (No. 2021124).

Institutional Review Board Statement: Not applicable.

Informed Consent Statement: Not applicable.

Data Availability Statement: All data and models generated or used during the study appear in the submitted article.

Conflicts of Interest: The authors declare no conflict of interest.

References

1. Kajiyama, S.; Wu, Y.; Hyodo, M. Experimental investigation on the mechanical properties of methane hydrate-bearing sand formed with rounded particles. *J. Nat. Gas Sci. Eng.* **2017**, *45*, 96–107. [CrossRef]
2. Zhou, S.W.; Li, Q.Y.; Lv, X. Key issues in development of offshore natural gas hydrate. *Front. Energy* **2020**, *14*, 433–442. [CrossRef]
3. Sloan, E.D. Fundamental principles and applications of natural gas hydrates. *Nature* **2003**, *426*, 353–359. [CrossRef]
4. Liu, J.J.; Shao, Z.L.; Wu, M.Y. Heat and mass transfer analysis of depressurization-induced hydrate decomposition with different temperatures of over-and under burden. *J. Nat. Gas Sci. Eng.* **2017**, *44*, 65–76. [CrossRef]
5. Feng, Y.; Chen, L.; Suzuki, A. Numerical analysis of gas production from layered methane hydrate reservoirs by depressurization. *Energy* **2019**, *166*, 1106–1119. [CrossRef]
6. Gambelli, A.M.; Rossi, F. Natural gas hydrates: Comparison between two different applications of thermal stimulation for performing CO2 replacement. *Energy* **2019**, *172*, 423–434. [CrossRef]
7. Xu, C.G.; Cai, J.; Yu, Y.S. Effect of pressure on methane recovery from natural gas hydrates by methane-carbon dioxide replacement. *Appl. Energy* **2018**, *217*, 527–536. [CrossRef]
8. Wang, Y.; Feng, J.C.; Li, X.S.; Zhang, Y. Experimental investigation of optimization of well spacing for gas recovery from methane hydrate reservoir in sandy sediment by heat stimulation. *Appl. Energy* **2017**, *207*, 562–572. [CrossRef]
9. Chen, C.; Pan, D.B.; Yang, L. Investigation into the water jet erosion efficiency of hydrate-bearing sediments based on the arbitrary Lagrangian-Eulerian method. *Appl. Sci.* **2019**, *9*, 182. [CrossRef]
10. Li, Z.Z.; Guo, X.Q.; Yang, L.Y. Exploitation of methane in the hydrate by use of carbon dioxide in the presence of sodium chloride. *Pet. Sci.* **2009**, *6*, 426–432. [CrossRef]
11. Zhou, S.W.; Zhao, J.Z.; Li, Q.Y. Optimal design of the engineering parameters for the first global trial production of marine natural gas hydrates through solid fluidization. *Nat. Gas Ind. B* **2018**, *5*, 118–131. [CrossRef]
12. Wang, L.Z.; Wang, G.R. Experimental and Theoretical Study on the Critical Breaking Velocity of Marine Natural Gas Hydrate Sediments Breaking by Water Jet. *Energies* **2020**, *13*, 1725. [CrossRef]

13. Zhou, S.W.; Chen, W.; Li, Q.P. The green solid fluidization development principle of natural gas hydrate stored in shallow layers of deep water. *China Offshore Oil Gas* **2014**, *26*, 1–7.
14. Wei, N.; Sun, W.T.; Meng, Y.T. Multiphase non-equilibrium pipe flow behaviors in the solid fluidization exploitation of marine natural gas hydrate reservoir. *Energy Sci. Eng.* **2018**, *6*, 760–782. [CrossRef]
15. Tang, Y.; Yao, J.X.; He, Y.; Sun, P.; Jing, X. Study on pressure-controlled sliding sleeve of jet breaking for natural gas hydrate mining based on throttle pressure drop principle. *Energy Sci. Eng.* **2020**, *8*, 1422–1437. [CrossRef]
16. Tang, Y.; Huang, S.X.; Yao, J.X.; Liu, Q.Y.; Wang, G.R.; Zhou, S.W. An Automatic Jet Breaking Tool for Solid-State Fluidized Exploitation of Natural Gas Hydrate. U.S. Patent No. 11,193,333, 17 January 2020.
17. Yi, X.Z.; Peng, Z.; Zhou, Y.H. Numerical simulation for erosion behavior of high-pressure fracturing fluids on JY-50 fracturing bend pipe. *Surf. Technol.* **2019**, *48*, 144–151.
18. Huang, Z.Q.; Xie, D.; Huang, X.B. Analytical and experimental research on erosion wear law of drill pipe in gas drilling. *Eng. Fail. Anal.* **2017**, *79*, 615–624. [CrossRef]
19. Huang, Y.; Yin, K.; Zhu, L.H. Numerical simulation of swivel bend wear in reverse circulation drilling. *J. Cent. South Univ.* **2013**, *44*, 2052059.
20. Wang, G.R.; Chu, F.; Tao, S.Y. Optimization design for throttle valve of managed pressure drilling based on CFD erosion simulation and response surface methodology. *Wear* **2015**, *338*, 114–121. [CrossRef]
21. Jafari, A.; Dehghani, K.; Bahaaddini, K. Experimental comparison of abrasive and erosive wear characteristics of four wear-resistant steels. *Wear* **2018**, *416*, 14–26. [CrossRef]
22. Habib, M.A.; Ben-Mansour, R.; Badr, H.M. Erosion and penetration rates of a pipe protruded in a sudden contraction. *Comput. Fluids* **2008**, *37*, 146–160. [CrossRef]
23. Alfonsi, G. Reynolds-averaged Navier-Stokes equations for turbulence modeling. *Appl. Mech. Rev.* **2009**, *62*, 4–20. [CrossRef]
24. Hinze, J.O. *Turbulence*; Mc Graw Hill: New York, NY, USA, 1975.
25. Chintala, V.; Subramanian, K.A. A CFD (computational fluid dynamics) study for optimization of gas injector orientation for performance improvement of a dual-fuel diesel engine. *Energy* **2013**, *57*, 709–721. [CrossRef]
26. Ji, M.K.; Utomo, T.; Woo, J.S.; Lee, Y.H.; Jeong, H.M.; Chung, H.S. CFD investigation on the flow structure inside thermo vapor compressor. *Energy* **2010**, *35*, 2694–2702. [CrossRef]
27. Dickenson, J.A.; Sansalone, J.J. Discrete phase model representation of Particulate Matter (PM) for simulating PM separation by hydrodynamic unit operations. *Environ. Sci. Technol.* **2009**, *43*, 8220–8226. [CrossRef]
28. Frawley, P.; Corish, J.; Niven, A.; Geron, M. Combination of CFD and DOE to analyse solid particle erosion in elbows. *Int. J. Comput. Fluid Dyn.* **2009**, *23*, 411–426. [CrossRef]
29. Grant, G.; Tabakoff, W. Erosion prediction in turbomachinery resulting from environmental solid particles. *J. Aircr.* **1975**, *12*, 471–478. [CrossRef]
30. Forder, A.; Thew, M.; Harrison, D. A numerical investigation of solid particle erosion experienced within oilfield control valves. *Wear* **1998**, *216*, 184–193. [CrossRef]

energies
MDPI

Article

Cenozoic Depositional Evolution and Stratal Patterns in the Western Pearl River Mouth Basin, South China Sea: Implications for Hydrocarbon Exploration

Entao Liu [1,*], Yong Deng [2,3,*], Xudong Lin [1], Detian Yan [4], Si Chen [4] and Xianbin Shi [5]

1 Hubei Key Laboratory of Marine Geological Resources, China University of Geosciences, Wuhan 430074, China
2 School of Geophysics, China University of Petroleum (Beijing), Beijing 102249, China
3 Zhanjiang Branch Company, China National Offshore Oil Corporation, Zhanjiang 524057, China
4 Key Laboratory of Tectonics and Petroleum Resources, Ministry of Education, China University of Geosciences, Wuhan 430074, China
5 Hubei Geological Survey Institute, Wuhan 430030, China
* Correspondence: liuentao@cug.edu.cn (E.L.); dengy@cnooc.com.cn (Y.D.); Tel./Fax: +86-27-67886151 (E.L.)

Abstract: Investigating the deposition evolution and stratal stacking patterns in continental rift basins is critical not only to better understand the mechanism of basin fills but also to reveal the enrichment regularity of hydrocarbon reservoirs. The Pearl River Mouth Basin (PRMB) is a petroliferous continental rift basin located in the northern continental shelf of the South China Sea. In this study, the depositional evolution process and stacking pattern of the Zhu III Depression, western PRMB were studied through the integration of 3D seismic data, core data, and well logs. Five types of depositional systems formed from the Eocene to the Miocene, including the fan delta, meandering river delta, tidal flat, lacustrine system, and neritic shelf system. The representative depositional systems changed from the proximal fan delta and lacustrine system in the Eocene–early Oligocene, to the tidal flat and fan delta in the late Oligocene, and then the neritic shelf system in the Miocene. The statal stacking pattern varied in time and space with a total of six types of slope break belts developed. The diversity of sequence architecture results from the comprehensive effect of tectonic activities, sediment supply, sea/lake level changes, and geomorphic conditions. In addition, our results suggest that the types of traps are closely associated with stratal stacking patterns. Structural traps were developed in the regions of tectonic slope breaks, whereas lithological traps occurred within sedimentary slope breaks. This study highlights the diversity and complexity of sequence architecture in the continental rift basin, and the proposed hydrocarbon distribution patterns are applicable to reservoir prediction in the PRMB and the other continental rift basins.

Keywords: depositional system; sequence architecture; stacking pattern; Pearl River Mouth Basin; hydrocarbon exploration

Citation: Liu, E.; Deng, Y.; Lin, X.; Yan, D.; Chen, S.; Shi, X. Cenozoic Depositional Evolution and Stratal Patterns in the Western Pearl River Mouth Basin, South China Sea: Implications for Hydrocarbon Exploration. *Energies* **2022**, *15*, 8050. https://doi.org/10.3390/en15218050

Academic Editors: Mohammad Sarmadivaleh and Shu Tao

Received: 18 September 2022
Accepted: 27 October 2022
Published: 29 October 2022

Publisher's Note: MDPI stays neutral with regard to jurisdictional claims in published maps and institutional affiliations.

1. Introduction

The continental rift basin is one of the most significant types of petroliferous basins. Approximately 31% of oil and gas fields in the world are distributed in continental rift basins with proven reserves of over 0.68×10^8 of oil equivalents [1]. Major petroleum systems sourced by Cenozoic sediments occur in a large number of sedimentary basins such as the Kachchh Basin (India), Congo Basin (Africa), Madrid Basin (Spain), Songliao Basin (China), and the basins in the South China Sea [1,2]. Investigation of the deposition evolution process in continental rift basins is critical to locating hydrocarbon reservoirs and predicting potential source rocks [2–6].

In sedimentary basins, stratigraphic and depositional architectures are fundamentally controlled by basin and tectonic evolution [7,8]. The dynamic process of basin filling is mainly controlled by the interaction between sea/lake-level changes, tectonic subsidence, sediment supply, and geomorphic conditions, but the main controlling factor varies in different types of sedimentary basins [9–13]. In continental rift basins, tectonics is the most significant factor controlling depositional processes, including episodic rifting, differential subsidence, and the evolution and linkage of normal faults [4,14–19]. It exerts significant control on sedimentary fills, sequence architecture, and sand dispersal patterns [8,20–24]. By contrast, in marine basins, sequence stratal patterns and depositional architecture are mainly controlled by eustatic sea-level changes [9,10,25–27]. Even in a single basin, different evolutionary stages are characterized by different stratigraphic and depositional architectures [28,29]. For example, in continental rift basins, episodic tectonic subsidence with short stages of accommodation creation in response to pulses of fault reactivation usually occurs in the early stage, while little accommodation is generated due to tectonic quiescence, and sediment supply consumes and fills the available space in the later stage [2,5,8,30,31].

The tectono-stratigraphic archive provides unique insights into the evolution of stratigraphic and depositional architectures in continental rift basins [5,31–35]. Various palaeogeographic features formed by syn-depositional faults are getting more concern since they play a significant effect in the transportation and deposition of sediments [36–38]. Structural slope breaks are important types of palaeogeographic features, controlling the distribution of deposition systems through changes in the accommodation [33,36,39,40]. Overall, the previous studies related to stratigraphic and depositional architecture mainly focused on marine basins. For the continental rift basins, attention has been mainly paid to the rifting stage of continental rift basins [32,33,41,42]. However, studies on the evolutionary features during the long-time process are rare, especially for the evolutionary process from lacustrine to marine environments [22].

The Pearl River Mouth Basin (PRMB) is a significant hydrocarbon-bearing basin located in the central region of the northern margin of the South China Sea continental shelf [43–45]. It is a focus point for intensive hydrocarbon resources exploration because of the high petroleum potential in the Eocene–Miocene interval revealed by recent drilling findings [46,47]. The previous studies show that PRMB has undergone complex structural evolution from the lacustrine environment in the Eocene to the marine environment in the Miocene [48,49]. During the evolution process, multiple types of depositional systems (e.g., lacustrine system, neritic system, fan delta) were developed with great variability in time and space [49,50]. Although previous studies [46,47,51] have investigated the tectonic evolution, depositional systems, and source rock potential of the PRMB, rare study has focused on the depositional evolution and sequence filling processes during the basin evolution from lacustrine to marine environments, and the controlling factors for the sequence architecture and depositional system distribution remain debatable [42]. Moreover, the characteristics, spatial and temporal distribution, and evolutionary process of stratal stacking patterns in different stages of basin evolution are not clear. Given the above disputes over the depositional evolution of the PRMB, this study was conducted with the following aims: (1) to describe the evolution of depositional systems from the Eocene to the Miocene within a sequence stratigraphic framework; (2) to identify and characterize stratal stacking patterns in different basin evolutionary stages; (3) to investigate the characteristics of stratigraphic and depositional architecture throughout the entire history of basin filling; (4) to discuss the controlling factors for basin evolution and sediment filling process. Our study can not only enhance the understanding of sequence filling and depositional evolution processes in the continental rift basins from lacustrine to marine environments, but also serve as a new example for further exploration in the PRMB and the other basins in the South China Sea.

2. Geological Setting

The Pearl River Mouth Basin (PRMB) is a passive margin Cenozoic basin located in the center area of the northern continental margin of the South China Sea [45,52,53]. It covers an area of approximately 26.7×10^4 km^2 and consists of four depressions from west to east, including the Zhu I Depression, Zhu II Depression, Zhu III Depression, and Chaoshan Depression (Figure 1A). The target study area, the Zhu III Depression, is located in the western region of the PRMB, covering a region of 12,180 km^2 with a water depth of 0–400 m. The Zhu III Depression is bounded by the South China Block to the north, Xisha Island to the south, Hainan Island to the west, and Taiwan Island to the east (Figure 1A). It consists of several subsidence centers (i.e., Wenchang A, B, C, D, and E Sags, Qionghai Sag, Yangjiang Sag) and structural highs (i.e., Qionghai Uplift, Shenhu Uplift, Yangjiang Uplift) (Figure 1B).

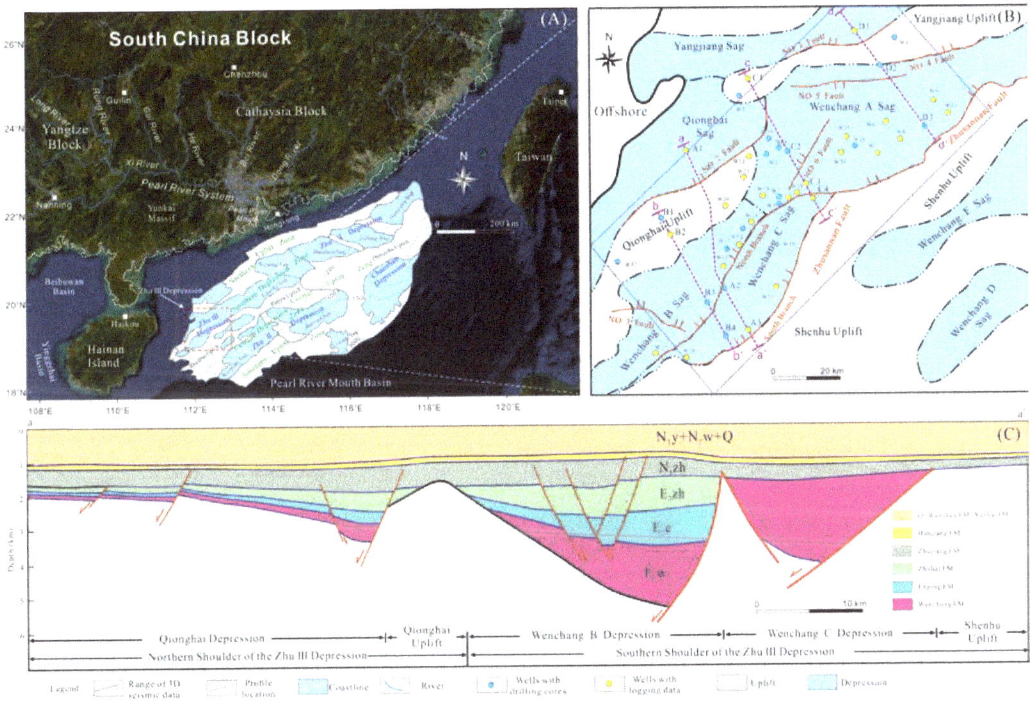

Figure 1. (**A**) Geological map of South China showing the regional setting and tectonic elements of the PRMB with major modern river systems that discharge into the northern South China Sea (modified after Liu et al. [50]). (**B**) Map showing tectonic elements of the Zhu III Depression and geographic locations of borehole samples. (**C**) Interpreted NE-trending profile in the Zhu III Depression and the location of the profile is shown in (**B**).

The Zhu III Depression has undergone a complex tectonic evolution associated with the continental rifting and evolution process of the South China Sea [52,54]. Its evolution history can be divided into four stages, including a pre-rifting stage (Shenhu Formation, 65–49 Ma), a rifting stage (Wenchang and Enping Formations, 39–30 Ma), a transition stage (Zhuhai Formation, 30–23.3 Ma), and a post-rifting stage (Zhujiang Formation, 23.3 Ma to present) (Figure 2). The Shenhu Formation (65–49 Ma) deposited during the pre-rifting stage and consisted of coarse-grained alluvial and volcanic deposits [49,52].

During the rifting stage (39–30 Ma), Wenchang (49–39 Ma) and Enping (39–30 Ma) Formations formed, and the depression is characterized by intense faulting and the development of several half-grabens [49,52,55]. The Wenchang and Enping Formations are represented by widespread terrestrial facies, including fan deltas and braided river deltas [55]. The boundary between the rifting stage and the transition stage was defined by the significant unconformity T_{70}, which is related to the regional Nanhai Event. The Oligocene Zhuhai Formation (30–23.3 Ma) that deposited during the transition stage with half-grabens coalesced is mainly composed of tidal flats, semi-enclosed bays, and deltas [49,50,52]. The Miocene Zhujiang Formation (23.3–16.3 Ma) formed during the post-rifting stage when the PRMB began to receive marine sediments from the Miocene [50]. The Zhujiang Formation consists of fan delta and neritic deposits [49,52,56]. The source rocks are deposited in the Wenchang and Enping Formations, and the main reservoirs are located in the stratum of the Zhuhai and Zhujiang Formations. Several commercial oil pools were discovered in the central and marginal areas of the Wenchang A Sag [57].

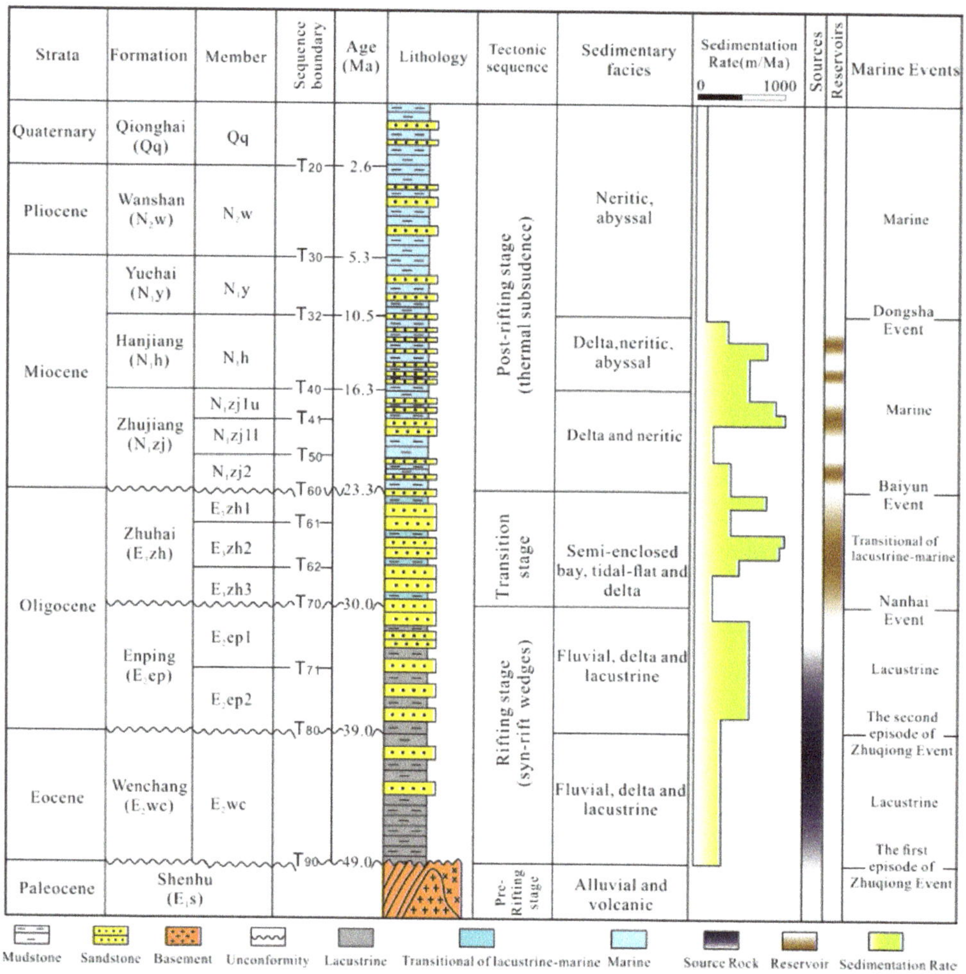

Figure 2. Generalized stratigraphic framework, sedimentary facies, tectonic evolution, and source-reservoir-cap assemblages of the PRMB. The sediment rate curve is from Xie et al. [49].

3. Material and Methods

3.1. Sequence Stratigraphic Framework

The combination of seismic data, drilling cores, logging data, and lithology data provided by CNOOC was used to conduct sedimentological and sequence stratigraphical studies. The seismic club comprises a region of approximately 2000 km^2, and its peak frequency is ~40 Hz. Seismic data were calibrated with well data using synthetic seismogram calibration. The sequence stratigraphic technique [9,58–60] was applied to establish a sequence stratigraphic framework across the depression and to identify different stratal stacking patterns in different structural units.

3.2. Analysis of Sedimentary System

The types and ranges of depositional systems were determined using geological and geophysical methods [61,62]. Over 600 m of drilling cores from 20 wells were described to determine sedimentary facies and depositional characteristics. As drilling cores in the Zhu III Depression are limited, the logging and lithological association data from more than 50 wells, including natural gamma, true resistivity, and acoustic logging data, were also employed to identify depositional facies and to constrain their distribution ranges [40,63]. In no-well regions, the ranges of depositional systems were determined by 3D seismic data using seismic reflection, horizontal slice, and attributive analysis. The analysis of structural slope breaks was conducted based on seismic interpretation of 3D seismic data and analysis of sedimentary systems.

4. Results

4.1. Characteristics of Depositional Systems

Based on the comprehensive sedimentological analysis, five types of depositional systems were identified in the Zhu III Depression, western PRMB, including the fan delta, meandering river delta, tidal flat, neritic shelf system, and lacustrine system (Figures 3–5).

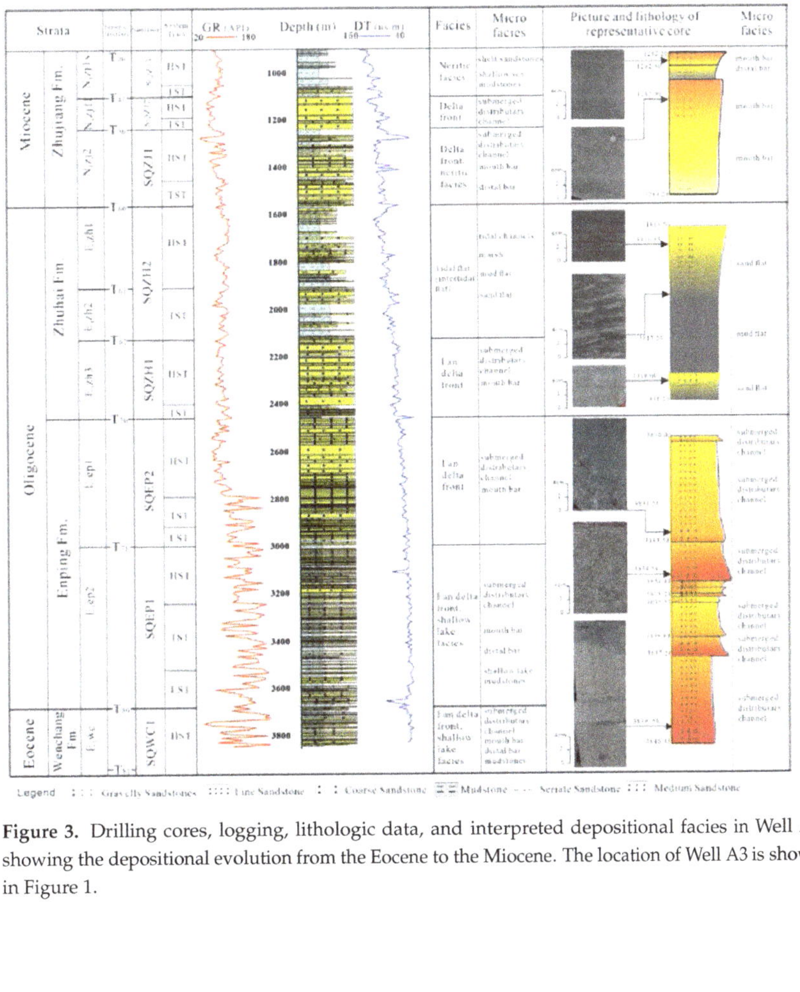

Figure 3. Drilling cores, logging, lithologic data, and interpreted depositional facies in Well A3 showing the depositional evolution from the Eocene to the Miocene. The location of Well A3 is shown in Figure 1.

Figure 4. Typical sedimentary characteristics of the deposits in the Enping (**a**–**f**) and Wenchang (**g**–**l**) Formations in the Zhu III Depression: (**a**) 4694.4 m in Well W1, gravelly sandstone with normally graded bedding; (**b**) 4698.1 m in Well W1, gravelly sandstone; (**c**) 4695.4 m in Well W1, gravelly sandstone with normally graded bedding; (**d**) 4693.2m in Well W1, gravelly sandstone with normally graded bedding; (**e**) 4694.6 m in Well W1, black mudstone; (**f**) 2.9 m (4696.8–4699.7 m) long core of the underwater distributary channels of the fan deltas; (**g**) 4042.9 m in Well W13, oil-bearing gravelly sandstone; (**h**) 3830.0 m in Well B3, gravelly sandstone with normally graded bedding; (**i**) 3041 m in Well W12, oil-bearing gravelly sandstone with normally graded bedding; (**j**) 3043 m in Well 13, oil-bearing gravelly sandstone; (**k**) 3581 m in Well W9, gravelly sandstone; (**l**) 2.4 m (3833.5–3835.9 m) long core of the underwater distributary channels of the fan deltas.

Figure 5. Typical sedimentary characteristics of the deposits in the Zhujiang (**a–f**) and Zhuhai (**g–l**) Formations in the Zhu III Depression: (**a**) 2092.5 m in Well W4, siltstone; (**b**) 2092.8 m in Well W4, siltstone; (**c**) 2090.2 m in Well W4, fine-grained sandstone; (**d**) 2091.3 m in Well W4, fine-grained sandstone and siltstone with parallel bedding; (**e**) 1220.3 m in Well W32, fine-grained sandstone with normally graded bedding; (**f**) 2.0 m (2091.9–2092.9 m and 2093.9–2094.9 m) long core of the shallow-marine detrital deposits from the Zhujiang Formation; (**g**) 3210.7 m in Well W3, siltstone interbedded with mudstone; (**h**) 3210.9 m in Well W3, siltstone interbedded with mudstone; (**i**) 1861.2 m in Well A3, tidal rhythmites; (**j**) 1816.4 m in Well A3, tidal rhythmites; (**k**) 3211.1 m in Well W3, fine-grained sandstone interbedded with mudstone; (**l**) 3.0 m (3207.3–3210.3 m) long core of the tidal flat deposits in the Zhuhai Formation.

4.1.1. Fan Delta

The fan delta system consists of three sub-facies, including the fan delta plain, fan delta front, and pro-fan delta with the fan delta front being the most extensively distributed (Figure 6). Three lithofacies are recognized in the fan delta deposits (Figures 3 and 4): (a) gravelly sandstones with erosional bases, normally graded bedding, and widespread sandstone gravels (Figure 4a–d,g,j,k); (b) normally graded bedded sandstones with mud laminae at the top (Figure 4i,f,l); (c) thin fine-grained sandstones, siltstones with coarsening upward successions (Figure 4h,f,l).

Lithofacies (a) and (b) are interpreted as submerged distributary channels, which are funnel-shaped on the GR log, while lithofacies (c) are interpreted as distal bar deposits, which have a funnel-shaped geometry on the GR log (Figure 3). In the seismic profiles, they are characterized by oblique progradational reflection configurations, low to medium continuous reflections, and medium to high amplitudes (Figures 7–9).

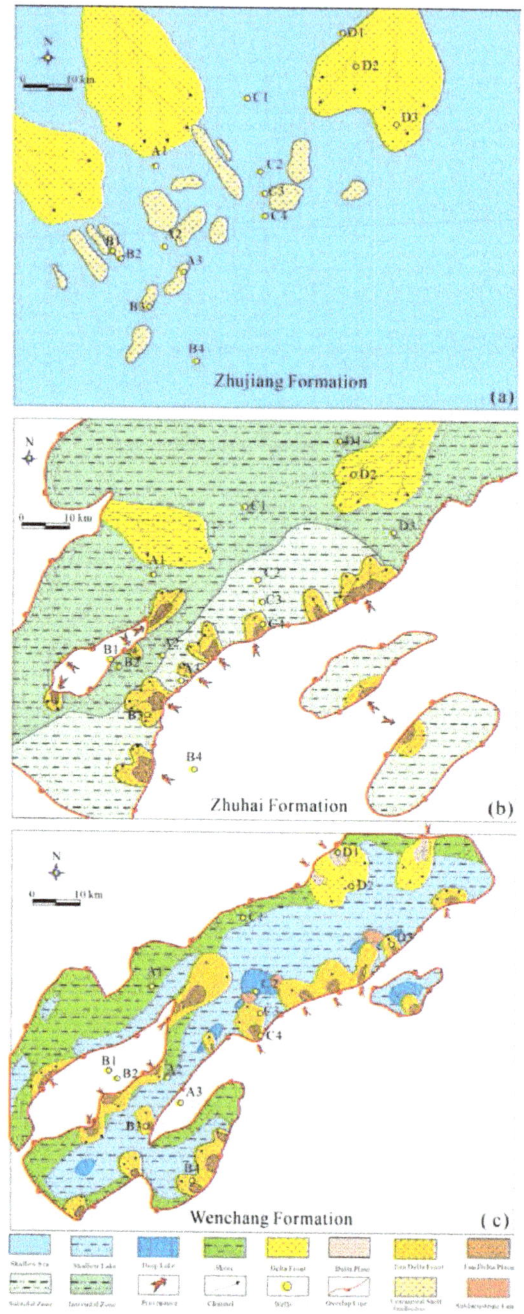

Figure 6. Deposition facies diagrams of the Zhujiang Formation (**a**), Zhuhai Formation (**b**), and Wenchang Formation (**c**) of the Zhu III Depression, PRMB, showing the depositional evolution from the Eocene Wenchang Formation to the Miocene Zhujiang Formation.

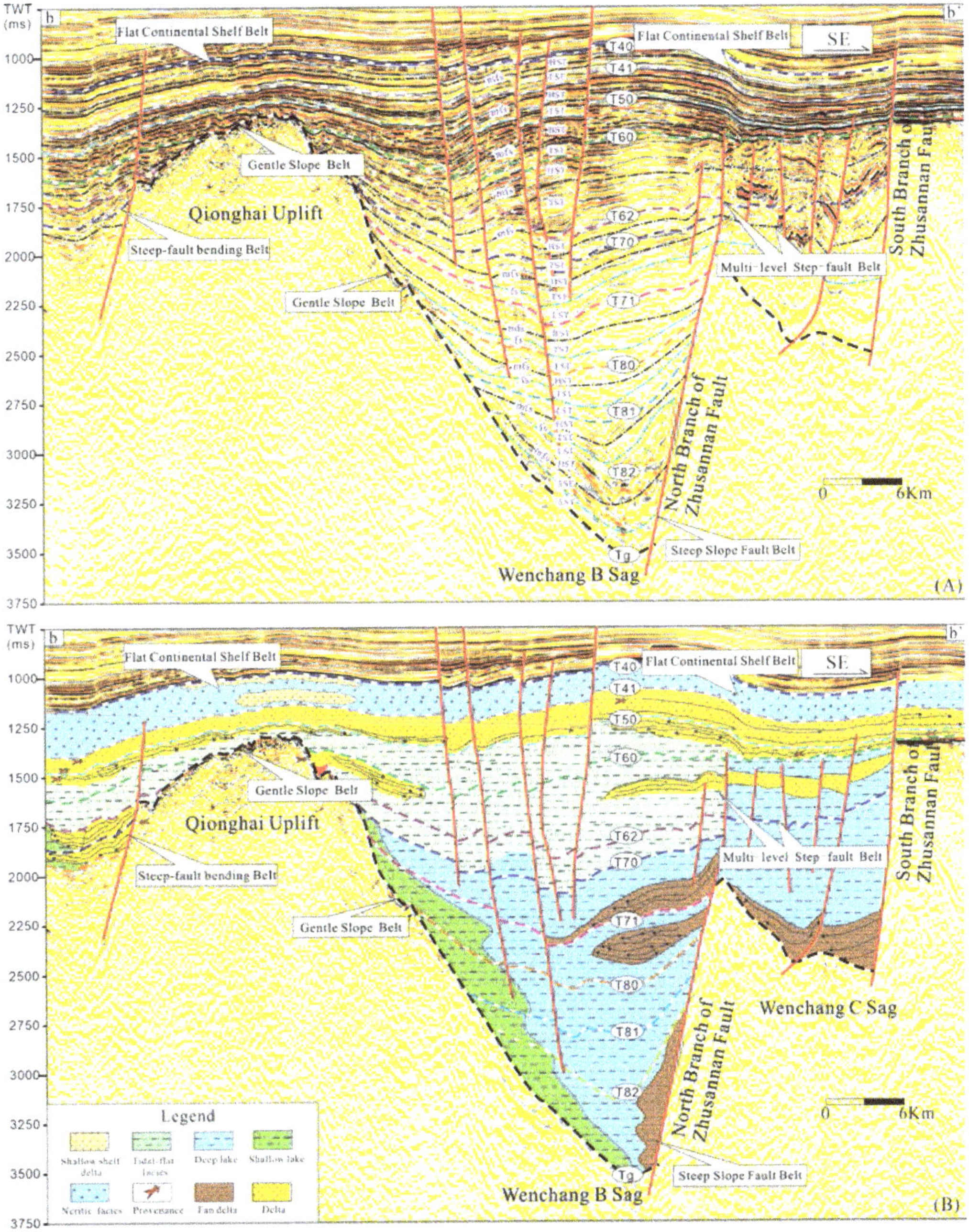

Figure 7. (**A**) interpreted seismic section bb′ showing seismic reflection characteristics and the distribution regions of different stratal stacking patterns; (**B**) the types of depositional facies and their distribution ranges in the section bb′ showing the depositional evolution from the Eocene to the Miocene. The location of section bb′ is shown in Figure 1.

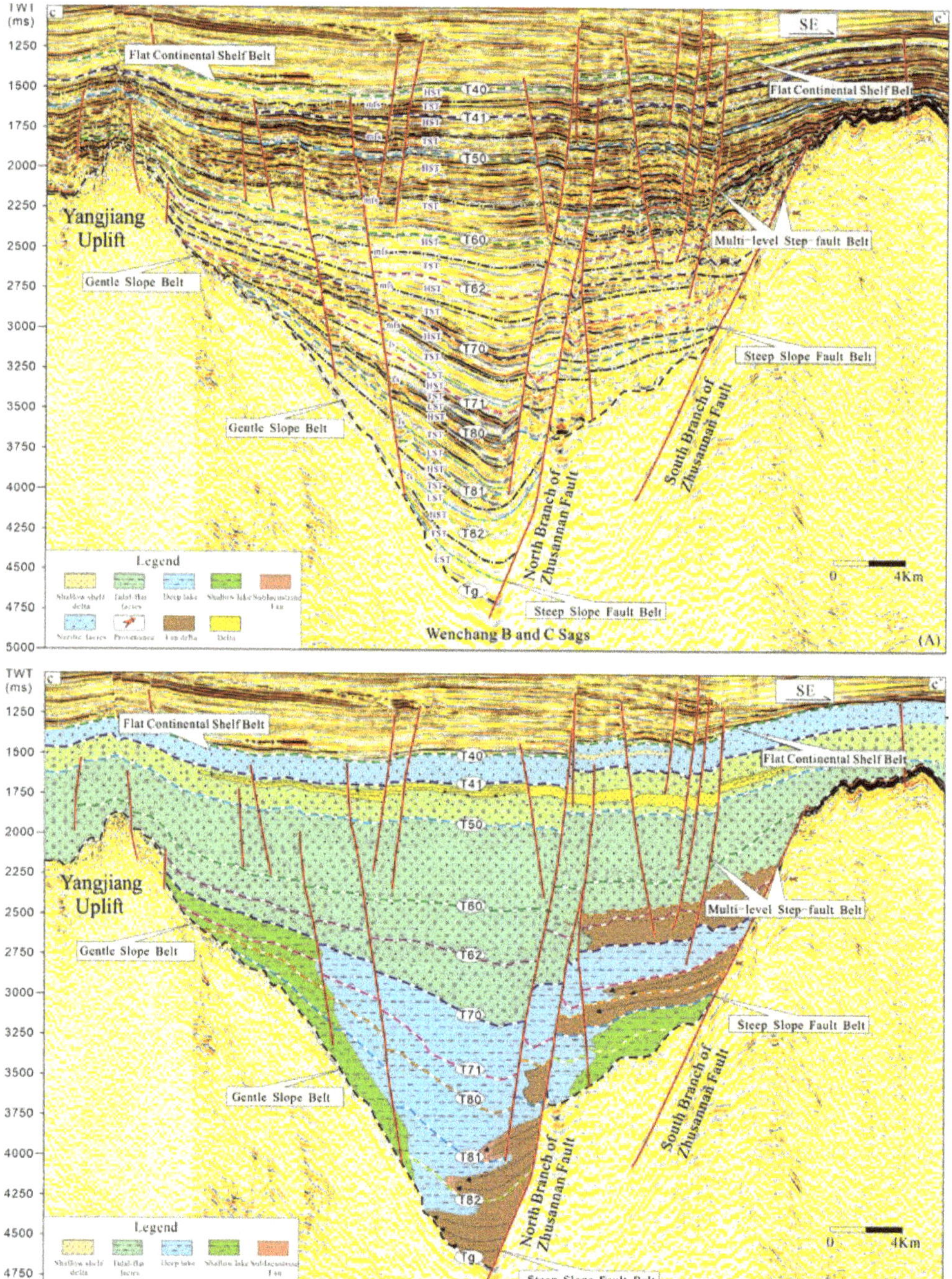

Figure 8. (**A**) interpreted seismic section cc′ showing seismic reflection characteristics and the distribution regions of different stratal stacking patterns; (**B**) the types of depositional facies and their distribution ranges in the section cc′ showing the depositional evolution from the Eocene to the Miocene. The location of section bb′ is shown in Figure 1.

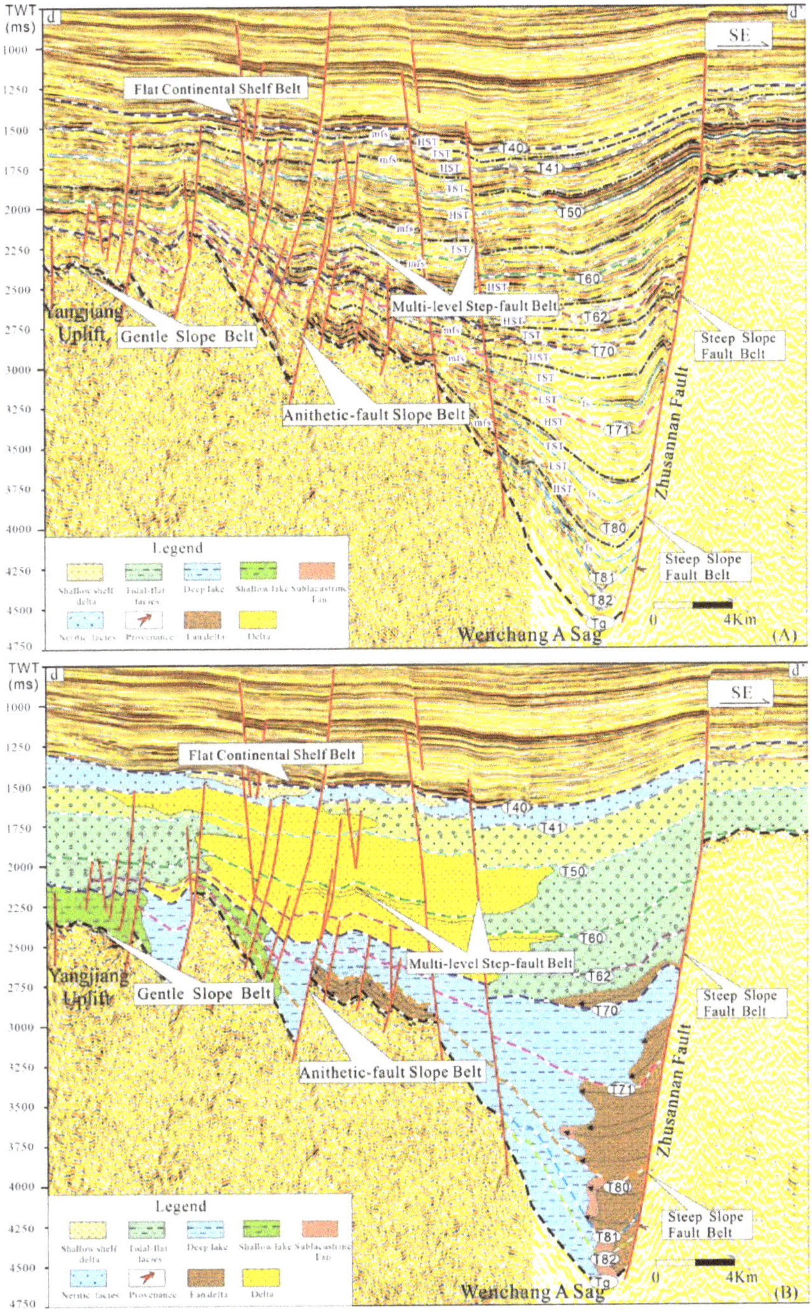

Figure 9. (**A**) interpreted seismic section dd′ showing seismic reflection characteristics and the distribution regions of different stratal stacking patterns; (**B**) the types of depositional facies and their distribution ranges in the section dd′ showing the depositional evolution from the Eocene to the Miocene. The location of section bb′ is shown in Figure 1.

Fan deltas are widely distributed in the Wenchang and Enping Formations, and are mainly located in the footwall blocks of the Zhusannan Fault as well as the adjacent region of the Qionghai Uplift in the Wenchang–Enping Formations (Figures 6 and 7). They exhibit fan-shaped plan view geometries and are distributed along the boundary faults.

4.1.2. Meandering River Delta

In contrast to fan delta deposits, meandering river delta deposits are significantly finer-grained with better sorting and roundness in detrital sediments (Figures 3 and 5a,b). The deposits are mainly composed of two lithofacies, including (a) normally graded bedded sandstones with mud laminae or mudstones at the top and (b) thin fine-grained sandstones, siltstones with coarsening upward successions (Figure 5b,c). Lithofacies (a) are interpreted as submerged distributary channels, whereas lithofacies (b) are interpreted as distal bars deposits.

Meandering river delta deposits are developed in the Zhu III Depression with their distribution ranges increasing from the Eocene to the Miocene (Figure 6). In the Wenchang–Enping Formations, meandering river delta deposits were confined to the northeast of the Zhu III Depression with their sediments from the Yangjiang Uplift (Figures 6, 8 and 9). In the later Oligocene Zhuhai Formation, meandering river delta deposits could be divided into two parts, including the northwestern delta system and the northeastern delta system (Figure 6). In the Miocene Zhujiang Formation, the distribution range of meandering river delta deposits further expanded (Figure 6).

4.1.3. Tidal Flat

The lithologies of tidal flat deposits are composed of siltstone, pelitic siltstone, fine-grained sandstone, and mudstone (Figures 3 and 5). Two types of lithofacies are developed including (a) siltstones interbedded with thin layers of mudstones (Figure 5g,h,k,l) and (b) tidal rhythmites that are characterized by packages of vertically accreted laminated to thinly bedded fine-grained sandstone and mudstone (Figure 5i,j,l).

Tidal flat deposits are widely distributed in the Oligocene Zhuhai Formation (Figures 6–9). Two types of sub-facies are identified in the Zhu III Depression, including the intertidal zone and the subtidal zone. The intertidal zone was developed in the northern depression ranging from Yangjiang Sag to Wenchang B Sag, while the subtidal zone was distributed in the southern depression (i.e., Wenchang C, D, and E Sags) (Figure 6).

4.1.4. Lacustrine System

Lacustrine deposits are widely distributed in the Wenchang and Enping Formations and are composed of two sub-facies (i.e., the shallow-lacustrine and the deep-lacustrine deposits) (Figure 6). The deep-lacustrine deposits developed primarily in the centers of half-grabens, and the lithofacies consists of thick layers of gray-black mudstones. On the seismic profiles, the deep-lacustrine deposits display continuous, high-amplitude, high-frequency, and unparallel reflections (Figures 7–9). By contrast, the shallow-lacustrine deposits consist of thick gray mudstones interbedded with thin layers of fine-grained sandstones with small-scale cross-bedding and waving bedding (Figure 4e). The shallow-lacustrine deposits are characterized by divergent reflector configurations on seismic sections (Figures 7–9).

4.1.5. Neritic Shelf System

Neritic shelf deposits are developed only in the Miocene Zhujiang Formation (Figure 6). They are composed of two types of lithofacies: (a) thick black marine mudstones that contain bivalve, brachiopod shells; (b) massive fine-grained sandstones that are well-sorted and well-rounded (Figures 3 and 4f). Lithofacies (a) is interpreted as offshore black marine mudstone, and Lithofacies (b) is interpreted as the deposits of neritic shelf sand bodies. On the seismic profiles, the neritic shelf deposits display continuous, high-amplitude, high-frequency, and parallel reflections (Figures 7–9).

4.2. Sequence Architecture and Stacking Patterns

4.2.1. Sequence Stratigraphic Framework

Sedimentological and sequence stratigraphy analysis show that the sequence architecture in the Zhu III Depression varies in time and space (Figures 7–9). The Eocene Wenchang Formation is composed of three sequences, and the Oligocene Enping Formation is composed of two sequences [42]. These sequences consist of lowstand, transgressive, and highstand systems tracts (Figures 7–9). The lowstand systems tract is located in the centers of the deep lake, which is characterized by thick wedge-shaped progradation seismic reflection. The transgressive systems tract is characterized by medium, continuous parallel reflections, whereas the highstand systems tract features low-angle oblique progradation reflections (Figures 7–9). By contrast, the Oligocene Zhuhai Formation and the Miocene Zhujiang Formation consist of two and three sequences, respectively, composed of only transgressive and highstand systems tracts [42]. In the Oligocene Zhuhai Formation, the highstand systems tract displays strong amplitude and continuous reflection characteristics, while the transgressive systems tract shows moderate–weak amplitude on the seismic reflection (Figures 7–9). Different from the overlying strata, the third sequence (SQZJ3) in the Miocene Zhujiang Formation is composed of shallow-marine detrital deposits, and the continental shelf sand bodies are developed only within the highstand systems tract (Figure 3).

4.2.2. Depositional Stacking Patterns

Six types of slope break belts are identified in the Zhu III Depression (Figure 10), and their features are described as follows.

(1) Steep slope fault belt

The steep slope fault belt is defined as a steep slope that is controlled by a syn-sedimentary boundary fault (Figure 10). Such a boundary fault usually has a large fault displacement and a straight plan pattern (Figure 10). A deep fault terrace is developed at the footwall of the boundary fault where clastic sediments are accumulated (Figure 10). In the Zhu III Depression, the steep slope fault belt is mainly located at the footwalls of the Zhusannan Fault, which constitute the southwest margins of the Wenchang A, B, and C Sags (Figures 7–9 and 11).

(2) Steep-fault bending belt

The steep-fault bending belt features a steep slope that is controlled by both a syn-sedimentary boundary fault and a pre-existing uplift (Figure 10). Compared with the steep slope fault belt, the slope of the steep-fault bending belt is much gentler, and fault displacement is smaller, which is influenced by the pre-existing uplift (Figure 10). The steep-fault bending belt in the Zhu III Depression developed mainly along the footwall of the No.2 Fault, the north margin of the Qionghai Uplift (Figures 7–9 and 11).

(3) Multi-level step-fault belt

The multi-level step-fault belt is characterized by multiple fault terraces jointly controlled by several syn-sedimentary faults (Figure 10). These faults usually have a ladder-shaped profile and are parallel to each other (Figure 10). Each fault terrace is controlled by a single syn-sedimentary fault, creating an accommodation zone for sediment deposition (Figure 10). The multi-level step-fault belts are mainly located at the footwalls of the Zhusannan Fault and No.5 Fault during the sedimentary period of the Zhuhai Formation (Figures 7–9 and 11).

(4) Antithetic-fault slope belt

The antithetic-fault slope belt is featured by multiple fault terraces formed by the development of antithetic faults (Figure 10). The dip direction of antithetic faults is opposite to the direction of sediment transport, which is significantly different from the multi-level step-fault belt (Figure 10). In the Zhu III Depression, the antithetic-fault slope belt is only located on the east margin of the Qionghai Uplift in the stratum of the Wenchang and Enping Formations (Figures 9 and 11).

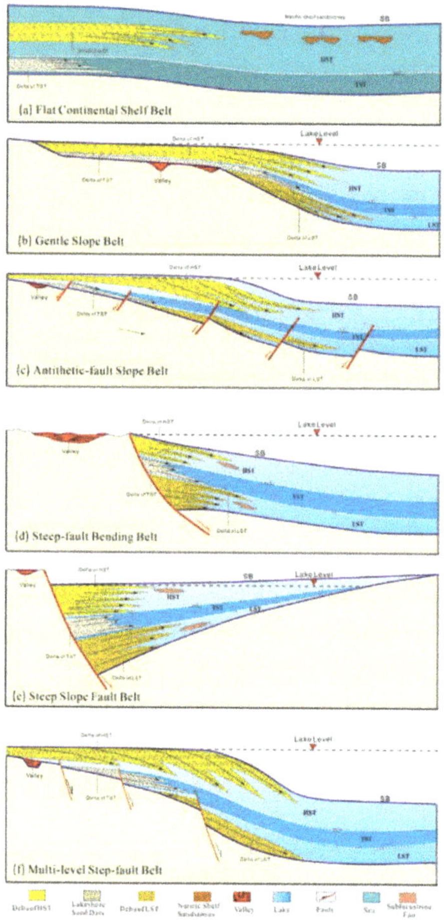

Figure 10. The types of slope break belts and their stratal stacking patterns in the Zhu III Depression, PRMB.

(5) Gentle slope belt

The gentle slope belt is characterized by limited change of slope gradient with a lack of intensive syn-sedimentary normal faults (Figure 10). The sediments are distributed and generally disappear along the slope (Figure 10). In the Zhu III Depression, the gentle slope belt is mainly located on the margins of the Qionghai and Yangjiang Uplifts during the depositional period from the Wenchang Formation to the Zhuhai Formation (Figures 7–9 and 11).

(6) Flat continental shelf Belt

The flat continental shelf belt is featured by a nearly horizontal stratum with a low angle of <2° (Figure 10). There is neither an intensive normal fault nor gentle slope (Figure 10). Long-distance transported sediments are deposited along the continental shelf as large-scale meandering river deltas (Figure 10). Neritic shelf sandstones are formed due to the modification of storms and tides (Figure 10). In the Zhu III Depression, the flat continental shelf belt is located almost in the whole depression during the depositional period of the Zhujiang Formation (Figures 7–9 and 11).

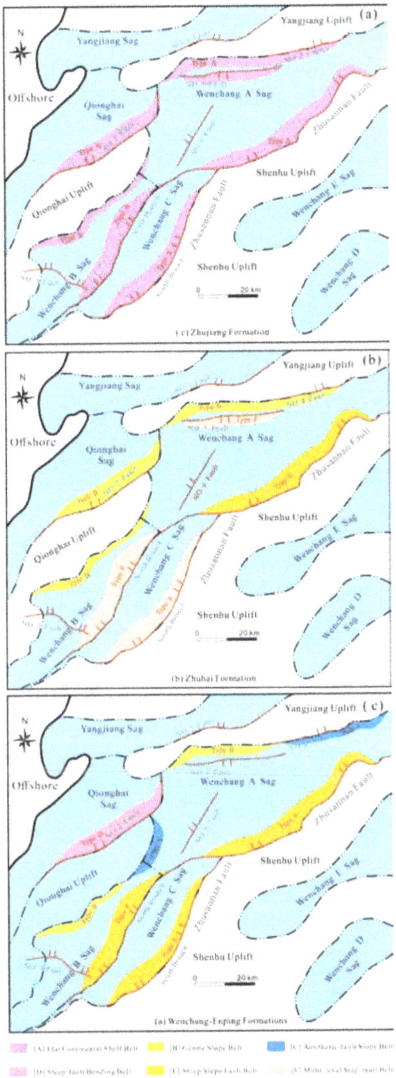

Figure 11. The spatial distributions of slope break belts of the Zhujiang Formation (**a**), Zhuhai Formation (**b**), and Wenchang-Enping Formations (**c**) in the Zhu III Depression, PRMB.

5. Discussion

5.1. Tectonic Controls on the Evolution of Depositional Systems

There are four stages in the tectonic evolution history of the Zhu III Depression, including the pre-rifting stage, rifting stage, transition stage, and post-rifting stage (Figure 2). Each evolutionary stage shows distinctive characteristics in basin configuration and stratigraphic architecture (Figures 6–9). During this evolution process, the representative depositional systems changed from proximal fan delta and lacustrine deposits in the Eocene Wenchang and Oligocene Enping Formations, to tidal flat and fan delta deposits in the Oligocene Zhuhai Formation, and then distal meandering river delta and neritic deposits in the Miocene Zhujiang Formation (Figure 6).

During the rifting stage, the basin is characterized by wedge-shaped geometries with the development of several NE trending half-grabens because of intense rifting (Figures 6–9). The structural highs developed along the basin margin and, among the half-grabens, acted as significant source areas for proximal fan delta deposits, which were accumulated mainly in the footwall scarps of boundary faults (Figures 7–9). During the transition stage, basin rifting was weakened, and the increase in subsidence of the PRMB led to large-scale transgression [56]. Tidal flats were widely developed during this stage, and the fan deltas were deposited at the footwalls of boundary faults. During the post-rifting stage, rifting was almost negligible, and the subsidence rate significantly weakened and gradually stabilized (Figures 6–9). As a result, the isolated sub-sages were gradually interconnected to form a unified depression, and the whole depression was covered by the shallow sea shelf (Figure 6). This consequently resulted in the formation of a structural foundation for the development of neritic shelf sandstones and long-distance transported deltas in the continental shelf (Figure 6). Therefore, it can be concluded that tectonic evolution controlled the basin configuration and evolution process of depositional systems.

Normal faults significantly control sequence architecture and syn-rift successions in extensional systems through the generation of accommodation space [33,41,64,65]. In the Zhu III Depression, the Zhusannan Fault was critical in controlling accommodation and the distribution of deposition systems (Figures 7–9). In different basin evolutionary stages, the Zhusannan Fault had different characteristics of fault activity and fault geometry, affecting the development position of sand bodies. During the rifting stage (Wenchang–Enping Formations), the Zhusannan Fault was marked by large displacement and a high activity rate (as high as 250 m/Ma) [42]. Because of the intensive faulting, the high steep slope topography was formed with the fault scarp directly adjoined to the deep lake (Figure 10). The root of the Zhusannan Fault was the largest accommodation space in the depression, providing a depositional site for the nearshore fan delta (Figures 7–9). In this stage, fan delta systems of different sizes were mainly distributed along the Zhusannan Fault with prograding stacking patterns (Figures 7–9).

By contrast, during the transition stage (Zhuhai Formation), the fault activity of the Zhusannan Fault significantly decreased to <100 m/Ma [42], and the subsidence centers gradually moved away from the footwalls of the Zhusannan Fault (Figures 7–9). In this stage, secondary syn-sedimentary faults that were parallel to the Zhusannan Fault were formed and maintained a tectonic low landform (Figures 7 and 8). These faults, together with the Zhusannan Fault, controlled the distribution ranges of fan delta systems (Figures 7 and 8). Compared with the fan deltas in the rifting stage, the distribution ranges of fan deltas are much smaller (Figures 6–9). During the post-rifting stage, the Zhusannan Fault exerted little effect on the distribution of deposition systems. The Shenhu Uplift was gradually submerged and unable to provide sediments to form fan deltas, while the large-scale meandering river delta sourced from the Pearl River in the north became the dominant deposition system [50]. In this stage, the distribution pattern of depositional systems was mainly controlled by sea-level change and sediment supply, which have been documented in previous studies [66].

In summary, the filling patterns and distribution of depositional systems were controlled by several factors (i.e., tectonic activities, sea-level change, and sediment supply), but the main controlling factor is not uniform during the basin evolution process from rifting stage to the post-rifting stage. In the rifting stage, the intense rifting of boundary faults significantly controlled the distribution of depositional systems, whereas sea-level change and sediment supply became significant controlling factors for stratigraphic architecture.

5.2. Sequence Architecture Evolution

Structural and sedimentary slope breaks have significant control over the distribution pattern of depositional systems [12,21,33,67]. Six types of slope break belts have been identified in the Zhu III Depression, exerting significant functions on stratal stacking patterns (Figure 10). The sequence architecture differs greatly in space and time (Figures 10 and 11).

For example, in the northern depression, the sequence architecture was mainly controlled by faulting, and two different faulting patterns (i.e., steep slope fault belt, multi-level step-fault belt) were developed during the evolving growth of the Zhusannan Fault (Figure 11). By contrast, the sequence architecture in the regions adjacent to the Qionghai Uplift in the western depression was mainly controlled by the gentle slope belt (Figure 11). Even in the regions along the footwall of the Zhusannan Fault, two different patterns, including the steep slope fault belt in the west and the multi-level step-fault belt in the east, were revealed in the depositional period of the Zhuhai Formation, indicating that the fault geometry (low-angle or high-angle) is also an important factor controlling the depositional patterns (Figure 11). Therefore, our results highlight the diversity and complexity of sequence architecture in the continental rift basin, which resulted from the comprehensive effect of sea/lake-level changes, faulting, palaeogeomorphology, and sediment supply.

The diversity and complexity of sequence architecture are closely associated with the tectonic evolution process [7,21,66]. The dynamic evolution process of sequence architecture can be divided into three stages: (1) During the rifting stage, the steep slope fault belt and steep-fault bending belt were developed in the footwalls of the Zhusannan and No.2 Fault, respectively (Figure 11). The antithetic-fault slope belt and gentle slope belt were developed around the uplifts (Figure 11). The depression was filled with widespread fan delta deposits on the gentle slopes and steep slopes (Figure 6). The distribution of fan deltas in the fault zones was mainly controlled by faulting, while that in the gentle slope was dominantly controlled by palaeogeomorphology (Figure 10). (2) During the transition stage, the steep slope fault belt developed in the footwall of the Zhusannan Fault converted into the multi-level step-fault belt, which controlled the distribution of sedimentary facies from the shallow to the deep lacustrine environment (Figures 7 and 8). By contrast, in the regions of the intrabasin uplifts (e.g., Qionghai Uplift), the previously developed filling patterns have changed to the gentle slope belt (Figure 11). (3) During the post-rifting stage, the end of tectonism caused the cessation of differential subsidence, and the previously developed sequence patterns were replaced by the flat continental shelf belt (Figures 6–9). The depositional systems and their stratal stacking patterns were dominantly controlled by sea-level change and sediment supply [66], and tectonic activities had little effect on the sequence architecture.

In summary, the spatial and temporal features of stratal stacking patterns varied with a single structural unit and between different structural units (Figure 11). From the rifting stage to the post-rifting stage, the dominant stacking patterns have changed from tectonic slope breaks (e.g., steep slope fault belt, steep-fault bending belt) to sedimentary slope breaks (e.g., flat continental shelf belt) (Figure 11). The diversity of sequence architecture resulted from the comprehensive effect of tectonic activities, sediment supply, sea/lake-level changes, and geomorphic conditions. The sequence architecture is mainly controlled by the tectonic activities in the rifting stages, whereas it is dominantly controlled by sediment supply and sea level changes in the post-rifting stage.

5.3. *Hydrocarbon Accumulation Rules and Exploration Direction*

The basin evolution and sequence architecture can control not only the depositional filling process but also the distribution of hydrocarbon resources and the types of traps [33,68–70]. Structural traps are widely developed in the regions of tectonic slope breaks (e.g., steep slope fault belt, steep-fault bending belt, multi-level step-fault belt) where fault traps and faulted anticlinal traps formed (Figures 12 and 13). By contrast, lithological traps have a close relationship with the development of the sedimentary slope breaks (e.g., gentle slope belt, flat continental shelf belt) (Figures 12 and 13). There are two types of lithological traps, including up-dip wedge-out traps and lithologic lenticular traps (Figure 12), but their formation regions are different. Up-dip wedge-out traps were formed in the stratum of Wenchang, Enping, and Zhuhai Formations, whereas lithologic lenticular traps were developed mainly in the Zhujiang Formation (Figure 12). In addition, stratigraphic onlap traps are developed by subsequent uplift and erosion, especially in the stratum of the Zhuhai Formation in the Qionghai Uplift (Figure 12).

Trap Type	Hydrocarbon Reservoir Type	Model Diagram	Development Region	Development Strata
Lithologic Trap	Updip Wedge-out Trap		Qionghai Uplift	Enping, Zhuhai and Zhujiang Formations
			Wenchang Sag B	Wenchang and Enping Formations
	Lithologic Lenticular Trap		Wenchang Sag A and B	Zhujiang Formation (T40-T41)
			Wenchang Sag A and B	Wenchang and Enping Formations
Stratigraphic Trap	Stratigraphic Onlap Trap		Qionghai Uplift	Zhuhai Formation
			Wenchang Sag A	Zhuhai Formation
Structural Trap	Fault Trap	fault	Zhusannan Fault No 2 Fault System No 6 Fault System No 5 Fault System	Wenchang, Enping, and Zhuhai Formations
	Faulted Anticlinal Trap	delta sand bodies fault	Wenchang Sag B	Zhuhai Formation

Figure 12. The types of traps and their development strata and regions in the Zhu III Depression, PRMB.

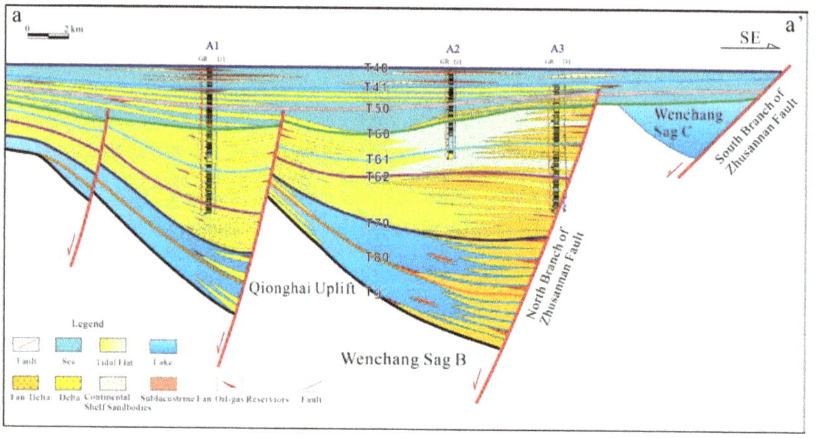

Figure 13. The type and distribution of oil and gas reservoirs across the representative section in the Zhu III Depression, PRMB. The location of section aa' is shown in Figure 1.

The sequence stacking pattern is closely associated with trap types and their distribution [68,69]. In the regions of steep slope fault belts and steep-fault bending belts, three different types of traps are developed, including fault traps at the fault terraces of the boundary faults, up-dip wedge-out traps in the north, and sublacustrine fan traps in the deep lake (Figure 13). In the multi-level step-fault belts, lobate sand bodies are deposited on each fault terrace where fault traps develop. In the gentle slope belts, up-dip wedge-out traps and lithologic lenticular traps are formed (Figure 13). By contrast, in the flat continental shelf belts, neritic shelf sand bodies developed in the Miocene Zhuhai Formation are favorable for forming lithologic lenticular traps (Figure 13).

In the last few decades, exploration achievements have been mainly achieved in the regions of structural traps [71]. The investigation of depositional evolution and stratal stacking patterns provides valuable insights into guiding the exploration and development [33,68,69]. According to exploration status and characteristics of trap distribution, the hydrocarbon exploration in the Zhu III Depression should be conducted at different levels: (1) The flat continental shelf belt developed in the Miocene Zhujiang Formation should be the most significant area for future exploration since the neritic shelf sand bodies distributed in the flat continental shelf are characterized by a high-quality reservoir property with average porosity and permeability values higher than 18% and 140 mD, respectively. The oils in the sandstone reservoirs are high maturity (Ro = 0.8~1.3%) and originated from the Enping Formation (Toc = 1.6~2.4%, HI = 233~332 mg/g TOC) [71]. Moreover, the neritic shelf sand bodies are favourable for forming "self-reservoir and self-coverage" lithologic reservoirs (Figure 13). Moreover, compared with the other traps, the hydrocarbon exploration for the neritic shelf sand bodies faces fewer difficulties, such as shallow burial, low exploration cost, and good seismic quality. These advantages profoundly influence the hydrocarbon enrichment and increase the success rate of hydrocarbon exploration, but determining the distribution range of neritic shelf sand bodies is primary. (2) The gentle slope belt in the adjacent region of the Qionghai Uplift should be an important area for future exploration (Figure 13). The up-dip wedge-out traps in this region are key exploration targets for seeking lithologic reservoirs. (3) The steep slope breaks (i.e., steep slope fault belt, steep-fault bending belt) located in the footwalls of the Zhusannan Fault should be scientifically investigated to achieve breakthroughs (Figure 13). The primary difficulty in these regions is poor seismic data quality, and more study is needed before locating an exploration target.

6. Conclusions

During the basin evolution process from the rifting stage to the post-rifting stage in the western Pearl River Mouth Basin (PRMB), in the northern South China Sea, five types of depositional systems were developed, including the fan delta, meandering river delta, tidal flat, lacustrine system, and neritic shelf system. The representative depositional systems changed from the proximal fan delta and lacustrine system in the Eocene–early Oligocene, to the tidal flat and fan delta in the late Oligocene, and then the neritic shelf system in the Miocene. The intense rifting of boundary faults significantly controlled the distribution of depositional systems in the rifting stage, whereas sea-level change and sediment supply became significant controlling factors for stratigraphic architecture in the post-rifting stage.

Six types of slope break belts (i.e., steep slope fault belt, steep-fault bending belt, multi-level step-fault belt, antithetic-fault slope belt, gentle slope belt, flat continental shelf belt) developed in the western PRMB, which exerted significant functions on stratal stacking patterns. Our results highlight the diversity and complexity of sequence architecture in the continental rift basin, which resulted from the comprehensive effects of sea/lake-level changes, faulting, palaeogeomorphology, and sediment supply.

Stratal stacking patterns are closely associated with trap types and their distribution. According to the exploration status and characteristics of trap distribution, the hydrocarbon exploration in the Zhu III Depression should be conducted at different levels: the flat continental shelf belt and the gentle slope belt are recommended as the key targets for exploration, whereas the steep slope break belt requires further investigations to achieve breakthroughs.

Author Contributions: Conceptualization, E.L. and Y.D.; methodology, S.C.; software, X.L.; validation, D.Y.; formal analysis, S.C.; investigation, X.S.; resources, Y.D.; data curation, X.S.; writing—original draft preparation, E.L.; writing—review and editing, S.C.; visualization, D.Y.; supervision, Y.D.; project administration, Y.D. and D.Y.; funding acquisition, E.L. All authors have read and agreed to the published version of the manuscript.

Funding: This work was supported by the National Natural Science Foundation of China (Nos. 42072142, 41702121, U19B2007).

Data Availability Statement: Not applicable.

Acknowledgments: We appreciate Zhanjiang Branch Company of China Offshore Oil Corporation for providing the data and permission to publish this paper. We are also very grateful to the reviewers and editors for their contributions to improving this paper.

Conflicts of Interest: The authors declared that they have no conflict of interest in this work. We declare that we do not have any commercial or associative interest that represents a conflict of interest in connection with the work submitted.

References

1. Dou, L.R.; Pan, X.H.; Tian, Z.J.; Xiao, K.Y.; Zhang, Z.W. Hydrocarbon formation and distribution of rift basins in Sudan: A comparative analysis of them with rift basins in East China. *Pet. Explor. Dev.* **2006**, *33*, 255–261.
2. Martins-Neto, M.A.; Catuneanu, O. Rift sequence stratigraphy. *Mar. Pet. Geol.* **2010**, *27*, 247–253. [CrossRef]
3. Maravelis, A.G.; Chamilaki, E.; Pasadakis, N.; Zelilidis, A.; Collins, W.J. Hydrocarbon generating potential of a Lower Permian sedimentary succession (Mount Agony Formation): Southern Sydney Basin, New South Wales, Southeast Australia. *Int. J. Coal Geol.* **2017**, *83*, 52–64. [CrossRef]
4. Wang, X.; He, S.; Wu, J. Tectonic controls on lacustrine source rock occurrence in the Huizhou Sag, Pearl River Mouth Basin, China. *Int. Geol. Rev.* **2020**, *62*, 72–93. [CrossRef]
5. Maravelis, A.G.; Catuneanu, O.; Nordsvan, A.; Landenberger, B.; Zelilidis, A. Interplay of tectonism and eustasy during the Early Permian icehouse: Southern Sydney Basin, southeast Australia. *Geol. J.* **2018**, *53*, 1372–1403. [CrossRef]
6. Shehata, A.A.; El Fawal, F.M.; Ito, M.; Abdel Aal, M.H.; Sarhan, M.A. Sequence stratigraphic evolution of the syn-rift early cretaceous sediments, west Beni Suef basin, the Western Desert of Egypt with remarks on its hydrocarbon accumulations. *Arab. J. Geosci.* **2018**, *11*, 313–331. [CrossRef]
7. Contreras, J.; Scholz, C.H. Evolution of stratigraphic sequences in multisegmented continental rift basins: Comparison of computer models with the basins of the East African rift system. *AAPG Bull.* **2001**, *85*, 1565–1581.
8. Martins-Neto, M.A.; Pedrosa-Soares, A.C.; Lima, S.A.A. Tectono-sedimentary evolution of sedimentary basins from Late Paleoproterozoic to Late Neoproterozoic in the Sao Francisco craton and Aracuai Fold belt, eastern Brazil. *Sediment. Geol.* **2001**, *141–142*, 343–370. [CrossRef]
9. Vail, P.R.; Mitchum, R.M.; Thompson, S. Global cycles of relative changes of sea level. *AAPG Bull.* **1977**, *26*, 99–116.
10. Posamentier, H.W.; Allen, G.P. Variability of the sequence stratigraphic model: Effects of local basin factors. *Sediment. Geol.* **1993**, *86*, 91–109. [CrossRef]
11. Bahroudi, A.; Koyi, H.A. Tectono-sedimentary framework of the Gachsaran Formation in the Zagros foreland basin. *Mar. Pet. Geol.* **2004**, *21*, 1295–1310. [CrossRef]
12. Lin, C.S.; Erikoson, K.; Li, S.T.; Wan, Y.X.; Ren, J.Y.; Zhang, Y.M. Sequence architecture, depositional systems and their controls of lacustrine sequence of the Erlian Basin. *AAPG Bull.* **2001**, *85*, 2017–2043.
13. Zecchin, M.; Catuneanu, O. High-resolution sequence stratigraphy of clastic shelves III: Applications to reservoir geology. *Mar. Pet. Geol.* **2015**, *62*, 161–175. [CrossRef]
14. Jackson, C.; Gawthorpe, R.L.; Carr, I.D.; Sharp, I.R. Normal faulting as a control on the stratigraphic development of shallow marine syn-rift sequences: The Nukhul and Lower Rudeis Formations, Hammam Faraun fault block, Suez Rift, Egypt. *Sedimentology* **2005**, *52*, 313–338. [CrossRef]
15. Escalona, A.; Mann, P. Tectonics, basin subsidence mechanisms, and paleogeography of the Caribbean-South American plate boundary zone. *Mar. Pet. Geol.* **2011**, *28*, 8–39. [CrossRef]
16. Wang, G.; Wang, H.; Gan, H.; Liu, E.; Xia, C.; Zhao, Y.; Chen, S.; Zhang, C. Paleogene tectonic evolution controls on sequence stratigraphic patterns in the Fushan Sag, Northern South China Sea. *J. Earth Sci.* **2016**, *27*, 654–669. [CrossRef]
17. He, J.; Wang, H.; Jiang, T.; Liu, E.; Chen, S.; Jiang, P. Sedimentary Characteristics of Lacustrine Beach-Bars and Their Formation in the Paleogene Weixinan Sag of Beibuwan Basin, Northern South China Sea. *Energies* **2022**, *15*, 3391. [CrossRef]
18. Shehata, A.A.; El Fawal, F.M.; Ito, M.; Aboulmagd, M.A.; Brooks, H.L. Senonian platform-to-slope evolution in the tectonically-influenced Syrian Arc sedimentary belt: Beni Suef Basin, Egypt. *J. Afr. Earth Sci.* **2020**, *170*, 103934. [CrossRef]
19. Sakran, S.; Shehata, A.A.; Osman, O.; El Sherbiny, M. Superposed tectonic regimes in West Beni Suef basin, Nile Valley, Egypt: Implications to source rock maturation and hydrocarbon entrapment. *J. Afr. Earth Sci.* **2019**, *154*, 1–19. [CrossRef]
20. Carroll, A.R.; Bohacs, K.M. Stratigraphic classification of ancient lakes: Balancing tectonic and climatic controls. *Geology* **1999**, *27*, 99–102. [CrossRef]
21. Feng, Y.L.; Li, S.T.; Lu, Y.C. Sequence stratigraphy and architectural variability in late Eocene lacustrine strata of the Dongying depression, Bohai Bay Basin, eastern China. *Sediment. Geol.* **2013**, *295*, 1–26. [CrossRef]
22. Zhang, Y.; Zhang, L.; Mi, L.; Lu, X.; Wu, S.; Tang, L.; Zhou, J.; Xiong, X.; Zhu, J. Quantitative Analysis of Cenozoic Extension in the Qiongdongnan Basin, South China Sea: Insight on Tectonic Control for Hydrocarbon Reservoir Accumulation and Formation. *Energies* **2022**, *15*, 4011. [CrossRef]
23. Shehata, A.A.; El Fawal, F.M.; Ito, M.; Abdel Aal, M.H.; Sarhan, M.A. Cenomanian–Turonian depositional history of a post-Gondwana rift succession in the west Beni Suef basin, Egypt. *J. Afr. Earth Sci.* **2019**, *150*, 783–798. [CrossRef]

24. Kassem, A.A.; Osman, O.; Nabawy, B.; Baghdady, A.; Shehata, A.A. Microfacies analysis and reservoir discrimination of channelized carbonate platform systems: An example from the Turonian Wata Formation, Gulf of Suez, Egypt. *J. Pet. Sci. Eng.* **2022**, *212*, 110272. [CrossRef]
25. Shanley, K.W.; McCabe, P.J. Perspectives on the sequence stratigraphy of continental strata. *AAPG Bull.* **1994**, *78*, 544–568.
26. Chen, S.; Steel, R.; Wang, H.; Zhao, R.; Olariu, C. Clinoform growth and sediment flux into late Cenozoic Qiongdongnan shelf margin, South China Sea. *Basin Res.* **2020**, *32*, 302–319. [CrossRef]
27. Xu, J.; Zhao, N.; Xu, S.; Zhao, X. Main controlling factors and development model of the Miocene marine source rocks in Yinggehai Basin. *Bull. Geol. Sci. Technol.* **2021**, *40*, 54–63.
28. Young, M.J.; Gawthorpe, R.L.; Sharp, I.R. Architecture and evolution of syn-rift clastic depositional systems towards the tip of a major fault segment, Suez Rift, Egypt. *Basin Res.* **2015**, *14*, 1–23. [CrossRef]
29. Willan, R.C.R.; Hunter, M.A. Basin evolution during the transition from continental rifting to subduction: Evidence from the lithofacies and modal petrology of the Jurassic Latady Group, Antarctic Peninsula. *J. S. Am. Earth Sci.* **2005**, *20*, 171–191. [CrossRef]
30. Elliott, W.S.J.; Suttner, L.J.; Pratt, L.M. Tectonically induced climate and its control on the distribution of depositional systems in a continental foreland basin, Cloverly and Lakota Formations (Lower Cretaceous) of Wyoming, U.S.A. *Sediment. Geol.* **2007**, *202*, 730–753. [CrossRef]
31. Liu, E.; Wang, H.; Feng, Y.; Pan, S.; Jing, Z.; Ma, Q.; Gan, H.; Zhao, J. Sedimentary architecture and provenance analysis of a sublacustrine fan system in a half-graben rift depression of the South China Sea. *Sediment. Geol.* **2020**, *409*, 105781. [CrossRef]
32. Nottvedt, A.; Gabrielsen, R.H.; Steel, R.J. Tectonostratigraphy and sedimentary architecture of rift basins, with reference to the northern North Sea. *Mar. Pet. Geol.* **1995**, *12*, 881–901. [CrossRef]
33. Liu, E.; Wang, H.; Li, Y.; Zhou, W.; Leonard, N.D.; Lin, Z.; Ma, Q. Sedimentary characteristics and tectonic setting of sublacustrine fans in a half-graben rift depression, Beibuwan Basin, South China Sea. *Mar. Pet. Geol.* **2014**, *52*, 9–21. [CrossRef]
34. Zhou, Y.; Ji, Y.L.; Pigott, J.D.; Meng, Q.A.; Wan, L. Tectonostratigraphy of lower cretaceous tanan sub-basin, tamtsag basin, mongolia: Sequence architecture, depositional systems and controls on sediment infill. *Mar. Pet. Geol.* **2014**, *49*, 176–202. [CrossRef]
35. Wang, H.; Chen, S.; Gong, T.; Yu, Z.; Huang, C.; Zhang, Y.; Zhao, R. Sedimentary process and accumulation mechanism of traction fluidization gravity flow: An example from Qikou Sag, Bohai Bay Basin. *Bull. Geol. Sci. Technol.* **2020**, *39*, 95–104.
36. Alfaro, E.; Holz, M. Seismic geomorphological analysis of deepwater gravity-driven deposits on a slope system of the southern Colombian Caribbean margin. *Mar. Pet. Geol.* **2014**, *57*, 294–311. [CrossRef]
37. Allen, P.A. From landscapes into geological history. *Nature* **2018**, *451*, 274–276. [CrossRef]
38. Andrić, N.; Sant, K.; Matenco, L.; Mandic, O.; Tomljenović, B.; Pavelić, D.; Ooms, J. The link between tectonics and sedimentation in asymmetric extensional basins: Inferences from the study of the Sarajevo-Zenica Basin. *Mar. Pet. Geol.* **2017**, *83*, 305–332. [CrossRef]
39. Prather, B.E. Controls on reservoir distribution, architecture and stratigraphic trapping in slope settings. *Mar. Pet. Geol.* **2003**, *20*, 529–545. [CrossRef]
40. Liu, E.; Wang, H.; Li, Y.; Leonard, N.D.; Feng, Y.; Pan, S.; Xia, C. Relative role of accommodation zones in controlling stratal architectural variability and facies distribution: Insights from the Fushan Depression, South China Sea. *Mar. Pet. Geol.* **2015**, *68*, 219–239. [CrossRef]
41. Khalil, S.M.; Mc Clay, K.R. Structural control on syn-rift sedimentation, northwestern Red Sea margin, Egypt. *Mar. Pet. Geol.* **2009**, *26*, 1018–1034. [CrossRef]
42. Yang, X.B.; Chang, L.; Xu, R.; Cheng, D.X.; Yin, Z.J.; You, J.J.; Liu, K.; Hu, G.W. Types and sand control effect of terrestrial slope breaks of Wenchang Sag inside the Pearl River Mouth Basin. *Earth Sci.* **2020**, *45*, 989–997, (In Chinese with English Abstract).
43. Pei, J.; Zhang, C.; Wang, Y.; Wang, K.; Liu, J.; Wang, S. Tectonic evolution and depositional response in southern continental marginal basins of South China Sea during period of rift- drift-foreland: A case study from the Liyue Basin. *Bull. Geol. Sci. Technol.* **2021**, *40*, 42–53.
44. Xie, Y. Sedimentary characteristics and hydrocarbon exploration potential of the upstream of the Central Canyon in the Yinggehai and Qiongdongnan basins. *Bull. Geol. Sci. Technol.* **2020**, *39*, 69–78.
45. Liu, Y.; Chen, H.; Wang, Y.; Han, J.; Li, Q. Diagenetic effect of mantle-derived CO_2 charge to clay minerals in the Baiyun-Liwan deepwater area of the Pearl River Mouth Basin, the South China Sea. *Bull. Geol. Sci. Technol.* **2021**, *40*, 85–95.
46. Huang, B.; Xiao, X.; Zhang, M. Geochemistry, grouping and origins of crude oils in the western Pearl River Mouth Basin, offshore south China sea. *Org. Geochem.* **2003**, *34*, 993–1008. [CrossRef]
47. Quan, Y.; Liu, J.; Zhao, D.; Hao, F.; Wang, Z.; Tian, J. The origin and distribution of crude oil in Zhu III sub-basin, Pearl River Mouth Basin. China. *Mar. Petrol. Geol.* **2015**, *66*, 732–747. [CrossRef]
48. Dong., D.D.; Zhang, G.C.; Zhong, K.; Yuan, S.Q.; Wu, S.G. Tectonic evolution and dynamics of deep water area of Pearl River Mouth Basin, Northern South China Sea. *J. Earth Sci.* **2009**, *20*, 47–159. [CrossRef]
49. Xie, H.; Zhou, D.; Li, Y.; Pang, X.; Li, P.; Chen, G.; Li, F.; Cao, J. Cenozoic tectonic subsidence in deepwater sags in the Pearl River Mouth Basin, northern South China sea. *Tectonophysics* **2014**, *615–616*, 182–198. [CrossRef]

50. Liu, E.; Chen, S.; Yan, D.; Deng, Y.; Wang, H.; Jing, Z.; Pan, S. Detrital zircon geochronology and heavy mineral composition constraints on provenance evolution in the western Pearl River Mouth basin, northern south China sea: A source to sink approach. *Mar. Pet. Geol.* **2022**, *145*, 105884. [CrossRef]
51. Tang, X.; Yang, S.; Zhu, J.; Long, Z.; Jiang, G.; Huang, S.; Hu, S. Tectonic subsidence of the Zhu 1 sub-basin in the Pearl River Mouth Basin, northern South China Sea. *Front. Earth Sci.* **2017**, *11*, 729–739. [CrossRef]
52. Gong, Z.; Li, S.; Xie, T.; Zhang, Q.; Xu, S.; Xia, K.; Yang, J.; Sun, Y.; Liu, L. *Continental Margin Basin Analyss and Hydrocarbon Accumulation of the Northern South China Sea*; China Science Press: Beijing, China, 1997.
53. Wang, Y.; Zhao, Y.; Ding, W.; Fang, P.; Li, J. Cenozoic propagated rifting in the dangerous grounds in response to the episodic seafloor spreading of the South China Sea. *J. Earth Sci.* **2022**, *33*, 1031–1046. [CrossRef]
54. Tian, L.X. Sedimentary-reservoir characteristics under control of transfer modeland implications for hydrocarbon exploration in huizhou depression, Pearl River Mouth Basin. *Earth Sci.* **2021**, *46*, 4043–4056.
55. Guo, W.; Xu, G.Q.; Liu, B.J.; Xiang, X.H.; Liu, D.Q.; Zhang, B. Structure-sedimentary response relationship of Wenchang Formation in Baiyun Sag, Pearl River Mouth Basin. *Earth Sci.* **2022**, *47*, 2433–2453.
56. Wu, J.; Zhang, X.; Bai, H.; Zheng, X.; Cai, G.; Li, Z. Miocene tidal control system and its exploration significance of lithologic trap in Yangjiang Sag, Pearl River Mouth Basin. *Earth Sci.* **2021**, *46*, 3673–3689.
57. He, Z.; Yin, X.; Jiang, S.; Lei, M.; Liu, Y.; Zhao, R.; Zhu, B. Source rock classification, maturity and their implications in paleoenvironment reconstruction in the Zhu III sub-basin, China. *J. Pet. Sci. Eng.* **2022**, *216*, 110799. [CrossRef]
58. Catuneanu, O. Model-independent Sequence Stratigraphy. *Earth-Sci. Rev.* **2019**, *188*, 312–388. [CrossRef]
59. Maravelis, A.; Boutelier, D.; Catuneanu, O.; Seymour, K.; Zelilidis, A. A review of tectonics and sedimentation in a forearc setting: Hellenic Thrace Basin, north Aegean Sea and mainland Greece. *Tectonophysics* **2016**, *674*, 1–19. [CrossRef]
60. Catuneanu, O.; Zecchin, M. Unique vs. non-unique stratal geometries: Relevance to sequence stratigraphy. *Mar. Pet. Geol.* **2016**, *78*, 184–195. [CrossRef]
61. Breckenridge, J.; Maravelis, A.G.; Catuneanu, O.; Ruming, K.; Holmes, E.; Collins, W.J. Outcrop analysis and facies model of an Upper Permian tidally influenced fluvio-deltaic system: Northern Sydney Basin, Southeast Australia. *Geol. Mag.* **2019**, *156*, 1715–1741. [CrossRef]
62. Melehan, S.; Botziolis, C.; Maravelis, A.G.; Catuneanu, O.; Ruming, K.; Holmes, E.; Collins, W.J. Sedimentology and Stratigraphy of an Upper Permian Sedimentary Succession: Northern Sydney Basin, Southeastern Australia. *Geosciences* **2021**, *11*, 273. [CrossRef]
63. Liu, E.; Wang, H.; Pan, S.; Qin, C.; Jiang, P.; Chen, S.; Yan, D.; Lü, X.; Jing, Z. Architecture and depositional processes of sublacustrine fan systems in structurally active settings: An example from Weixinan Depression, northern South China Sea. *Mar. Pet. Geol.* **2021**, *134*, 105380. [CrossRef]
64. Gawthorpe, R.L.; Leeder, M.R. Tectono-sedimentary evolution of active extensional basins. *Basin Res.* **2000**, *12*, 195–218. [CrossRef]
65. Schlische, R.W. Half-graben filling models: Implications for the evolution of continental extensional basins. *Basin Res.* **1991**, *3*, 123–141. [CrossRef]
66. Ge, J.; Zhao, X.; Tan, M.; Zhuo, H.; Liu, C.; Jones, B.G. Sequence stratigraphy and depositional evolution of the north-eastern shelf (33.9–10.5 Ma) of the Pearl River Mouth basin, South China Sea. *Mar. Pet. Geol.* **2022**, *141*, 105697. [CrossRef]
67. Ravnås, R.; Steel, R.J. Architecture of marine rift-basin successions. *AAPG Bull.* **1998**, *82*, 110–146.
68. Cai, Q.; Hu, M.; Ngia, N.R.; Hu, Z. Sequence stratigraphy, sedimentary systems and implications for hydrocarbon exploration in the northern Xujiaweizi Fault Depression, Songliao Basin, NE China. *J. Pet. Sci. Eng.* **2017**, *152*, 471–494. [CrossRef]
69. Khalifa, M.K.; Mills, K.J. Predicting sequence stratigraphic architecture and its implication for hydrocarbon reservoir potential of the uppermost Silurian through Lower Devonian Winduck Interval, central Darling Basin of western New South Wales, SE Australia. *Mar. Pet. Geol.* **2022**, *142*, 105725. [CrossRef]
70. Li, H.; Wu, J.; Huang, J.; Wang, Y.; Li, Z. Quantitative analysis of fault vertical sealing ability and its application in an Oilfield of Bohai Bay Basin. *Bull. Geol. Sci. Technol.* **2020**, *39*, 125–131.
71. Cheng, P.; Tian, H.; Huang, B.; Wilkins, R.W.T.; Xiao, X. Tracing early-charged oils and exploration directions for the Wenchang a sag, western Pearl River Mouth Basin, offshore south China sea. *Org. Geochem.* **2013**, *61*, 15–26. [CrossRef]

energies

Article

New Insights into the Genetic Mechanism of the Miocene Mounded Stratigraphy in the Qiongdongnan Basin, Northern South China Sea

Litao Xu [1,2], Wanzhong Shi [1,2,*], Ren Wang [1,2], Jinfeng Ren [3], Yulin He [3], Hao Du [1,2], Tingna Zuo [1,2], Jin Huang [1,2] and Yang Dong [1,2]

1 Key Laboratory of Tectonics and Petroleum Resources, Ministry of Education, China University of Geosciences, Wuhan 430074, China
2 School of Earth Resources, China University of Geosciences, Wuhan 430074, China
3 Guangzhou Marine Geological Survey, China Geological Survey, Guangzhou 510075, China
* Correspondence: shiwz@cug.edu.cn

Abstract: The origin of deep-water mounds has been a topic of debate in recent years. In this study, newly collected seismic data were employed to characterize the mounds within the Meishan Formation in the Qiongdongnan Basin and a novel model was proposed. The result showed that pervasive mounds and 'V'-shaped troughs were alternately distributed at the top of the Meishan Formation. They appeared as elongated ridges flanked by similarly elongated gullies, with the trending parallel with the strike of the basinward slope. The mounded features were considered to be formed in response to the tectonically induced seabed deformation. The differential subsidence steepened the slope that was equivalent to the top of the Meishan Formation (ca. 10.5 Ma), which offered sufficient driving forces triggering the slope's instability. Correspondingly, the uppermost deposits glided along a bedding-parallel detachment surface, creating a number of listric detachment faults that ceased downward to this surface. The uppermost layer was cut into a range of tilted fault blocks with tops constituting a seemingly mounded topography. Some of the downfaulted troughs between mounds steered the gravity flows and were filled by sand-rich lithologies. The differential subsidence played a decisive role in the formation of a mounded stratigraphy, which in turn acted as clues to the important tectonic phase since the late Miocene.

Keywords: mounded stratigraphy; genetic mechanism; slope instability; gravity gliding; detachment faults; Qiongdongnan Basin

Citation: Xu, L.; Shi, W.; Wang, R.; Ren, J.; He, Y.; Du, H.; Zuo, T.; Huang, J.; Dong, Y. New Insights into the Genetic Mechanism of the Miocene Mounded Stratigraphy in the Qiongdongnan Basin, Northern South China Sea. *Energies* **2022**, *15*, 9478. https://doi.org/10.3390/en15249478

Academic Editor: Dameng Liu

Received: 16 November 2022
Accepted: 12 December 2022
Published: 14 December 2022

1. Introduction

Extensive attention has been paid to the formation of deep-water mounds in recent decades [1,2]. The presence of mounds within the strata can cause dome-like bedforms with mounded seismic reflections [3,4]. Various origins of deep-water mounds in debates mainly involved the carbonate reefs [5,6], mud diapirs or mud volcanoes [4,7], deep-water wave sediments [8], sand bars related to differential compaction [9–11], and remnant mounds originated from channel incision [12–15].

In recent years, pervasive mounded seismic reflections were detected in the Miocene Meishan Formation (Fm.) in the Qiongdongnan Basin (QDNB) [13,16]. Numerous models were proposed over the years to interpret their origins but there is no unified conclusion [4,5,14–18]. A widely accepted view in prior studies was that the middle Miocene was the dominant reef-building period in the northern South China Sea, and the mounded reflections were then thought to be the carbonate reefs [5,16,19]. Nevertheless, recent drilling revealed that some mounds were mainly composed of mud-rich lithology rather than reefs [18,20]. Channel incision has also been considered the possible controlling factor in the formation of the Miocene mounds in the study area [4,15]. However, most of the mounds trend roughly parallel with the inferred strike of the slope, which contradicts the principle that remnant mounds generated by the incision have their axes pointing downslope in most situations [14,16,21]. Additionally, these mounds were interpreted as sediment waves generated by bottom currents, but lacked the overwhelming proof supporting the existence of the bottom currents at that time [13]. Therefore, the models developed previously might not be readily applicable to the study area, and other mechanisms are required to explain the formation of mounds. It is worth noting that much of the topographic relief can appear in the structurally disturbed strata [3,22]. For the QDNB, an important tectonic transformation occurred at the turn of the middle and late Miocene, accompanied by intense tectonic activation corresponding to the time equal to the top of the Meishan Fm. [23–25]. Whether the formation of the mounds within the uppermost layer of the middle Miocene is related to such tectonic change has not been noticed previously.

In this study, we applied 2D and 3D seismic data to describe the characteristics of the mounds within the uppermost layer of the Miocene Meishan Fm. Combined with the tectonic settings, a novel model was developed to explain the formation process of the mounds. This model accentuates the decisive role played by tectonic activities in the creation of the mounds. The research results may offer a new idea in the interpretation of similar geological phenomena in other areas around the world.

2. Geological Setting

The QDNB is located at the west of the NE-trending extensional system in the northern South China Sea [26–29] and comprises the Northern Depression, Northern Uplift, Central Depression, and Southern Uplift from the north to south [30–32] (Figure 1). The Central Depression is located in the deep-water area, including the Ledong, Lingshui, Songnan, Ganquan, Beijiao, Baodao, and Changchang Sags, as well as the Lingnan and Songnan highs [33,34]. The water depths vary from 500 to 1500 m, with the maximum depth reaching 2500 m [35]. The tectonic evolution of the basin is consistent with a typical passive margin containing rifting and thermal subsidence. The unconformity of T6 that formed at ~21 Ma divided the Cenozoic strata of the basin into the syn-rift and post-rift layers. Faults were extremely developed in the syn-rift stage while seldom active in the post-rift stage [13,27]. Since the late Miocene, tectonic subsidence accelerated, accompanied by intense reactivation of faults in the eastern basin [23,24,36]. Correspondingly, the water depth of the basin increased [16,37]. In addition, since the cessation of submarine expansion (~16 Ma), magma has been widely active in the South China Sea and its adjacent areas [25,38]. There is a high heat flow zone in the Baodao to Changchang Sags, which was closely related to the thermal events involving massive magmatic intrusions and eruptions from the Miocene to Pliocene [37].

The main sequence boundaries of the basin including Tg, T8, T7, T6, T5, T4, T3, and T2 divided the Cenozoic strata into Lingtou (E_2l), Yacheng (E_3y), Lingshui (E_3l), Sanya (N_1s), Meishan (N_1m), Huangliu (N_1h), Yinggehai (N_2y) and Ledong (Ql) Fms. (Figure 2). The depositional environment transformed from a coast and shallow sea to a semi-deep and deep sea from the middle to late Miocene [4,18]. The lower member of the Meishan Fm. is composed of mudstones, sandstones, and calcareous sandstones; its upper member mainly consists of mudstones mixed with thin argillaceous siltstones [15,20,39]. The lower member of the Huangliu Fm. is dominantly fine sandstones interbedded with thin mudstones and its upper member primarily comprised sandy limestones and fine sandstones [15]. A large submarine canyon (i.e., Central Canyon) was developed in the late Miocene, with an 'S'-shaped axis parallel to the shelf break of the basin [18]. The mounded stratigraphy within the Miocene Meishan Fm. is restricted to the Southern Slope (between the Central Depression and Southern Uplift) which has the topography dipping from SE to NW, and is bounded by the Central Canyon to the north and the Xisha Islands to the south [5,18].

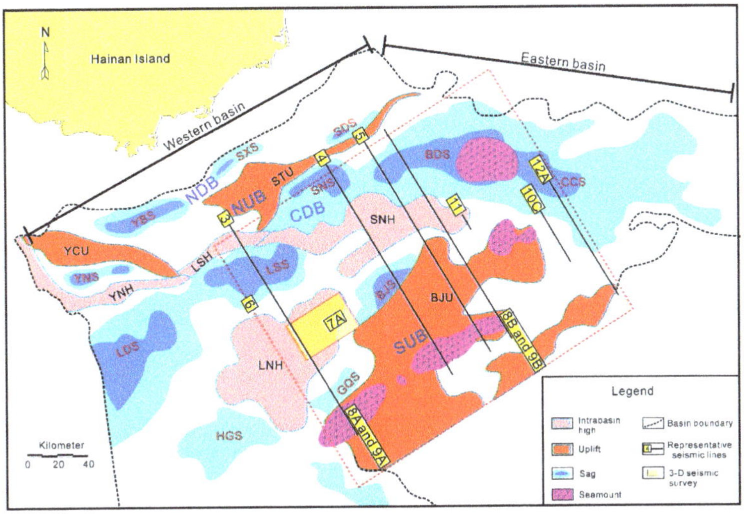

Figure 1. Tectonic units of the Qiongdongnan Basin (modified from [30,31]). The Qiongdongnan basin is divided into the eastern and western basin according to the direction change of the basin axis. Seamounts composed of basaltic volcanic rocks developed in the south eastern uplifts and eastern sags. NDB = Northern Depression Belt. NUB = Northern Uplift Belt. CDB = Central Depression Belt. SUB = Southern Uplift Belt. YCU = Yacheng Uplift. STU = Songtao Uplift. BJU = Beijiao Uplift. YNH = Yanan High. LSH = Lingshui High. LNH = Lingnan High. SNH = Songnan High. YNS = Yanan Sag. YBS = Yabei Sag. SXS = Songxi Sag. SDS = Songdong Sag. LDS = Ledong Sag. LSS = Lingshui Sag. BJS = Beijiao Sag. SNS = Songnan Sag. BDS = Baodao Sag. CCS = Changchang Sag. GQS = Ganquan Sag. HGS = Huangguang Sag.

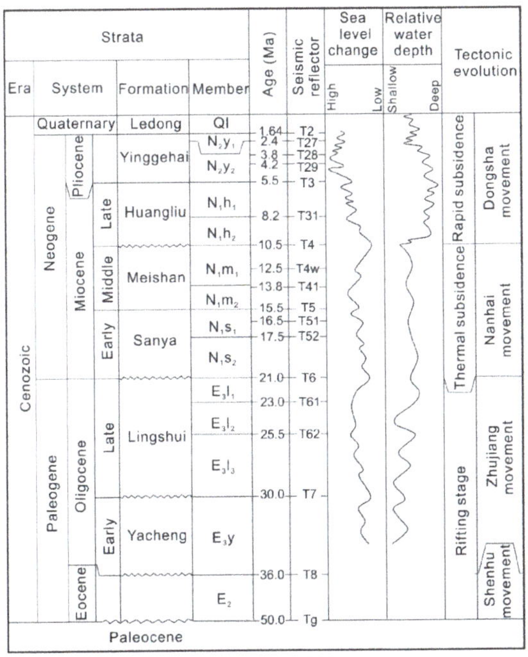

Figure 2. Generalized stratigraphic column of the Qiongdongnan Basin (modified from [28,29]).

3. Data and Methods

The 2D and 3D seismic data used in this study were acquired over the past several years by Guangzhou Marine Geological Survey (GMGS). The G Gun II type air-gun sources were employed to acquire the seismic lines [40]. The seismic streamer used the Seal 24-digit cable, with a submerged depth of 8 m, and the receiver streamer had 240 traces and 60-fold coverage [41]. The 5700 km of 2D seismic lines covering the QDNB have a dominant frequency of 30 HZ on average, with a vertical resolution of 20 m and a trace interval of 6.25 m. In particular, new multi-channel 3D seismic volumes acquired in 2018 cover an area of ~800 km^2 located in the Lingnan High, with the dominant frequency ranging from 30 to 70 Hz. The sampling interval for the 3D seismic data is 1 ms, with a bin size of 3.125 m $\times$ 18.75 m. The seismic data were processed by the Institute of Processing in the GMGS using the GeoCluster (a product of CGG®) processing system. The processing procedures including data input, trace editing, static correction, prestack noise attenuation, amplitude compensation, prestack deconvolution, CMP sorting, velocity analysis, dynamic correction, residual static correction, dip moveout (DMO) correction, prestack time migration, quality control in processing, poststack noise attenuation, and poststack time migration have been described by Wang et al. (2010) and Zhang (2020) [41,42].

The interpretation of seismic data was carried out by the Geoframe® software (a product of Schlumberger®) to construct the seismic stratigraphic framework. The seismic data were zero-phased and displayed normal polarity, indicating a positive event on the seismic profiles which is shown as a red reflection [25]. The identification of seismic facies differences and contact relationships between strata supported the tracking of sequence boundaries [43]. Eight main seismic sequence boundaries including Tg, T8, T7, T6, T5, T4, T3, and T2 were identified (Figures 3–5). Faults can be identified in seismic profiles in the form of the offset of seismic events or zones of poor seismic signal across which reflectors are offset [3,44]. Fault interpretation was conducted by adopting appropriate geometric and kinematic models, e.g., the listric normal fault, domino-type fault, and flower structure [45]. Based on this, the seismic interpretation emphasized describing the geometric characteristics of the mounds within the uppermost layer of the Meishan Fm. The width and height of the mounds and their distribution range were determined based on the 2D seismic profiles. In addition, the isopach map of the Huangliu Fm., prepared based on seismic surfaces T3 and T4, was used to recover the paleogeomorphology corresponding to the end of the Meishan deposition. Moreover, the seismic attribute map of the curvature was extracted by the Geoframe® to reveal the plan-view characters of the mounded topography. Curvature calculates the bending degree of the stratum, which can effectively reflect geomorphic features such as valleys, ridges, and domes [46,47].

To determine the tectonic histories, backstripping was employed to reconstruct the amount and rate of tectonic subsidence in different stages [48,49]. Back-stripping employs the tectonic subsidence equation presented by Steckler and Watts (1978) [50]. The method is based on the crustal isostatic principle to backstrip the present stratigraphy layer by layer and to obtain the true basement subsidence caused by the tectonic driving forces [48]. Tectonic subsidence reconstruction was performed by dividing and dating the sequence boundary of the seismic profiles. The ages of the sequence boundary are shown in Figure 2. In the process, the seismic sequence boundaries were time-depth conversed by using the formula provided by the GMGS. There is a decrease in the porosity of sediment with an increase in overlying strata thickness during the compaction process [49]. To remove the effect of compaction, the original amount of subsidence in each stratum can be recovered by the relationship between porosity and depth that fits an exponential function to recover the original thicknesses of strata [51]. Additionally, considering that the depositional thickness is not equivalent to the subsidence when the sedimentary interface remains below the water surface, the accurate paleo-water depths and eustatic variations were taken into consideration [48,52]. The paleowater depth data were collected from the previous research results of Zhai et al. (2013) and Zuo et al. (2022) (Table 1) [32,53]. The eustatic variations are shown in Figure 2.

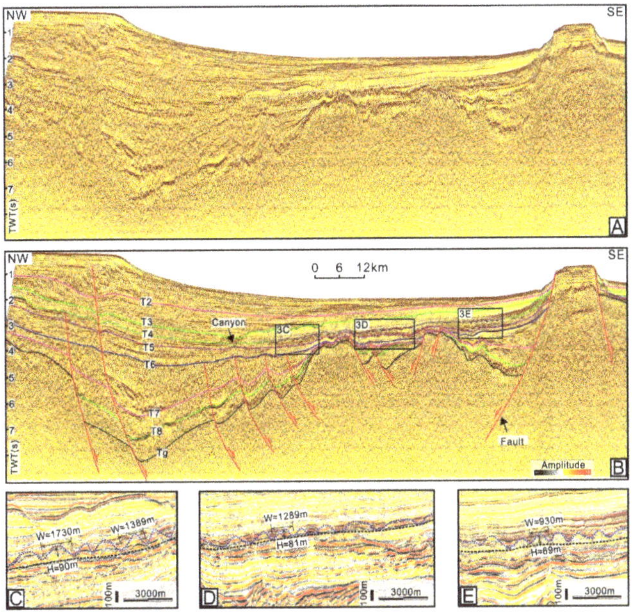

Figure 3. (**A**,**B**) Uninterpreted and interpreted seismic section traversing the Lingshui Sag, showing the stratigraphic and structural characteristics (for location, see Figure 1). (**C**–**E**) Seismic profiles and their interpretations showing the mounded topography at the top of Meishan Fm (for locations, see Figure 3A). W = width; H = height; TWT = two-way traveltime.

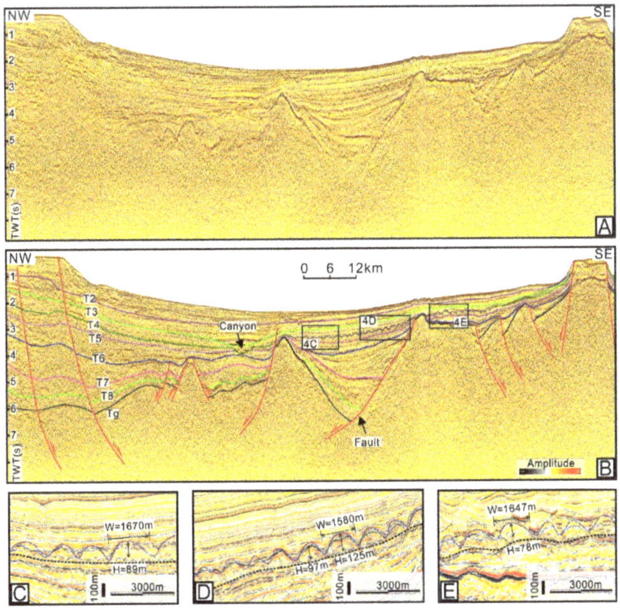

Figure 4. (**A**,**B**) Uninterpreted and interpreted seismic section traversing the Beijiao Sag, showing the stratigraphic and structural characteristics (for location, see Figure 1). (**C**–**E**) Seismic profiles and their interpretations showing the mounded topography at the top of Meishan Fm (for locations, see Figure 4A). W = width; H = height; TWT = two-way traveltime.

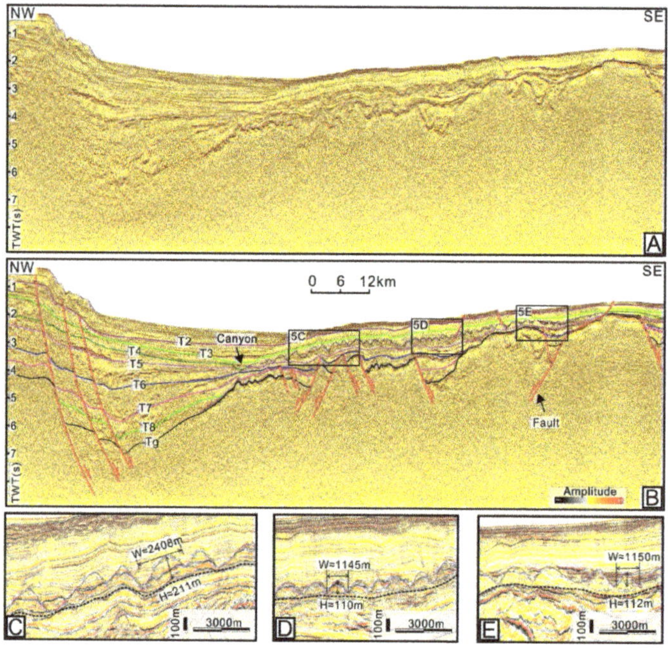

Figure 5. (**A**,**B**) Uninterpreted and interpreted seismic section traversing the Baodao Sag, showing the stratigraphic and structural characteristics (for location, see Figure 1). (**C–E**) Seismic profiles and their interpretations showing the mounded topography at the top of Meishan Fm (for locations, see Figure 5A). W = width; H = height; TWT = two-way traveltime.

Table 1. Reference value of paleo heat flow and water depth during different periods (referenced from [32,53]). HF = Heat flow; PWD = Paleowater depth. HF1, HF2, and PWD1 apply to the northern slope; HF3, HF4, and PWD2 apply to the central depression; HF5, HF6, and PWD3 apply to the southern uplift.

Age (Ma)	HF1 (mW/m^2)	HF2 (mW/m^2)	HF3 (mW/m^2)	HF4 (mW/m^2)	HF5 (mW/m^2)	HF6 (mW/m^2)	Age (Ma)	PWD1 (m)	PWD2 (m)	PWD3 (m)
0	58	65	72	77	84	92	1.9	120	1600	700
2	61	68	76	80	88	96	5.5	90	1100	500
5.5	54	58	64	69	75	80	10.5	50	700	300
10.5	55	59	66	71	76	81	15.5	60	500	150
17.5	57	61	70	75	79	82	17.5	40	400	120
23	58	62	70	74	78	81	21	20	200	30
30	50	53	58	63	66	69	22	15	100	20
36	44	47	52	57	60	62	23	40	100	50
38	45	48	53	57	60	62	25.5	20	80	30
50	42	44	46	48	50	52	33	10	50	20
							50	20	20	25

4. Results

4.1. Seismic Stratigraphy

The seismic reflector T4 corresponds to the regional unconformity which represented the Dongsha movement in the South China Sea [16]. In the Southern Slope, the reflector T4 truncated the underlying strata and was onlapped by overlying sediments [4,15], as shown in Figures 3–5. The underlying Meishan Fm. can be divided into upper and lower members by an obvious interface. The upper member is characterized by strong-moderate amplitudes, moderate frequency, and mounded reflections, and the lower member shows subparallel sheet-shaped reflections with moderate-strong amplitudes and moderate frequency (Figures 3–5). The overlying Huangliu Fm. is dominated by moderate amplitude, moderate-weak frequency, and parallel-subparallel reflections (Figures 3–5). The Huangliu Fm. gradually thins onto the southern margin of the basin. A large submarine canyon (Central Canyon) was developed in the post-rift sequence of the QDNB, which is distributed in the Central Depression from west to east (Figures 3–6). The large magmatic body between Songnan and Baodao Sags divides the canyon into two sections, the west section trending NE and the east section trending nearly EW (Figure 6).

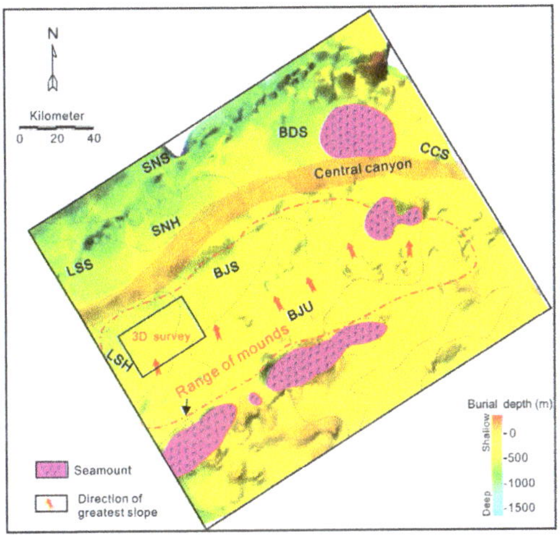

Figure 6. Paleogeomorphology at the beginning of the deposition of the Huangliu Fm., showing the directions of the greatest slope in the southern area. The red dashed line indicates the range of the mounds. LSS = Lingshui Sag. BJS = Beijiao Sag. SNS = Songnan Sag. BDS = Baodao Sag. CCS = Changchang Sag. LNH = Lingnan High. SNH = Songnan High. BJU = Beijiao Uplift.

4.2. Characteristics of Mounded Stratigraphy

The NW–SE oriented seismic profiles show that the top of the Meishan Fm. is characterized by the elevations alternating with 'V'-shaped troughs, displaying undulated morphologies (Figures 3–5). The elevations tend to be present as lenticular seismic reflectors that are thickest in the middle part and thin toward both wings, showing dome-like bedforms, which are common not only in the QDNB but also in the Pearl River Basin in the northern South China Sea [5,16,18]. Similar elevations were described as 'mounds' previously, and strata characterized by the presence of mounds are commonly said to have 'mounded topography' [3,18]. The mounded stratigraphy was constrained between the T4 and the boundary of the upper and lower member of the Meishan Fm. (i.e., T5) (Figures 3–5). The mounds distributed in the Lingnan High have heights varying from 70 to 90 m, with the widths ranging from 0.9 to 1.7 km (Figure 3C–E). The heights of a single mound in the Beijiao area vary from 90 to 125 m, and their average width is greater than 1.5 km (Figure 4C–E). The mounds in the south of the Baodao Sag have heights of about 110~210 m, with widths ranging from 1.0 to 2.4 km (Figure 5C–E). On the whole, the scale of the single mound in the eastern area seems larger. The internal reflectors of the mounds show strong-moderate amplitudes and moderate frequency, which are characterized by obvious fold structures and disturbed bedding (Figures 3–5). The seismic reflections within the troughs vary greatly in different areas, displaying strong-moderate amplitudes and moderate frequency in the Lingnan High (Figures 3C–E and 7B,C), and moderate-weak amplitudes and low frequency in the Beijiao area and the south of the Baodao Sag (Figures 4 and 5). The reflector T4 commonly truncated the flanks of the mounds and was onlapped, filled, and leveled by subsequent seismic strata corresponding to the Huangliu Fm. (Figures 3–5). Spatially, the mounds were most developed in the middle of the Southern Slope and disappeared in the Central Depression and South Uplift (Figures 3–6). The mounded stratigraphy covers an area restricted to the east of the Huaguang Sag, west of the Changchang Sag, north of the Beijiao Uplift, and south of the Central Canyon (Figure 6). Additionally, the curvature attribute map calculated for the top of the Meishan Fm. shows that the mounded topography is characterized by coupled elongated ridges and gullies, with most of their axes predominantly oriented SWW–NEE and SW-NE (Figure 7A). These elongated ridges and gullies show a subparallel configuration with a few intersections.

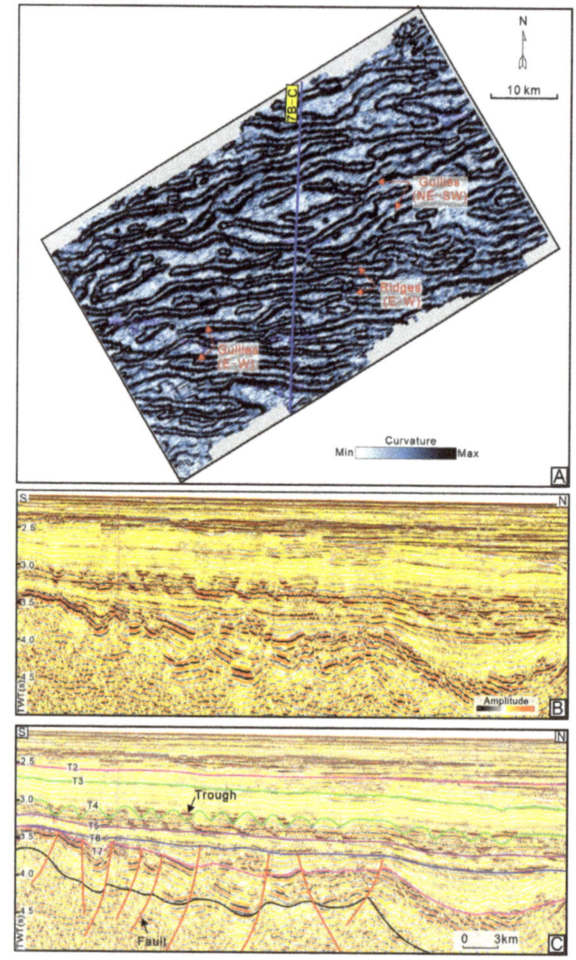

Figure 7. (**A**) Characteristics of mounded topography revealed by curvature analysis based on 3D seismic data (See Figure 6 for the location of the 3D survey). The mounded topography is characterized by the alternating distribution of nearly EW-trending elongate troughs and ridges. (**B**,**C**) The top of the Miocene Meishan Formation has a mounded topography characterized by the mounds alternating with troughs (for location, see Figure 7A). TWT = two-way traveltime.

4.3. Tectonic Subsidence

The balanced cross-sections were created by using the backstripping technique to determine the tectonic subsidence in different stages [48,49]. The result shows that the sedimentary centers were controlled by the boundary faults in the Paleogene, during which the sedimentary thickness and subsidence rate in the depression centers are larger than those in the uplift belts (Figures 8 and 9). The Eocene is relatively thin and the subsidence rate is low. The sedimentary thickness and subsidence rate in the early Oligocene significantly increased. The sedimentary thickness in the late Oligocene is relatively thick while the subsidence rate decreased. The faults were almost inactive when the basin entered the post-rift stage and its evolution was mainly controlled by thermal subsidence (Figure 8). During the depositional period of Sanya and Meishan Fms., the sedimentary thickness is relatively thin and the subsidence rate was extremely low, with an average value of less than 150 m/myr (Figure 9). In addition, there was little difference in tectonic subsidence between the Central Depression and the Southern Slope. During the deposition of the Huangliu Fm., the sedimentary thickness increased slightly and the subsidence rate in the Central Depression was ~200 m/myr and gradually decreased southward (Figure 9). During the depositional period of the Yinggehai Fm., sedimentary thickness and subsidence increased sharply, with a maximum cumulative rate approaching ~1100 m/myr in the Lingshui Sag and ~1600 m/myr in the Baodao Sag (Figures 8 and 9). The difference in subsidence between the Central Depression and the Southern Slope and Southern Uplift became more obvious. During the deposition of the Ledong Fm., the deposition thickness was still large, and the Baodao and Lingshui Sags had a subsidence rate of ~800 m/myr and ~950 m/myr, respectively (Figure 9). At the same time, the subsidence of the Southern Uplift was much less than that of the Central Depression. Overall, the most obvious feature of the basin is that since 10.5 Ma, the subsidence began to accelerate, showing higher subsidence in the depressions and lower subsidence in the Southern Uplift.

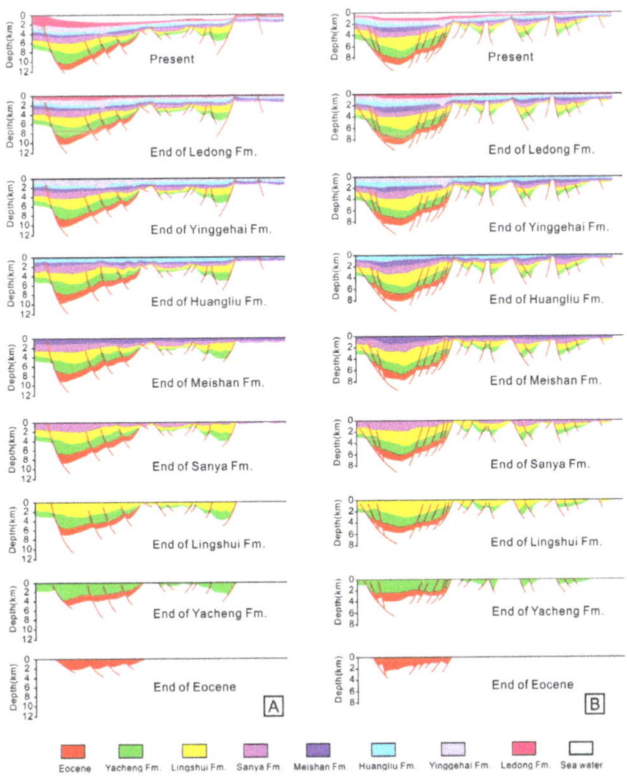

Figure 8. (**A**,**B**) Balanced cross-sections created based on the two profiles traversing the Lingshui and the Baodao Sag, respectively (for location, see Figure 1).

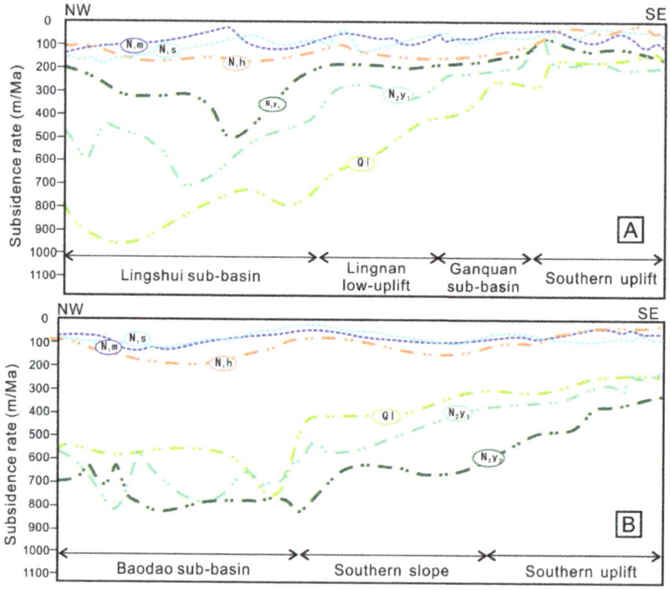

Figure 9. (**A**,**B**) Subsidence curves created based on the two sections traversing the Lingshui and the Baodao Sags, respectively (for location, see Figure 1). The subsidence rate was low in the period of the Sanya and Meishan deposition, then increased slightly during the period of the Huangliu Deposition, and increased sharply from the period of the Yinggehai Deposition. Moreover, the subsidence rate of the depressions is higher than that of the Southern Uplift after the period of the Huangliu Deposition. N_1s = Sanya Fm., N_1m = Meishan Fm.; N_1h = Huangliu Fm.; N_2y_1 = Lower Yinggehai Fm.; N_2y_2 = Upper Yinggehai Fm.; Ql = Ledong Fm. Ma = Megaannus.

5. Discussion

The mounded topography in the study area was previously supposed to be the result of the incision of bottom currents or channels [14,15]. The main evidence supporting this view is that some troughs between the mounds are characterized by high amplitude reflections marking the sand-prone lithologies which may be a clue to the incision [4,14,15], as shown in Figures 3C–E and 7B,C. However, quite a few troughs distributed over a wider area are characterized by weak amplitudes, probably marking the bathyal–abyssal deposits [5,18] (Figures 4C and 5C). Additionally, the internal reflectors within most of the mounds have folded seismic patterns, which might be related to other forces rather than the incision (Figures 3–5). Moreover, paleogeomorphology indicated that the seabed over the Southern Slope roughly inclined toward NNW after the deposition of the Meishan Fm (Figure 6). If the troughs were indeed derived from the incision, their axes should have tended to the downdip of slope (NNW) in most situations [21]. However, it is not consistent with the result shown in Figure 7, with most of the troughs tending to be parallel to the strike of the slope (SWW–NEE and SW-NE). These doubts implied that the origin of the mounded topography cannot be simply explained solely by incision. In the examples from the Gulf of Mexico, the Brazilian margin, and the Central North Sea, the undulated morphologies can appear over tectonically active slopes [3,21,22,54–57]. Similar geological phenomena were observed at the seabed of the Shenhu Sea (Figure 10A) [40] and the south of the QDNB (Figure 10B), featured by seemingly mounded topography related to the disturbance of strata by detachment faults. These phenomena are commonly explained by the gravity gliding of sediments down a slope, which can generate a type of listric detachment fault (flattening downward) with associated rotated blocks to cause seabed relief [21,56–59]. Such a process can be triggered by rapid deposition, earthquakes, seafloor steepening, and excess pore pressure [58,60]. These studies provided an important insight into discovering the origin of a mounded topography in the study area.

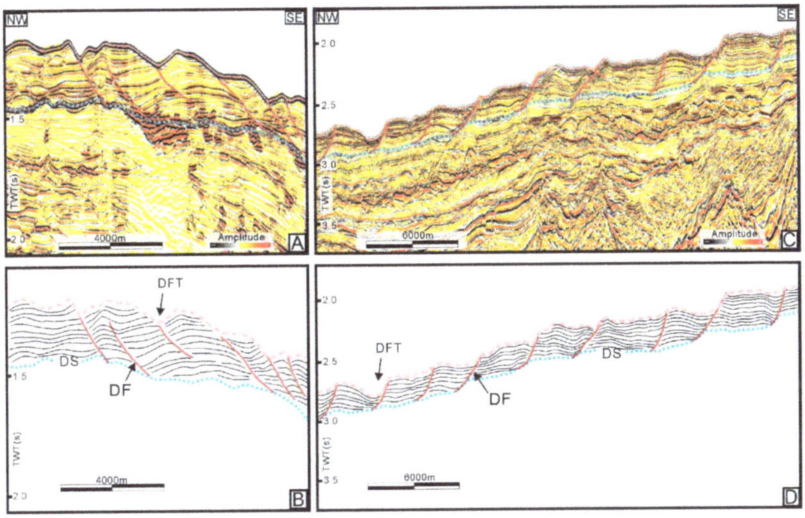

Figure 10. (**A**,**B**) Seismic profile and its geological interpretation showing the seemingly mounded topography on the seabed caused by submarine landslides in the Shenhu Sea in the northern South China Sea (profile quoted from [40]). (**C**,**D**) Seismic profile and its interpretation showing the seemingly mounded topography on the seabed in the Southern Slope (for location, see Figure 1). DS = detachment surface; DF = detachment fault; DFT = downfaulted trough; TWT = two-way traveltime.

In this study, the mounded features were interpreted to result from the gravity gliding that was induced by tectonic activities. Firstly, detailed examinations of seismic data revealed that the uppermost layer of the Meishan Fm. was dismembered into a series of moderately deformed blocks by a number of faults with listric geometries which ceased downward to a unified surface (Figure 11A–C). The blocks resting upon the fault planes have shown clockwise rotation, with their internal beds perpendicular to the concave upward fault planes (Figure 11A–C). Secondly, soft-sediment deformation associated with gravity gliding might have occurred in the mounds (blocks) with their internal reflections showing convex-up fold structures and disturbed bedding (Figures 3–5 and 11A–C) [61]. Such phenomenon can be related to the rollover anticlines unique to the listric growth faults or associated with the local compressive stresses pervasive in slide blocks which cause internal beds to buckle [22,62,63]. These features are consistent with the gravity slide tectonics shown in some important examples documented previously [3,21,22,54–57]. It can be inferred that, after the end of Meishan deposition, the uppermost unconsolidated deposits composed of mudstones mixed with thin argillaceous siltstones glided along an underlying shear plane which is probably the top of deep lithified strata [3,15,20,21,39]. Finally, the unbalanced subsidence corresponding to the time equal to the top of the Meishan Fm. was likely to trigger the gravity gliding of sediments which created the mounds. Specifically, the subsidence of the basin shows a decreasing trend from the Central Depression to the southern margin since 10.5 Ma, which immediately steepened the Southern Slope with an NNW inclination (Figures 6 and 9). Studies indicated that the occurrence of a steepening slope induced by the basement-driven differential subsidence was an important aspect of creating sufficient driving forces to drive the gravity gliding [3]. Some scholars suggested that the differential subsidence triggered an intense erosion of the turbidity currents, which created the embryo of the Central Canyon in the late Miocene [64–68]. Therefore, it can be inferred that the differential subsidence across the

Southern Slope since 10.5 Ma might have generated sufficient forces to drive the gravity gliding that created the mounds.

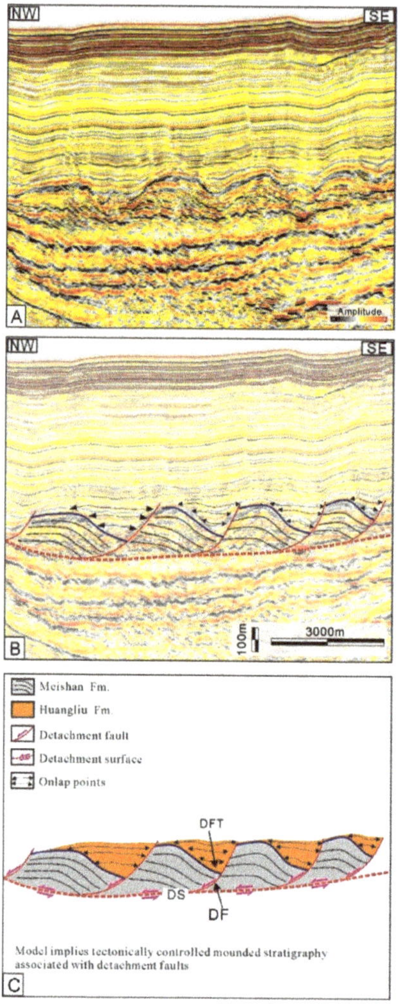

Figure 11. (**A**) Uninterpreted seismic profiles, (**B**) interpretations, and (**C**) their geological models showing the formation of the mounded stratigraphy (for locations, see Figure 1). The tilted fault blocks rest on the fault plane of detachment faults, with their tops constituting the seemingly mounded topography. DS = detachment surface; DF = detachment fault; DFT = downfaulted trough; TWT = two-way traveltime.

Further analysis indicated a time–space coupling relationship between the formation of mounds and the tectonic events including accelerated subsidence, magmatic activity, and fault reactivation. The accelerated subsidence in the post-rift stage has been observed in many rifted basins, which was considered to be related to the deep thermal anomaly [23,69–71]. Detailed examination of seismic data revealed a large number of magmatic intrusions distributed near the deeply rooted faults (Figure 12B). These faults likely extend deeply into magma chambers beneath the Central Depression to provide primary pathways for magma migration [25]. Studies indicated that massive magmas invading up to the cold upper crust through the faults might contribute to the decay of a deep thermal source and the cooling of the asthenosphere, which would cause rapid subsidence [72–74]. The samples from adjacent areas indicated that the magmatism typically occurred in the Miocene and more recently [25,37]. Therefore, the accelerated subsidence since 10.5 Ma was likely related to the magmatic activities. In addition, such subsidence showed a higher rate at the crustal thinning zone beneath the Central Depression and decreased to the north and south of the basin (Figure 9). Under the influence of differential subsidence, the stratum of the basin was folded into a syncline, which induced extensional strain at the flanks of the syncline [74] (Figure 13A). Many Neogene faults extended upward to the interface T4 (Figure 12C,D), indicating that their formation age dated 10.5 Ma. It can be inferred the extensional strain induced by differential subsidence was accommodated by the generation of faults in the deep rigid layers, as well as the gravity slide tectonics in the shallow soft layers (Figures 12D and 13A) [75]. Although it is a speculation, local seismic disturbance caused by magmatic activity and fault reactivation might also have triggered slope instability and gravitational sliding [57,60,76–79].

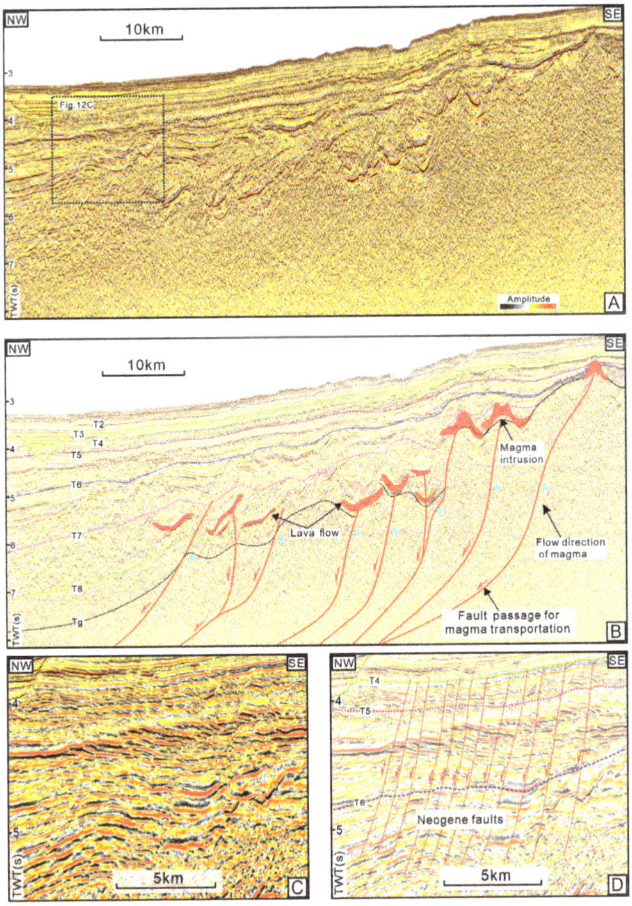

Figure 12. (**A**) Seismic profile and (**B**) its interpretation showing the magmatic intrusions associated with the deeply rooted faults (for location, see Figure 1). (**C**,**D**) Seismic profile and its interpretation show the dense post-rift faults developed in the eastern depression (for location, see Figure 12A). TWT = two-way traveltime.

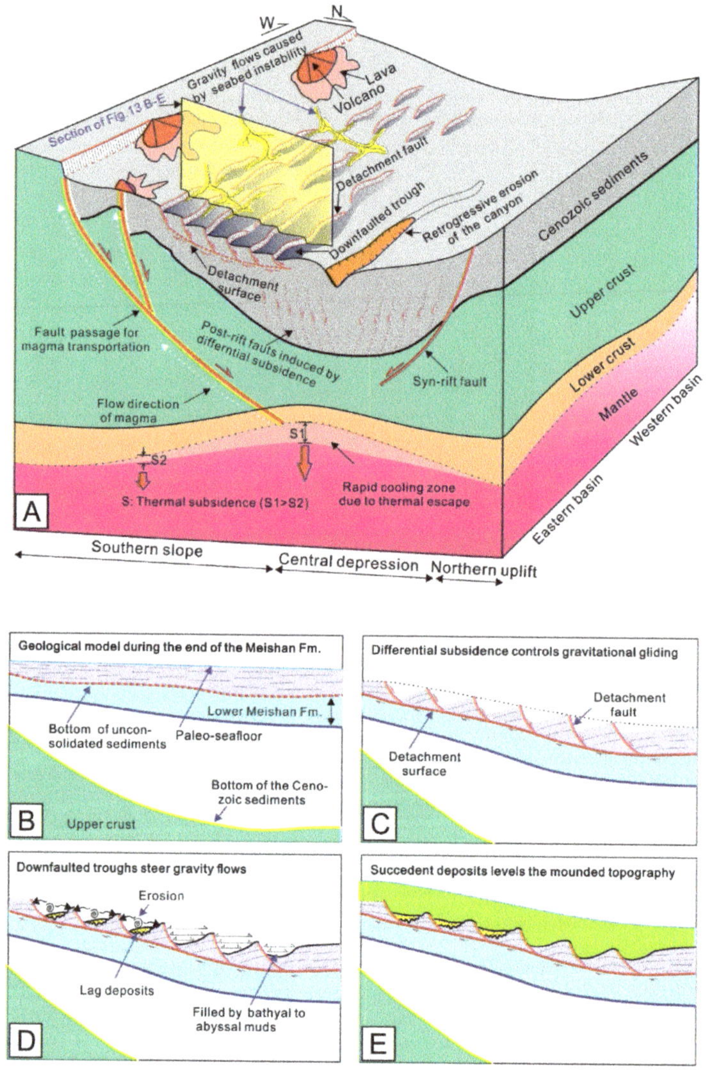

Figure 13. (**A**) Model to explain the formation of mounded morphologies in the study area of the Qiongdongnan Basin. Escape of massive magmas through the faults contributed to the cooling of the hot mantle material beneath the crustal thinning zone, resulting in quicker subsidence in the overlying Central Depression. The differential subsidence caused the gravitational gliding of the uppermost layers and the generation of faults in the deep rigid layers. The downfaulted troughs steered the gravity flows along their axes. (**B–E**) Diagrams illustrating the geological processes responsible for the creation of mounded stratigraphy.

Here, the processes responsible for the formation of mounded stratigraphy are summarized. Firstly, differential subsidence associated with magmatic activities steepened the slope at the end of the Meishan deposition, leading to the gravity gliding of the uppermost unconsolidated layer along a detachment surface (Figure 13B,C). Such a process generated a series of listric detachment faults with numerous tilted fault blocks resting on their fault planes (Figure 13C). The arcuate fault planes and the flanks of blocks together formed the undulated topography. Studies indicated that the denudation of some structural highs in the southern margin at the turn of the middle and late Miocene generated sedimentary sources, which were transported farther into the basin by gravity flows [15]. Then, there might have been some downfaulted troughs steering the gravity flows along their axes, with the unloading of sand-rich deposits at the bottom (Figure 13D). Deposition of other troughs without passing of gravity flows was predominantly from the succeeding bathyal–abyssal muds (Figure 13D). Finally, the Huangliu Fm. onlapped onto and leveled the mounded topography with its final shape adjusted by compaction (Figure 13E). The results of this study do not deny the existence of the erosion of gravity flows, which is also one of the processes that enhanced the relief amplitude of the mounded geometries. It is worthwhile to mention that gravity gliding derived from differential subsidence is decisive for the creation of mounded topography, while erosion and other processes are considered to be of secondary importance and will embellish such topography rather than create it. The differential subsidence associated with magmatic activities led to the development of mounds, which in turn became important clues implying revolutionary changes in tectonic movement that took place at the time equivalent to T4.

6. Conclusions

Detailed examination of seismic data revealed a unique mounded topography corresponding to the top of the Meishan Fm. in the Southern Slope of the QDNB. Such topography is characterized by the mounds alternating with 'V'-shaped troughs, which display elongated ridges flanked by similar gullies trending roughly parallel with the strike of the slope (SWW–NEE and SW-NE). The height of the mounds varies from about 70~210 m, and they are about 0.9~2.4 km in width. The internal reflectors within the mounds were truncated by the reflector T4 which is in turn overlain by seismic strata corresponding to the Huangliu Fm. Since the late Miocene (ca. 10.5 Ma), under the influence of magmatism, rapid subsidence occurred, which was manifested by the decreasing of subsidence from the center to the south of the depression. Such differential subsidence induced fault reactivation and steepened the slope. These tectonic events triggered the gliding of the uppermost deposits, which generated a series of listric detachment faults that ceased downwards a unified detachment surface. The top of tilted fault blocks that were constrained by these faults formed the rudiment of mounded topography. The downfaulted troughs steered the gravity flows and leveled by the lag deposits or bathyal–abyssal muds. The genetic model of the mounds established in this study accentuates the decisive contribution of tectonic activities to the creation of the mounded topography.

Author Contributions: Methodology, L.X. and T.Z.; investigation, L.X. and J.R.; formal analysis, L.X., J.R. and H.D.; writing—original draft preparation, L.X.; supervision, W.S., R.W. and Y.H.; conceptualization, R.W.; writing—review and editing, R.W.; software, Y.H., H.D., J.H. and Y.D.; funding acquisition, R.W. All authors have read and agreed to the published version of the manuscript.

Funding: This research was funded by the National Natural Science Foundation of China (No. 42202121).

Data Availability Statement: The data relevant with this study can be accessed by contacting the corresponding author.

Acknowledgments: This study was financially supported by the National Natural Science Foundation of China (No. 42202121). The data acquired in the research process is primarily attributed to Guangzhou Marine Geological Survey, Guangzhou. Thanks to the editor and reviewers for the suggestions and other assistance in improving this manuscript.

Conflicts of Interest: The authors declare no conflict of interest.

References

1. Hart, P.E.; Hutchinson, D.R.; Gardner, J.; Carney, R.S.; Fornari, D. A photographic and acoustic transect across two deep-water seafloor mounds, Mississippi Canyon, northern Gulf of Mexico. *Mar. Pet. Geol.* **2009**, *25*, 969–976. [CrossRef]
2. Meredyk, S.P.; Edinger, E.; Piper, D.J.W.; Huvenne, V.A.I.; Hoy, S.; Ruffman, A. Enigmatic deep-water mounds on the Orphan Knoll, Labrador Sea. *Front. Mar. Sci.* **2020**, *6*, 744. [CrossRef]

3. Koša, E. Differential subsidence driving the formation of mounded stratigraphy in deep-water sediments; Palaeocene, central North Sea. *Mar. Pet. Geol.* **2007**, *24*, 632–652. [CrossRef]
4. Li, Y.; Pu, R.; Niu, N.; Li, B. Characteristics and origins of middle Miocene mounds and channels in the northern South China Sea. *Acta Oceanol. Sin.* **2021**, *40*, 65–80. [CrossRef]
5. Wu, S.; Yuan, S.; Zhang, G.; Ma, Y.; Mi, L.; Xu, N. Seismic characteristics of a reef carbonate reservoir and implications for hydrocarbon exploration in deepwater of the Qiongdongnan Basin, northern South China Sea. *Mar. Pet. Geol.* **2009**, *26*, 817–823. [CrossRef]
6. Schindler, E.; Wehrmann, A. Genesis and internal architecture of the Middle to Upper Devonian Gwirat Al Hyssan reef-mound (Western Sahara). *Palaeogeogr. Palaeoecol.* **2011**, *304*, 184–193. [CrossRef]
7. Zhao, F.; Wu, S.G.; Sun, Q.L.; Huuse, M.; Li, W.; Wang, Z.J. Submarine volcanic mounds in the Pearl River Mouth Basin, northern South China Sea. *Mar. Geol.* **2014**, *355*, 162–172. [CrossRef]
8. Wynn, R.B.; Stow, D.A.V. Classification and characterisation of deep-water sediment waves. *Mar. Geol.* **2002**, *192*, 7–22. [CrossRef]
9. Purvis, K.; Kao, J.; Flanagan, K.; Henderson, J.; Durant, D. Complex reservoir geometries in a deep water clastic sequence, Gryphon Field, UKCS: Injection structures, geological modelling and reservoir simulation. *Mar. Pet. Geol.* **2002**, *19*, 161–179. [CrossRef]
10. Xu, S.; Hao, F.; Xu, C.; Wang, Y.; Zou, H.; Gong, C. Differential compaction faults and their implications for fluid expulsion in the northern Bozhong Subbasin, Bohai Bay Basin, China. *Mar. Pet. Geol.* **2015**, *63*, 1–16. [CrossRef]
11. Ward, N.I.P.; Alves, T.M.; Blenkinsop, T.G. Differential compaction over Late Miocene submarine channels in SE Brazil: Implications for trap formation. *GSA Bull.* **2018**, *130*, 208–221. [CrossRef]
12. Eschard, R.; Albouy, E.; Deschamps, R.; Euzen, T.; Ayub, A. Downstream evolution of turbiditic channel complexes in the Pab Range outcrops (Maastrichtian, Pakistan). *Mar. Pet. Geol.* **2003**, *20*, 691–710. [CrossRef]
13. Zhao, T.; Pu, R.; Qu, H.; Zhang, G.; Liang, J.; Feng, Y. An origin discussion of mound-shaped reflections in Miocene, southern Qiongdongnan Basin. *Acta Oceanol. Sin.* **2013**, *35*, 112–120, (In Chinese with English Abstract).
14. Sun, Q.; Cartwright, J.; Wu, S.; Zhong, G.; Wang, S.; Zhang, H. Submarine erosional troughs in the northern South China Sea: Evidence for Early Miocene deepwater circulation and paleoceanographic change. *Mar. Pet. Geol.* **2016**, *77*, 75–91. [CrossRef]
15. Xiong, P.; Jiang, T.; Kuang, Z.; Cheng, C.; Ren, J.; Lai, H. Sedimentary characteristics and origin of moundes in Meishan Formation, southern Qiongdongnan Basin. *Bull. Geol. Sci. Technol.* **2021**, *40*, 11–21, (In Chinese with English Abstract). [CrossRef]
16. Feng, Y.; Qu, H.; Zhang, G.; Pu, R. Seismic interpretation and hydrocarbon accumulations implication of the Miocene Meishan Formation reefs in southern Qiongdongnan Basin, northern South China Sea. *J. Palaeogeogr.* **2017**, *6*, 206–218. [CrossRef]
17. Feng, Y.; Ren, Y.; Li, Z.; Jin, L. Geological interpretation and hydrocarbon exploration potential of three types of mound-shaped reflectors in the Meishan Formation, Southern Qiongdongnan Basin. *Acta Geol. Sin.* **2021**, *95*, 167–176. [CrossRef]
18. Tian, J.; Wu, S.; Lv, F.; Wang, D.; Wang, B.; Zhang, X.; Ma, B. Middle Miocene mound-shaped sediment packages on the slope of the Xisha carbonate platforms, South China Sea: Combined result of gravity flow and bottom current. *Deep-Sea Res. Part II* **2015**, *122*, 172–184. [CrossRef]
19. Wu, S.; Yang, Z.; Wang, D.; Lü, F.; Lüdmann, T.; Fulthorpe, C.; Wang, B. Architecture, development and geological control of the Xisha carbonate platforms, northwestern South China Sea. *Mar. Geol.* **2014**, *350*, 71–83. [CrossRef]
20. Lei, C.; Ren, J.Y. Hyper-extended rift systems in the Xisha Trough, northwestern South China Sea: Implications for extreme crustal thinning ahead of a propagating ocean. *Mar. Pet. Geol.* **2016**, *77*, 846–864. [CrossRef]
21. Cobbold, P.R.; Szatmari, P. Radial gravitational gliding on passive margins. *Tectonophysics* **1991**, *188*, 249–289. [CrossRef]
22. Alves, T.M. Submarine slide blocks and associated soft-sediment deformation in deep-water basins: A review. *Mar. Pet. Geol.* **2015**, *67*, 262–285. [CrossRef]
23. Xie, X.; Müller, R.D.; Li, S.; Gong, Z.; Steinberger, B. Origin of anomalous subsidence along the Northern South China Sea margin and its relationship to dynamic topography. *Mar. Pet. Geol.* **2006**, *23*, 745–765. [CrossRef]
24. Yin, X.; Ren, J.; Lei, C.; Wang, S. Postrift rapid subsidence characters in Qiongdongnan Basin, South China Sea. *J. Earth Sci.* **2011**, *22*, 273–279. [CrossRef]
25. Sun, Q.; Alves, T.M.; Zhao, M.; Sibuet, J.; Calvès, G.; Xie, X. Post-rift magmatism on the northern South China Sea margin. *GSA Bull.* **2020**, *132*, 2382–2396. [CrossRef]
26. Zhang, C.; Wang, Z.; Sun, Z.; Liu, J.; Wang, Z. Structural differences between the western and eastern Qiongdongnan Basin: Evidence of Indochina block extrusion and South China Sea seafloor spreading. *Mar. Geophys. Res.* **2013**, *34*, 309–323. [CrossRef]
27. Hu, B.; Wang, L.S.; Yan, W.B.; Liu, S.W.; Cai, D.S.; Zhang, G.C.; Zhong, K.; Pei, J.X.; Sun, B. The tectonic evolution of the Qiongdongnan Basin in the northern margin of the South China Sea. *J. Asian Earth Sci.* **2013**, *77*, 163–182. [CrossRef]
28. Shi, W.; Xie, Y.; Wang, Z.; Li, X.; Tong, C. Characteristics of overpressure distribution and its implication for hydrocarbon exploration in the Qiongdongnan Basin. *J. Asian Earth Sci.* **2013**, *66*, 150–165. [CrossRef]
29. Xu, Q.; Shi, W.; Xie, Y.; Wang, Z.; Li, X.; Tong, C. Identification of low-overpressure interval and its implication to hydrocarbon migration: Case study in the Yanan sub-basin of the Qiongdongnan Basin, South China Sea. *PLoS ONE* **2017**, *12*, e0183676. [CrossRef]
30. Zhao, Z.; Sun, Z.; Liu, J.; Pérez-Gussinyé, M.; Zhuo, H. The continental extension discrepancy and anomalous subsidence pattern in the western Qiongdongnan Basin, South China Sea. Earth Planet. *Sci. Lett.* **2018**, *501*, 180–191. [CrossRef]

31. Yang, G.; Yin, H.; Gan, J.; Wang, W.; Zhu, J.; Jia, D.; Xiong, X.; Xu, W. Explaining structural difference between the eastern and western zones of the Qiongdongnan Basin, northern South China Sea: Insights from scaled physical models. *Tectonics* **2022**, *41*, e2021tc006899. [CrossRef]
32. Zuo, T.; Wang, R.; He, Y.; Shi, W.; Liang, J.; Xu, L.; Du, H.; Deng, Y.; Xu, X. Natural gas migration pathways and their influence on gas hydrate enrichment in the Qiongdongnan Basin, South China Sea. *Geofluids* **2022**, *2022*, 1954931. [CrossRef]
33. Lei, C.; Ren, J.; Pei, J.; Lin, H.; Yin, X.; Tong, D. Tectonic Framework and Multiple Episode Tectonic Evolution in Deepwater Area of Qiongdongnan Basin, Northern Continental Margin of South China Sea. *Earth Sci.* **2011**, *36*, 151–162, (In Chinese with English Abstract). [CrossRef]
34. Xia, Z.; Wan, Z.; Wang, X.; Shi, Q.; Cai, S.; Xia, B. The tectonic differences between the east and the west in the deep-water area of the northern South China Sea. *Acta Oceanol. Sin.* **2016**, *35*, 86–95. [CrossRef]
35. Ren, J.; Xu, L.; Shi, W.; Yang, W.; Wang, R.; He, Y.; Du, H. Shallow overpressure formation in the deep water area of the QiongdongnanBasin, China. *Front. Earth Sci.* **2022**, *10*, 922802. [CrossRef]
36. Wang, L.; Sun, Z.; Yang, J.; Sun, Z.; Zhu, J.; Zhuo, H.; Stock, J. Seismic characteristics and evolution of post-rift igneous complexes and hydrothermal vents in the Lingshui sub-basin (Qiongdongnan basin), northwestern South China Sea. *Mar. Geol.* **2019**, *418*, 106043. [CrossRef]
37. Shi, X.; Jiang, H.; Yang, J.; Yang, X.; Xu, H. Models of the rapid post-rift subsidence in the eastern Qiongdongnan Basin, South China Sea: Implications for the development of the deep thermal anomaly. *Basin Res.* **2017**, *29*, 340–362. [CrossRef]
38. Mao, K.; Xie, X.; Xie, Y.; Ren, J.; Chen, H. Post-rift tectonic reactivation and its effect on deep-water deposits in the Qiongdongnan Basin, northwestern South China Sea. *Mar. Geophys. Res.* **2015**, *36*, 227–242. [CrossRef]
39. Sun, Z.; Zhai, S.; Xiu, C.; Liu, X.; Zong, T.; Luo, W.; Liu, X.; Chen, K.; Li, N. Geochemical characteristics and their significances of rare-earth elements in deep-water well core at the Lingnan Low Uplift Area of the Qiongdongnan Basin. *Acta Oceanol. Sin.* **2014**, *33*, 81–95. [CrossRef]
40. Cheng, C.; Jiang, T.; Kuang, Z.; Yang, C.; Zhang, C.; He, Y.; Cheng, Z.; Tian, D.; Xiong, P. Characteristics of gas chimneys and their implications on gas hydrate accumulation in the Shenhu area, northern South China Sea. *J. Nat. Gas Sci. Eng.* **2020**, *84*, 103629. [CrossRef]
41. Wang, X.; Wu, S.; Yuan, S.; Wang, D.; Ma, Y.; Yao, G.; Gong, Y.; Zhang, G. Geophysical signatures associated with fluid flow and gas hydrate occurrence in a tectonically quiescent sequence, Qiongdongnan Basin, South China Sea. *Geofluids* **2010**, *10*, 351–368. [CrossRef]
42. Zhang, W.; Liang, J.; Wei, J.; Lu, J.; Su, P.; Lin, L.; Huang, W.; Guo, Y.; Deng, W.; Yang, X.; et al. Geological and geophysical features of and controls on occurrence and accumulation of gas hydrates in the first offshore gas-hydrate production test region in the Shenhu area, Northern South China Sea. *Mar. Pet. Geol.* **2020**, *114*, 104191. [CrossRef]
43. Yuan, H.; Wang, Y.; Wang, X. Seismic Methods for Exploration and Exploitation of Gas Hydrate. *J. Earth Sci.* **2021**, *32*, 839–849. [CrossRef]
44. Gao, M.; Xu, S.; Zhuo, H.; Wang, Y.; Wu, S. Coupling relationship between shelf-edge trajectories and slope morphology and its implications for deep-water oil and gas exploration: A case study from the passive continental margin, East Africa. *J. Earth Sci.* **2020**, *31*, 820–833. [CrossRef]
45. Gui, B.; He, D.; Zhang, Y.; Sun, Y.; Zhang, W. 3D geometry and kinematics of the Niudong Fault, Baxian Sag, Bohai Bay Basin, Eastern China—Insights from high-resolution seismic data. *J. Struct. Geol.* **2021**, *146*, 104307. [CrossRef]
46. Marfurt, K.J.; Kirlin, R.L.; Farmer, S.L.; Bahorich, M.S. 3-D seismic attributes using a semblance-based coherency algorithm. *Geophysics* **1998**, *63*, 1150–1165. [CrossRef]
47. Cao, L.; Yao, Y.; Liu, D.; Yang, Y.; Wang, Y.; Cai, Y. Application of seismic curvature attributes in the delineation of coal texture and deformation in Zhengzhuang field, southern Qinshui Basin. *AAPG Bull.* **2020**, *104*, 1143–1166. [CrossRef]
48. Fang, P.; Ding, W.; Fang, Y.; Zhao, Z.; Feng, Z. Cenozoic tectonic subsidence in the southern continental margin, South China Sea. *Front. Earth Sci.* **2017**, *11*, 427–441. [CrossRef]
49. Fang, P.; Ding, W.; Zhao, Y.; Lin, X.; Zhao, Z. Detachment-controlled subsidence pattern at hyper-extended passive margin: Insights from backstripping modelling of the Baiyun Rift, northern South China Sea. *Gondwana Res.* **2022**, in press. [CrossRef]
50. Steckler, M.S.; Watts, A.B. Subsidence of the Atlantic-type continental margin off New York. Earth Planet. *Sci. Lett.* **1978**, *41*, 1–13. [CrossRef]
51. Long, H.; Flemings, P.B.; Germaine, J.T.; Saffer, D.M. Consolidation and overpressure near the seafloor in the Ursa Basin, Deepwater Gulf of Mexico. *Earth Planet. Sci. Lett.* **2011**, *305*, 11–20. [CrossRef]
52. Fang, P.; Ding, W.; Lin, X.; Zhao, Z.; Fang, Y.; Li, C. Neogene subsidence pattern in the multi-episodic extension systems: Insights from backstripping modelling of the Okinawa Trough. *Mar. Pet. Geol.* **2020**, *111*, 662–675. [CrossRef]
53. Zhai, P.; Chen, H.; Xie, Y.; Wang, Z.; Tong, C. Modelling of evolution of overpressure system and hydrocarbon migration in deepwater area of Qiongdongnan basin, South China Sea. *J. Cent. South Univ.* **2013**, *44*, 4187–4201, (In Chinese with English Abstract).
54. Petrobras, P. Campos and Espirito Santo Basins, offshore Brazil. In *Seismic Expression of Structural Styles: A Picture and Work Atlas*; Bally, A.W., Ed.; American Association of Petroleum Geologists: Tulsa, OK, USA, 1983; Volume 2, pp. 51–58.
55. Buffler, R.T.; Locker, S.D.; Bryant, W.R.; Hall, S.A.; Pilger, R.H. Gulf of Mexico. In *Ocean Margin Drilling Program, Regional Atlas Ser., Atlas 6*; Marine Science International: Woods Hole, MA, USA, 1984; 26p.

56. Alves, T.M.; Lourençob, S.D.N. Geomorphologic features related to gravitational collapse: Submarine landsliding to lateral spreading on a Late Miocene–Quaternary slope (SE Crete, eastern Mediterranean). *Geomorphology* **2010**, *123*, 13–33. [CrossRef]
57. Alves, T.M. Scale-relationships and geometry of normal faults reactivated during gravitational gliding of Albian rafts (Espírito Santo Basin, SE Brazil). *Earth Planet. Sci. Lett.* **2012**, *331–332*, 80–96. [CrossRef]
58. Mourgues, R.; Cobbolda, P.R. Sandbox experiments on gravitational spreading and gliding in the presence of fluid overpressures. *J. Struct. Geol.* **2006**, *28*, 887–901. [CrossRef]
59. Carter, G.D.O.; Cooper, R.; Gafeira, J.; Howe, J.A.; Long, D. Morphology of small-scale submarine mass movement events across the northwest United Kingdom. *Geomorphology* **2020**, *365*, 107282. [CrossRef]
60. Biancardi, C.A.; Alves, T.M.; Martins-Ferreira, M.A.C. Unpredictable geometry and depositional stacking patterns of mass-transport complexes in salt minibasins. *Mar. Pet. Geol.* **2020**, *120*, 104522. [CrossRef]
61. Alsop, G.I.; Weinberger, R.; Marco, S.; Levi, T. Identifying soft-sediment deformation in rocks. *J. Struct. Geol.* **2019**, *125*, 248–255. [CrossRef]
62. Jackson, C.A.L. Three-dimensional seismic analysis of megaclast deformation within a mass transport deposit; implications for debris flow kinematics. *Geology* **2011**, *39*, 203–206. [CrossRef]
63. Imber, J.; Childs, C.; Nell, P.A.R.; Walsh, J.J.; Hodgetts, D.; Flint, S. Hanging wall fault kinematics and footwall collapse in listric growth fault systems. *J. Struct. Geol.* **2003**, *25*, 197–208. [CrossRef]
64. Li, X.; Fairweather, L.; Wu, S.; Ren, J.; Zhang, H.; Quan, X.; Jiang, T.; Zhang, C.; Su, M.; He, Y.; et al. Morphology, sedimentary features and evolution of a large palaeo submarine canyon in Qiongdongnan basin, Northern South China Sea. *J. Asian Earth Sci.* **2013**, *62*, 685–696. [CrossRef]
65. Su, M.; Xie, X.; Xie, Y.; Wang, Z.; Zhang, C.; Jiang, T.; He, Y. The segmentations and the significances of the Central Canyon System in the Qiongdongnan Basin, northern South China Sea. *J. Asian Earth Sci.* **2014**, *79*, 552–563. [CrossRef]
66. Su, M.; Xie, X.; Wang, Z.; Jiang, T.; Zhang, C.; He, Y. Sedimentary evolution of the Central Canyon System in the Qiongdongnan Basin, northern South China Sea. *Pet. Res.* **2016**, *1*, 81–92. [CrossRef]
67. Su, M.; Wu, C.; Chen, H.; Li, D.; Jiang, T.; Xie, X.; Jiao, H.; Wang, Z.; Sun, X. Late Miocene provenance evolution at the head of Central Canyon in the Qiongdongnan Basin, Northern South China Sea. *Mar. Pet. Geol.* **2019**, *110*, 787–796. [CrossRef]
68. Su, M.; Lin, Z.; Wang, C.; Kuang, Z.; Liang, J.; Chen, H.; Liu, S.; Zhang, B.; Luo, K.; Huang, S.; et al. Geomorphologic and infilling characteristics of the slope-confined submarine canyons in the Pearl River Mouth Basin, northern South China Sea. *Mar. Geol.* **2020**, *424*, 106166. [CrossRef]
69. Ziegler, P.A.; Cloetingh, S. Dynamic processes controlling evolution of rifted basins. *Earth Sci. Rev.* **2004**, *64*, 1–50. [CrossRef]
70. Ding, W.; Franke, D.; Li, J.; Steuer, S. Seismic stratigraphy and tectonic structure from a composite multi-channel seismic profile across the entire Dangerous Grounds, South China Sea. *Tectonophysics* **2013**, *582*, 162–176. [CrossRef]
71. Franke, D.; Savva, D.; Pubellier, M.; Steuer, S.; Mouly, B.; Auxietre, J.; Meresse, F.; Chamot-Rooke, N. The final rifting evolution in the South China Sea. *Mar. Pet. Geol.* **2014**, *58*, 704–720. [CrossRef]
72. Zhao, Z. The deep mantle upwelling beneath the northwestern South China Sea: Insights from the time-varying residual subsidence in the Qiongdongnan Basin. *Geosci. Front.* **2021**, *12*, 101246. [CrossRef]
73. Davies, J.H.; Bunge, H.P. Are splash plumes the origin of minor hotspots? *Geology* **2006**, *34*, 349–352. [CrossRef]
74. Wang, X.; Li, Z.; Li, X.; Li, J.; Xu, Y.; Li, X. Identification of an ancient mantle reservoir and young recycled materials in the source region of a young mantle plume: Implications for potential linkages between plume and plate tectonics. *Earth Planet. Sci. Lett.* **2013**, *377*, 248–259. [CrossRef]
75. Berra, F.; Carminati, E. Differential compaction and early rock fracturing in high-relief carbonate platforms: Numerical modelling of a Triassic case study (Esino Limestone, Central Southern Alps, Italy). *Basin Res.* **2012**, *24*, 598–614. [CrossRef]
76. Alsop, G.I.; Weinberger, R.; Marco, S.; Levi, T. Distinguishing coeval patterns of contraction and collapse around flow lobes in mass transport deposits. *J. Struct. Geol.* **2020**, *134*, 104013. [CrossRef]
77. Conforti, M.; Ietto, F. Influence of Tectonics and Morphometric Features on the Landslide Distribution: A Case Study from the Mesima Basin (Calabria, South Italy). *J. Earth Sci.* **2020**, *31*, 393–409. [CrossRef]
78. Wang, R.; Shi, W.; Xie, X.; Zhang, W.; Qin, S.; Liu, K.; Busbey, A.B. Clay mineral content, type, and their effects on pore throat structure and reservoir properties: Insight from the Permian tight sandstones in the Hangjinqi area, north Ordos Basin, China. *Mar. Pet. Geol.* **2020**, *115*, 104281. [CrossRef]
79. He, X.; Xu, C.; Qi, W.; Huang, Y.; Cheng, J.; Xu, X.; Yao, Q.; Lu, Y.; Dai, B. Landslides Triggered by the 2020 Qiaojia Mw5.1 Earthquake, Yunnan, China: Distribution, Influence Factors and Tectonic Significance. *J. Earth Sci.* **2021**, *32*, 1056–1068. [CrossRef]

energies

Article

The Facies Analysis, Evolution, and Coal-Bearing Source Rock Features of the Middle–Late Triassic Shallow-Water Delta in the North Carnarvon Basin, Northwest Shelf of Australia

Zhiwei Zeng [1,2,3,*], Wei Wang [2,3,*], Hongtao Zhu [2,3], Xianghua Yang [2,3] and Dan Li [4]

1 School of Geophysics and Geomatics, China University of Geosciences, Wuhan 430074, China
2 School of Earth Resources, China University of Geosciences, Wuhan 430074, China
3 Key Laboratory of Tectonics and Petroleum Resources, China University of Geosciences, Wuhan 430074, China
4 China National Offshore Oil Corporation Research Institute, Beijing 100028, China
* Correspondence: zwzeng@cug.edu.cn (Z.Z.); wang.wei@cug.edu.cn (W.W.)

Abstract: The sedimentary facies, architecture, and depositional mechanism of deltaic systems have been one of the global research hotspots in recent decades; however, the detailed distribution, sedimentary evolution, source rock potential, and major control factors of the coal-bearing shallow-water delta are still unclear. A typical shallow-water braided delta with coal-bearing source rocks developed in the Middle–Late Triassic Mungaroo Formation of the North Carnarvon Basin, which can be a good study area for an analysis of coal-bearing source rocks. In this study, the sedimentary facies, distribution and evolution, and coal-bearing source rock features of the Triassic strata were analyzed based on the integrated study of wireline logs, drilled cores, thin sections, seismic facies and attributes, and geochemical data. A range of shallow-water delta sedimentary facies was identified, including the proximal delta plain channel/interdistributary bay, distal delta plain channel/interdistributary bay, and the delta front. The coal-bearing shallow-water delta system of the Middle–Late Triassic Mungaroo Formation was characterized by the largest scale delta system with relatively broad proximal and distal delta plains and relatively narrow delta front subfacies. The scale of the delta system showed a trend of increasing from the Early Triassic Locker Shale to the Middle–Late Triassic Mungaroo Formation and then decreasing to the Late Triassic Brigadier Formation. The distal delta plain subfacies of the Mungaroo Formation should have the highest potential coal-bearing source rock, and the proximal delta plain also can be a favorable target for source rock evaluation. The major control factors of the coal-bearing source rocks of the Mungaroo shallow-water delta mainly included the Triassic megamonsoon climate, the topographic features, eustatic changes, and provenance supply. The proximal and distal delta plains of the shallow-water delta system with thin coal seams, carbonaceous mudstone, and dark mudstone lithologies' association could be a favorable source rock exploration facies for the next stage of natural gas field exploration.

Keywords: braided river delta; delta plain; Middle–Late Triassic; Mungaroo Formation; coal-bearing source rock; depositional model; North Carnarvon Basin

Citation: Zeng, Z.; Wang, W.; Zhu, H.; Yang, X.; Li, D. The Facies Analysis, Evolution, and Coal-Bearing Source Rock Features of the Middle–Late Triassic Shallow-Water Delta in the North Carnarvon Basin, Northwest Shelf of Australia. *Energies* **2023**, *16*, 2265. https://doi.org/10.3390/en16052265

Academic Editor: Jiafei Zhao

Received: 21 January 2023
Revised: 17 February 2023
Accepted: 25 February 2023
Published: 27 February 2023

1. Introduction

The sedimentary facies, stratal architecture, depositional evolution, and mechanism of modern and ancient deltaic deposits have been one of the global research hotspots in recent decades, including the continental shelf marginal delta system [1–10] and lake basin delta system [11–16]. The shallow-water delta is a special type of deltaic system, which was first proposed by Fisk et al. (1954) [17] with a relatively gentle slope, a broad and shallow water environment or low accommodation, a stable tectonic background, and an abundant sediment supply [17–20], based on the outcrop analysis and seismic-based interpretation [8,11,12,21–24]. Shallow-water deltas obviously differ from the Gilbert deltas [25] and are generally characterized by a different scale of braided distributary channels with a large-scale delta plain and a small range delta front [8,12,26]. The delta area in the dry period differs greatly from that in the flood period [11,12]. Previous studies have shown that the topographic features [11], climatic conditions (humid, semiarid, or arid) [27–30], river inflow, and tectonic subsidence [21,31,32] would be the significant controlling factors for the sedimentary architecture and depositional process of the shallow-water delta.

Coal-bearing source rocks are widely developed in the coastal delta environment, with many natural gas fields in the global Tethys region [33], such as the Northwest Shelf Basins of Australia [8,34], the Central Sumatra Basin of Indonesia [35], the Brunei–Sabah Basin of Malaysia [36], the East China Sea Shelf Basin [37], and the Yinggehai Basin [38] and Pearl River Mouth Basin [15,39] of the northern South China Sea. Generally, the coal-bearing source rocks are closely related with a coal seam and carbonaceous and dark mudstone with a high content of organic matter (rich with terrigenous higher plants) [37]. This special type of source rock shows a significant relationship with the natural gas exploration, and more than 70% of the gas fields in China were located in coal-related source rock [40]. However, the maceral composition in the coal-bearing source rocks of different delta subfacies or microfacies are obviously different, which could have a significant impact on the evaluation and prediction of the hydrocarbon generation potential. Meanwhile, the development characteristics, detailed distribution, sedimentary evolution, and major control factor of the favorable coal-bearing source rock of a shallow-water delta are still unclear. This might seriously restrict further exploration in the related oil and gas fields. Therefore, it is urgent to select a favorable research area to carry out these studies.

The Northwest Shelf (NWS) of Australia has become an important oil and gas exploration region in recent years [41–43]. The Middle–Late Triassic Mungaroo Formation of the large-scale and coal-bearing shallow-water delta in the North Carnarvon Basin (NCB) also has been a hot research target in this area, which was affected by the Pangaea megamonsoon [8,44], especially during the Carnian Pluvial Episode (CPE) [45]. The Carnian stage witnessed major changes in both the marine and terrestrial ecosystems and underwent a climate change event associated with global warming, the eruption of the Wrangellia large igneous province (LIP) [44], and the strong enhancement of the hydrological cycle [46]. Based on the previous studies [8,47], a typical shallow-water braided delta with coal-bearing source rocks was developed in the NCB during the Mungaroo Formation. The NCB has experienced many years of oil and gas exploration, with a large amount of drilling and 3D seismic data, which makes it a good study area to carry out the comprehensive analysis of the sedimentary facies, depositional evolution, and hydrocarbon generation potential of the coal-bearing source rocks of the shallow-water delta system.

Therefore, based on the integrated analysis of wireline logs, drilled cores, thin sections, seismic facies and attributes, and a series of geological and geochemical data in the NCB, the motivations of this research mainly include: (1) to characterize the stratal architectural features of various sedimentary subfacies and microfacies in the shallow-water delta system, especially during the Middle to early Late Triassic, (2) to analyze the source rock types, maceral compositions' characteristics, and hydrocarbon generation potential of different delta subfacies, (3) to investigate the sedimentary evolution of the deltaic system from the Early Triassic to the Late Triassic and to establish a representative depositional model of the Mungaroo Formation, (4) to further discuss the major control factor of the development of coal-bearing source rock, and to predict the favorable source rock distribution in different Triassic sequences.

2. Geological Setting

The North Carnarvon Basin (NCB) is located at the southern part of the Northwest Shelf of Australia (Figure 1a); the basin is a giant oil and gas bearing basin formed by continuous rifting and subsidence from the Late Paleozoic to the Cenozoic [41,43]. Since the first oil field discovery in the NCB in 1954, more than fifty oil and gas fields have been put into hydrocarbon production, including the Jansz, Gorgon, North Rankin, Perseus, Goodwyn, Sarborough, Pluto, Wheatstone, Geryon, and Clio gas fields (Figure 1b). The land and sea regions of the basin cover approximately 115,000 km^2 and 535,000 km^2, respectively, and the maximum water depth is approximately 3500 m [8,47].

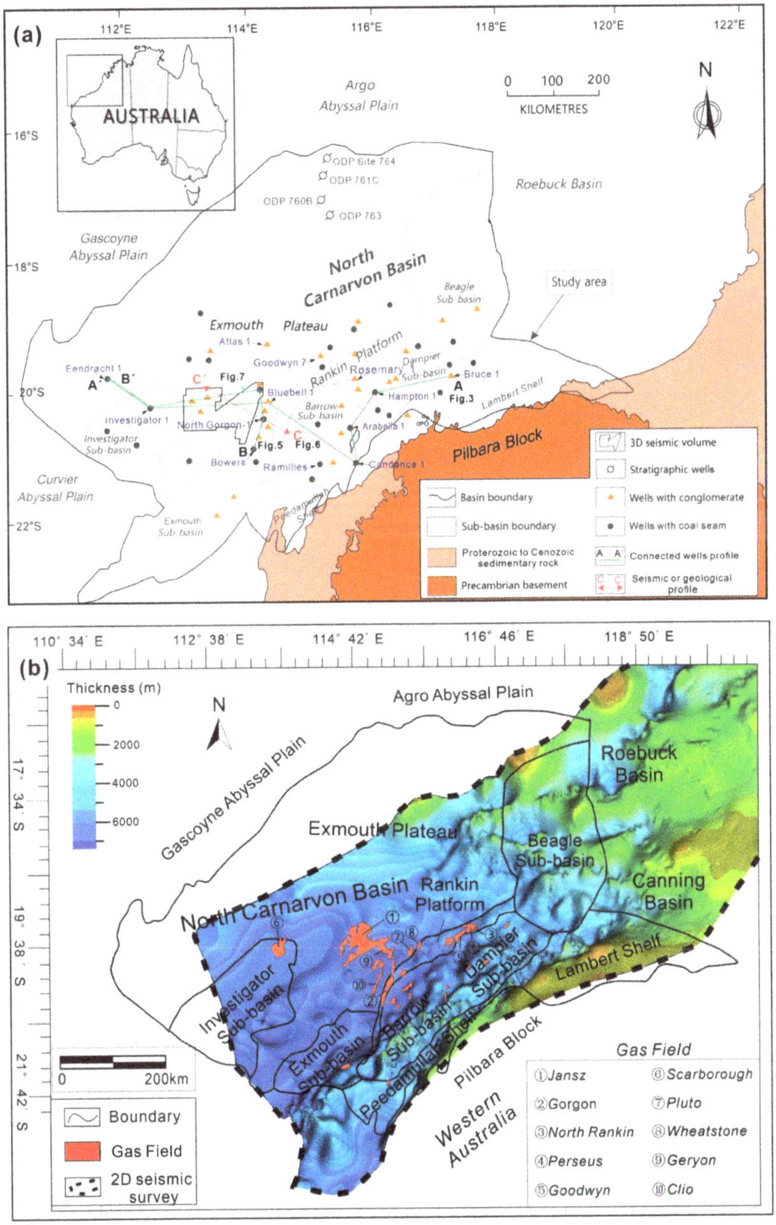

Figure 1. (**a**) Regional geological maps of the North Carnarvon Basin in the Northwest Shelf of Australia, including the drilled wells which penetrated the Triassic conglomerate or coal seam and the location of the connecting-well sections and seismic profile used in this study. (**b**) Thickness map of the Triassic strata in the North Carnarvon Basin and a series of significant gas field in this region.

Six significant evolution stages developed in the NCB, including: (1) the pre-rift stage (Silurian to Toarcian), (2) the early syn-rift stage (Toarcian to earliest Callovian), (3) the main syn-rift stage (earliest Callovian to Berriasian), (4) the late syn-rift stage (Berriasian to Valanginian), (5) the post-breakup subsidence stage (Valanginian to mid-Santonian) and (6) the passive margin stage (mid-Santonian to present) [48]. In this study, the thick Triassic strata was deposited in the pre-rifting stage (Late Permian to Triassic) with typical siliciclastic sediment deposits (Figure 2).

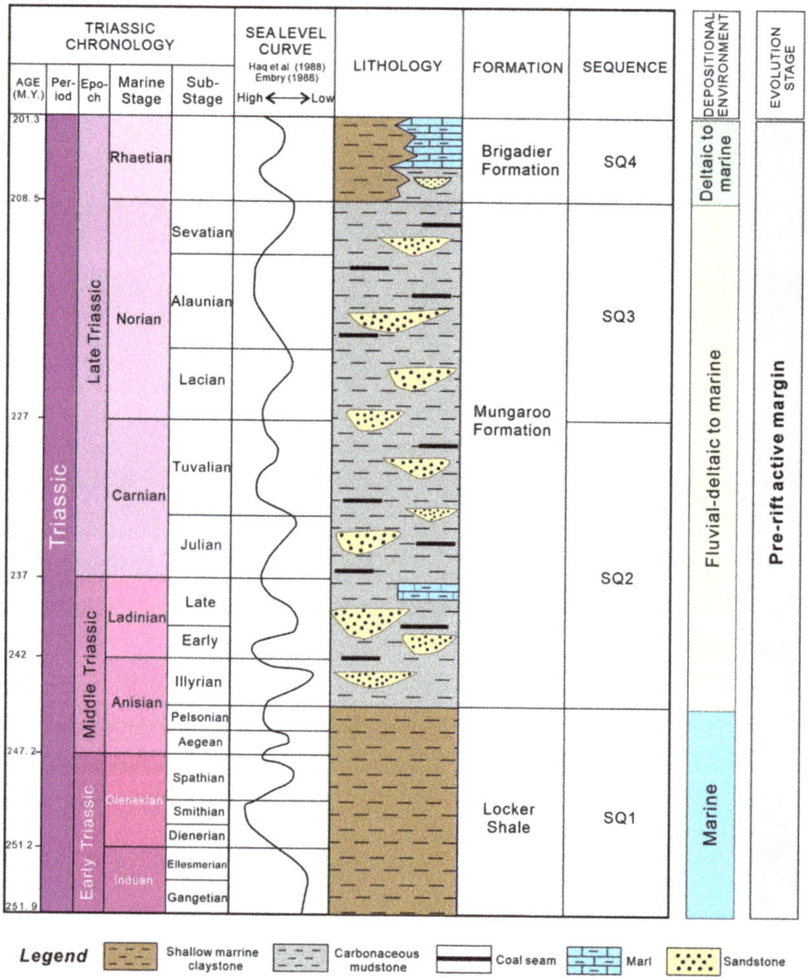

Figure 2. Stratigraphic map of the NCB, showing the Triassic chronology, sea-level curve, lithology, formation, sequences, depositional facies, and evolution stage. The major interval is the Triassic strata (especially the Mungaroo Formation) with a typical fluvial–deltaic to marine environment.

The study intervals are the Triassic Locker Shale (SQ1), the Mungaroo Formation (SQ2 and SQ3), and the Brigadier Formation (SQ4) (Figure 2). The Early Triassic Locker Shale was characterized by marine claystone and siltstone with shelfal limestone. The Middle–Late Triassic Mungaroo Formation was characterized by typical fluvial–deltaic to marine environments, and the large-scale coal-bearing shallow water delta covered much of the offshore NCB [47]. The palynologic associations published in previous studies [49,50] indicated typical fluvial–deltaic facies deposition. During the Latest Triassic to Early Jurassic, rapid subsidence caused the deposition of the transgressive Brigadier Formation, which was characterized by thin and interbedded claystone, marl, and sandstone. After the drowning of the Mungaroo delta, the depositional environment of the Brigadier Formation was deltaic to marine deposits.

3. Dataset and Methods

This study focuses mainly on the comprehensive research of wireline logs, drilled cores, thin sections, seismic facies and attributes, and a series of geological and geochemical data to promote a comprehensive understanding of the sedimentary facies, evolution, and the related coal-bearing source rock potential of the Triassic delta, especially for the Mungaroo Formation.

There were over 50 wells with detailed drilled wireline logs that penetrated the Triassic stratum, and approximately 1100 m of drilled cores were available from six of these wells. The cored samples were mainly derived from the Mungaroo Formation. A 3D seismic volume covered an area of over 8200 km^2, and the NCB covered 2D seismic data over 24,000 km (Figure 1a). The 3D seismic data were represented by inline and xline spacings of 12.5 m and 18.75 m, respectively. The vertical sample interval was 4 ms, and the signal-to-noise ratio was relatively high at ~8.3 with the dominant frequency of 40–50 Hz (measured velocity of ~2520 m/s). All geological data applied in this comprehensive study were provided by the CNOOC. All the drilled wells used in this research revealed the Mungaroo and Brigadier formations, and only few of them (such as Bruce-1, Hampton-1) penetrated the Locker Shale strata. The detailed sedimentary facies interpretations were based on the analysis of the well wireline logs, cores features, and microscopic thin sections. The thin sections and scanning electron microscope (SEM) micro-photos provided favorable records for the depositional environment. The seismic facies and seismic attributes (RMS attribute used in this study) were also useful to support the analysis of the facies' dispersal pattern and depositional evolution.

In addition, a series of geochemical analyses of mudstone, carbonaceous mudstone, and coal seam source rock samples (>200) from different sedimentary facies or subfacies of the shallow-water delta system were also carried out in this study. The test items of hydrocarbon source rock mainly included the total organic carbon (TOC), hydrocarbon generation potential (S_1 + S_2), hydrogen index (HI), maximum pyrolysis peak temperature (T_{max}), and organic maceral compositions in various sedimentary facies. These geochemical elemental analyses were carried out at the State Key Laboratory of Biogeology and Environmental Geology, China University of Geosciences (Wuhan). Based on replicate analyses of the Chinese national standard GB/T 19145-2003 [51], the analytical precision of the TOC measurements was better than 0.5 wt%.

4. Results

4.1. Microscopic Sedimentological Characteristics

The thin sections of the Middle–Late Triassic Mungaroo Formation showed different petrographic features with characteristics of various mineral compositions and assemblages (Figure 3). The sandstone samples of the Mungaroo Formation were characterized as quartz sandstone and subarkose with poor and medium sorting and angular round quartz grains (Figure 3a,b). Some primary pores, secondary pores, siderite, organic matter filled, and quartz overgrowth were also identified in the thin section graphs. In addition, kaolinite cements, banded organic matter, rootlet debits (Figure 3c,d), and coal seams (Figure 3e,f) also developed within the delta plains of the Mungaroo delta system.

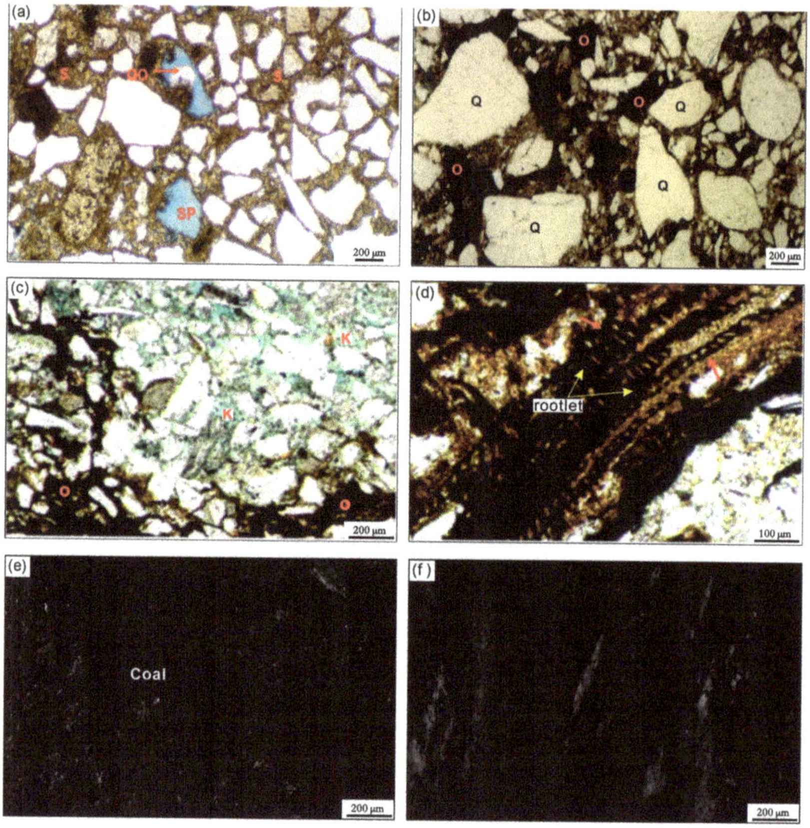

Figure 3. Photomicrographs of different petrographic features of the Mungaroo Formation in the NCB. (**a**,**b**) show the shapes and distribution patterns of the quartz (Q), feldspar (F), primary pores (P), secondary pores (SP), and a small amount of siderite (S) and quartz overgrowth (QO), 4050.6–4051.0 m, well North orgon-1. (**c**,**d**) show the microscopic characteristics of quartz (Q), kaolinite (K), organic matter (O), and rootlets, 4035.4–4035.7 m, well Bluebell-1. (**e**,**f**) display the features of the coal seam, SEM, 4159.0 m, well Bluebell-1. See the location of the wells in Figure 1a.

4.2. Well-Based Sedimentary Facies' Characteristics

The facies and vertical associations in this study were identified based on the analysis of the well-based wireline-log signatures, lithologies, grain-size, and sedimentary structures/fabrics (Figures 4–6). In this study, a range of shallow-water delta sedimentary environments was recognized, including the proximal delta plain channel/interdistributary bay, distal delta plain channel/interdistributary bay, delta front, beach bar, carbonate platform, and prodelta or littoral neritic sea. In the study area, the distributary channel sandstone of the delta plain was characterized by box or bell shapes in the GR logging curve and was mainly composed of thick gravel coarse sandstone with a typical scouring surface that was shown in the cores (Figure 4b). The interdistributary bay in the delta plain was mainly characterized by the frequent interbedding of sand and mudstone with typical rhythmic bedding (Figure 4a), and the GR logging curve was characterized by toothed features and interbedded with typical coal seams. The delta front was characterized by the interbedding of fine sandstone, siltstone, and mudstone with a certain reverse cycle feature (upward coarsening) and a visible crossbedding and bioturbated structure (Figure 4c).

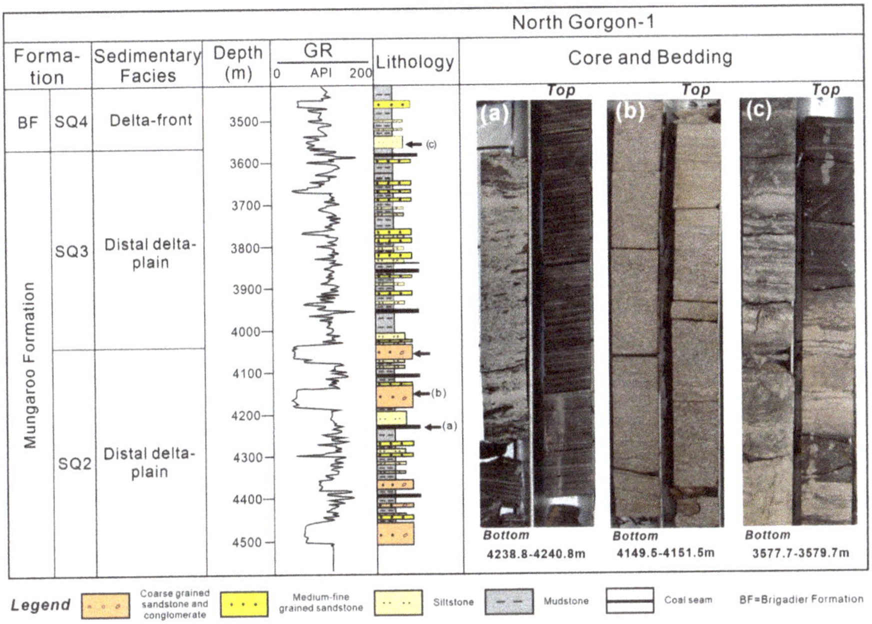

Figure 4. The gamma ray (GR) wireline log, lithologies and vertical evolution, delta–marine sedimentary facies, and the related cores and crossbedding sedimentary structures from North Gorgon-1. The depositional strata are related to the Mungaroo and Brigadier Formation (BF) from SQ2 to SQ4. The drilled cores' photos indicate the depositional features of the (**a**) coal seam and carbonaceous mudstone, (**b**) coarse grained sandstone and conglomerate of the distributary channel sandstone, and (**c**) interbedding of the distributary channel sandstone and mudstone of the interdistributary channel bay.

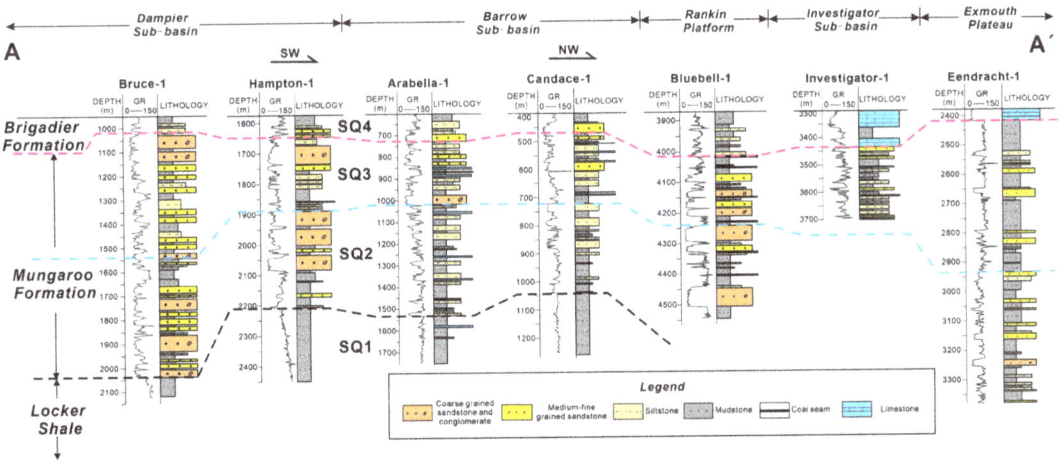

Figure 5. The cross-wells' stratigraphic profile A–A′ of the Triassic Locker Shale (SQ1), Mungaroo Formation (SQ2 and SQ3), and Brigadier Formation (SQ4), showing the characteristics of the gamma ray (GR) wireline log, lithologies, lithological association, and vertical evolution from bottom to top. See Figure 1a for the location of the cross-wells profile A–A′.

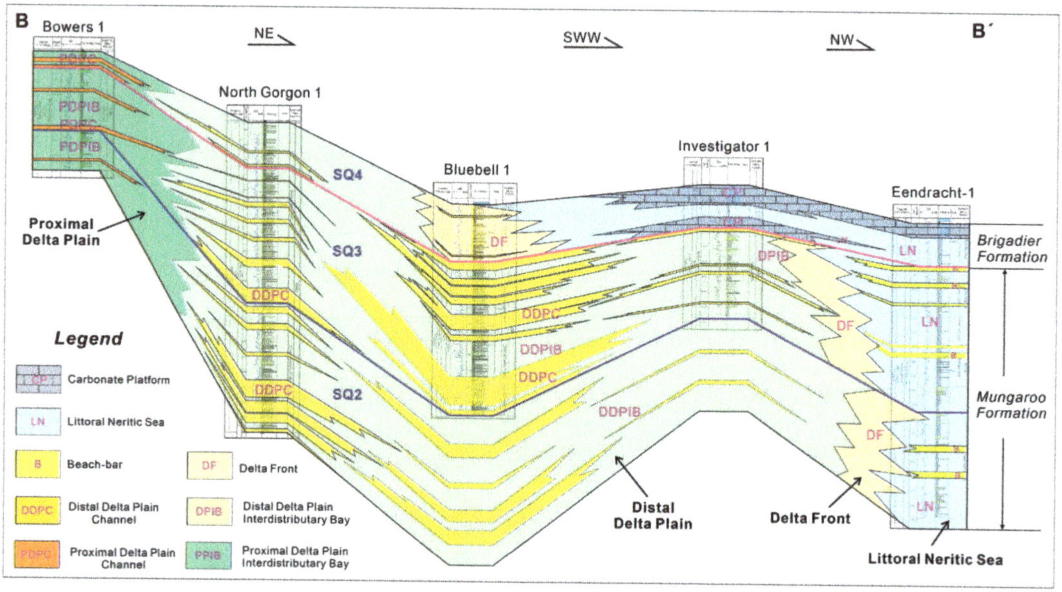

Figure 6. Cross-well profile B–B′ showing the sedimentary facies (proximal delta plain channel/interdistributary bay, distal delta plain channel/interdistributary bay, delta front, beach bar, carbonate platform, and littoral neritic sea) of the shallow-water delta system of the Middle–Late Triassic Mungaroo and Brigadier Formation (BF) from SQ2 to SQ4. The scale of the delta is gradually decreasing accompanied by vertical changes in the sedimentary facies and lithologic association. See Figure 1a for the location of the cross-wells profile B–B′.

In addition, the cross-wells stratigraphic profile A–A′ showed that the Early Triassic Locker Shale (SQ1) consisted of thick (40–200 m) mud-rich lithofacies integrated with siltstone (0.3 m–2.0 m) (Figure 5). The thin sheet siltstones were characterized by low gamma (GR) peaks.

The Middle–Late Triassic Mungaroo Formation (SQ2 and SQ3) was dominated by coal-bearing and braided delta with broad proximal and distal delta plains and a relatively narrow range delta front (Figure 5). The proximal and distal delta plains were characterized by deltaic distributary channels with thick (12–30 m) channels of conglomeratic sandstones (with box or bell shapes in the GR curve) interbedded with coal seams and carbonaceous mudstone (Figures 4 and 5). The delta plains also displayed multiple fining upward fluvial aggradational cycles (FACs; 0.1–1.2 m) with deposition of a braided distributary channel and an interdistributary bay (Figure 4). The deposition of the thin coal beds (Figure 1a) may be the consequence of the redeposit of coals that developed at the delta system. In addition, there was a certain reverse grain sequence in the delta front of the Mungaroo Formation (Figures 4 and 6). It is worth noting that the scale, thickness, and grain-size of the sediment deposits of the SQ2 were obviously larger than that of the SQ3, especially in the sedimentary facies of the proximal and distal delta plain distributary channels (Figure 4).

The Late Triassic Brigadier Formation (SQ4) was characterized by thin (2.2 m–4.0 m) fine stone, mudstone, and limestone with low gamma characteristics (Figures 4–6). Compared with the SQ2 and SQ3, the delta-scale and grain-size of the deltaic deposits obviously decreased, which was well indicated by North Gorgon-1 (Figure 4). During the deposition of the Brigadier Formation, the features of the FACs also decreased from bottom to top.

4.3. Seismic Facies and Attribute Features of the Coal-Bearing Shallow-Water Delta

The thickness map of the Triassic strata in the North Carnarvon Basin (Figure 1b) shows that the topography of the study area had a relatively gentle slope with a gradient of approximately 1.2° to 2.3°. The cross-well seismic profile C–C′ (Figure 7) presented a typical low-angle shingled progradational seismic reflection configuration of the SQ2 and SQ3, while the underlying Locker Shale (SQ1) and overlying Brigadier Formation (SQ4) were characterized by a parallel or subparallel seismic configuration (Figure 7). Integrated with the drilled-well based sedimentary facies analysis (Figures 4–6), the shingled progradational reflection in this research corresponded to the large-scale proximal and distal delta plain, with typical distributary channel gravel-rich or sand-rich deposits. In contrast, the parallel or subparallel seismic reflection could be linked to the interdistributary bays with mud-rich facies, carbonaceous mudstone, or coal seam deposits (Figure 7).

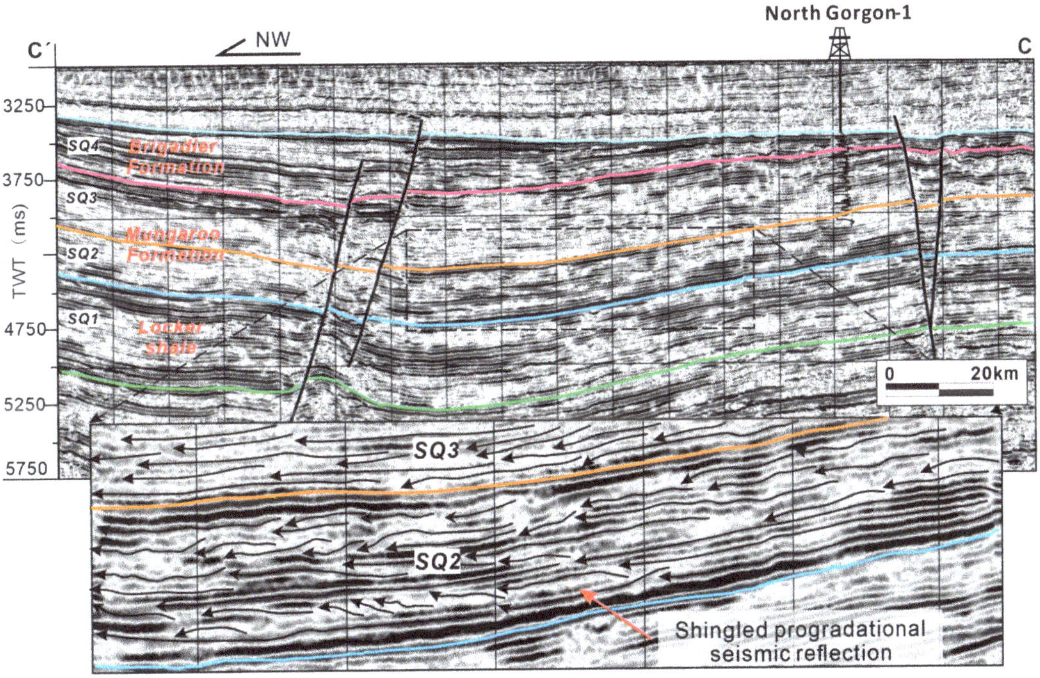

Figure 7. Cross-well seismic profile C–C′ presents a typical progradational seismic reflection of the SQ2 and SQ3. See Figure 1a for the position of the seismic profile C–C′. Abbreviation: TWT = two-way travel time.

The root mean square (RMS) seismic amplitude attribute map (Figure 8) also can be used to depict the planar distribution features of the various-scale channel sandstone geobodies of the shallow-water delta system. The distributary channel belts of the Mungaroo Formation were obviously imaged, showing the dominant paleoflow direction (SE–NW orientation), characteristics, and dispersal pattern. The features of the distributary channels also further supported the analysis of the delta system of the Mungaroo Formation in the study area.

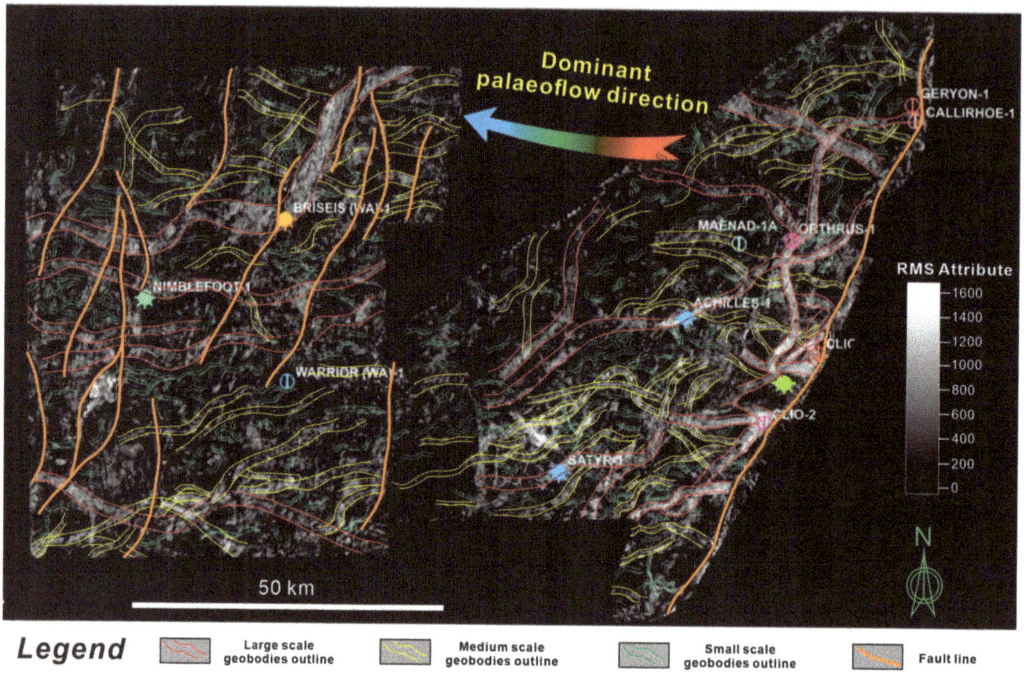

Figure 8. Root mean square (RMS) seismic amplitude attribute map of a 40 ms sculpt from the SQ2 of the Mungaroo Formation, showing the dominant paleoflow direction, characteristics, and dispersal pattern of the shallow-water braided delta distributary channels. See the location of this seismic attribute map in Figure 1a.

5. Discussion

5.1. Depositional Evolution and Dispersal Pattern of the Triassic Coal-Bearing Delta System

Based on the integrated analysis of the well-based lithologies, seismic facies and attributes, and microscopic sedimentary characteristics, in this study, the sedimentary facies distribution and evolution from the Early Triassic Locker Shale (SQ1) to the Middle–Late Triassic Mungaroo Formation (SQ2 and SQ3) and Late Triassic Brigadier Formation (SQ4) were well characterized (Figure 9a).

The Early Triassic Locker Shale (SQ1) was characterized by a relatively small-scale delta system that was mainly distributed along the coast of the Northwest Shelf of Australia. The feature of the Locker Shale sedimentary facies distribution was consistent with the characteristics of a relatively mud-rich lithologic association revealed by a large number of existing wells (Figures 5 and 6). The lower Mungaroo Formation (SQ2) corresponded to the depositional stage of the largest coal-bearing shallow-water delta systems that developed in the NCB (Figure 9a). It was characterized by a large-scale delta system with relatively broad proximal and distal delta plains and relatively narrow delta front subfacies. The feature of the lower Mungaroo Formation (SQ2) was also well validated by the drilled wells with thick distributary channel pebbly sandstone and widely developed thin coal seams (Figures 4–6). Based on the previous studies [8,44,46], it is believed that the large-scale delta system of the Mungaroo Formation might have a good inner relationship with the impact of the Triassic megamonsoon and the influence of Carnian Pluvial Episode (CPE). During the upper Mungaroo Formation (SQ3), the delta system was characterized by the interbedding of medium–fine channel sandstone and siltstone, mudstone, and thin coal beds. During the Brigadier Formation (SQ4), the delta system's scale decreased rapidly

(Figure 9a) and was characterized by thick limestone and dolomite and interbedded with thin siltstone and mudstone.

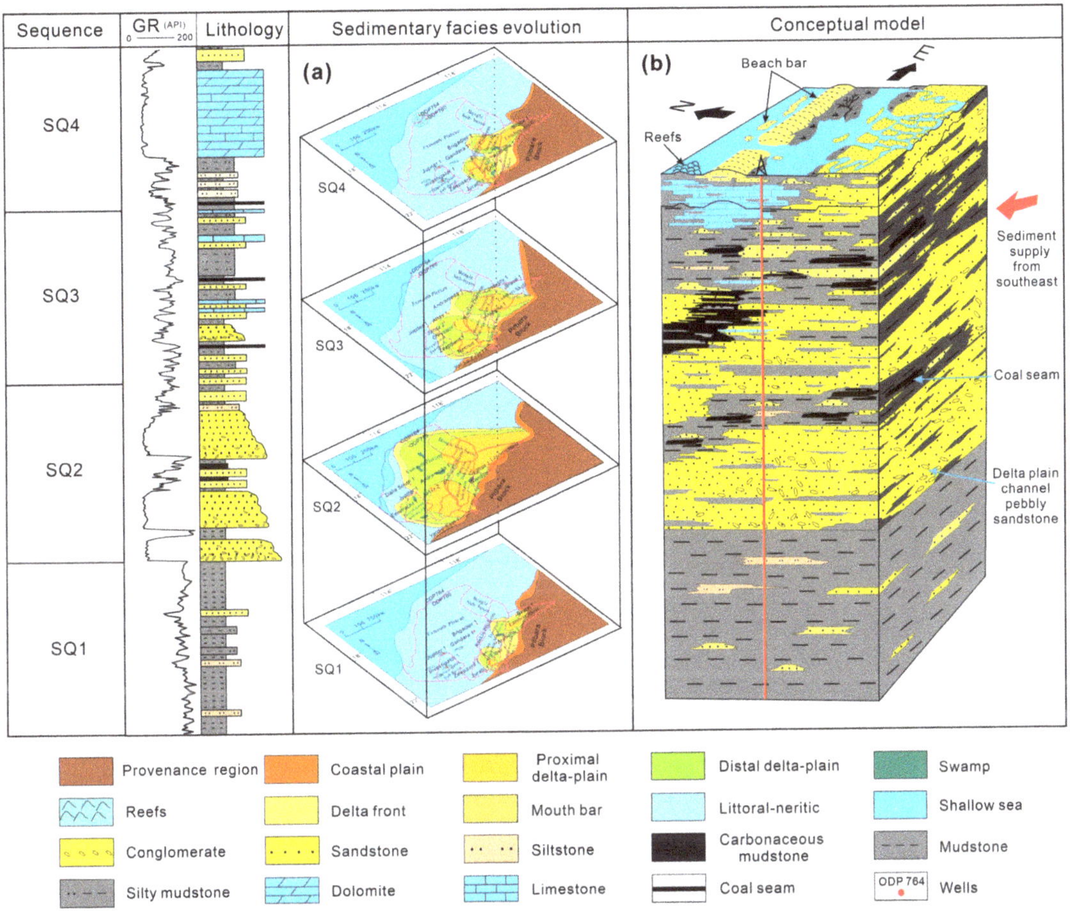

Figure 9. The lithology, GR wireline log feature, facies distribution and evolution, and the related three-dimensional conceptual model of the Triassic strata from the Locker Shale to the Brigadier Formation in the NCB. (**a**) The sedimentary facies maps of the SQ1 to SQ4 indicate the distribution pattern of the shallow water delta system. (**b**) The three-dimensional conceptual model of the the Locker Shale to the Brigadier Formation with typical lithologic distribution.

Therefore, based on the sedimentary facies' distribution and evolution of the Triassic sequences, a three-dimensional (3D) conceptual model from bottom to top is also provided (Figure 9b). The 3D model obviously displays the dispersal pattern of the sedimentary facies or subfacies association, lithologies distribution patterns, and the spatiotemporal distribution characteristics. Overall, the scale of the delta system showed a trend of increasing from SQ1 to SQ2 and then decreasing from SQ2 to SQ4. At the same time, the lithologic association and distribution also showed a matching evolution trend.

5.2. Coal-Bearing Source Rock Characteristics and the Potential of Different Sedimentary Subfacies

In this study, the Mungaroo Formation was the major study interval for detailed coal-bearing source rock analysis with different sedimentary facies or delta subfacies (Figures 10 and 11; Tables 1 and 2). The organic matter of the Mungaroo Formation was mainly represented by thin coal seams, carbonaceous mudstones, and mudstone-rich terrigenous organic matter. Among them, the thin coal seams and carbonaceous mudstone were mainly concentrated in the distal delta plain. Due to the distributary channels being well-developed in the proximal delta plain, the effect of river scouring was strong, and the fine sediment such as peat swamp deposited in the early stage was difficult to preserve. Most of the organic matter in the sediment was oxidized and decomposed in the Mungaroo Formation. The oxidized vitrinite was also seen in some sample analysis. In addition, the delta front was mainly enriched in silty mudstone and siltstone with typical dispersed organic matter.

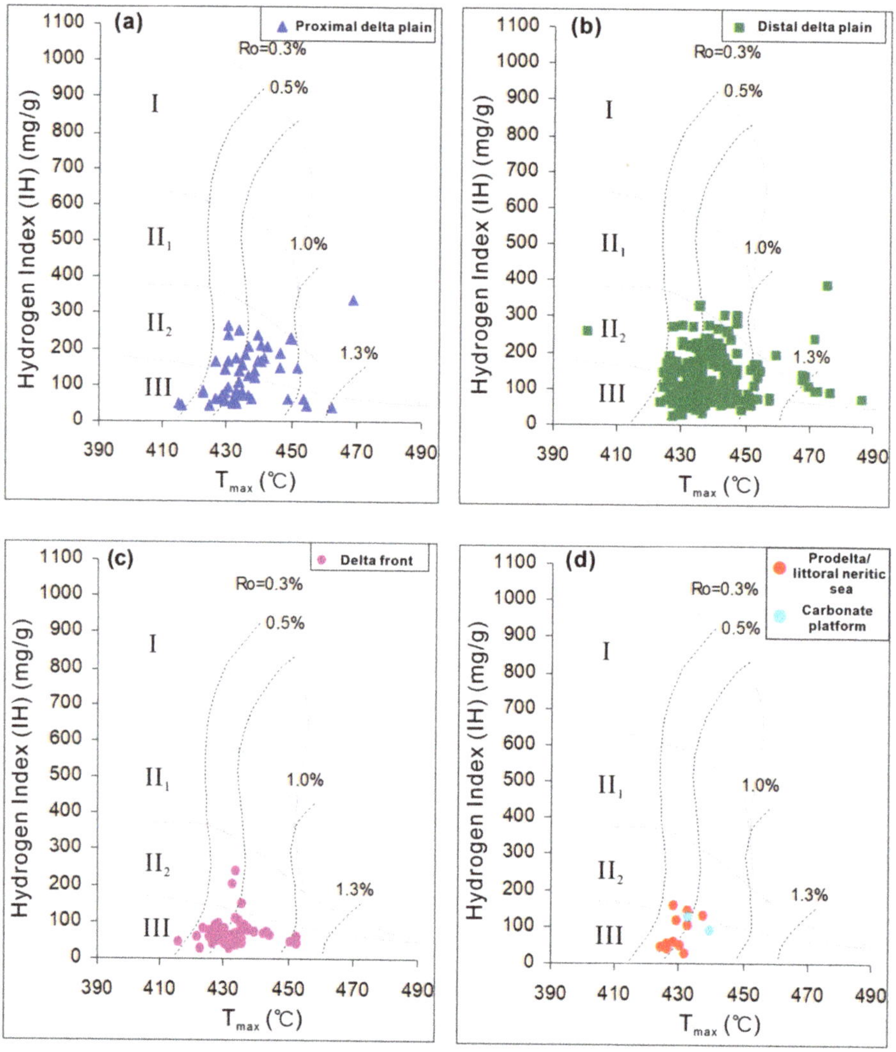

Figure 10. Organic matter types, maximum pyrolysis peak temperature (T_{max}), and hydrogen index (IH) characteristics of the Mungaroo Formation source rocks in different sedimentary facies or subfacies, including the (**a**) proximal delta plain, (**b**) distal delta plain, (**c**) delta front and (**d**) prodelta, littoral neritic sea, and carbonate platform.

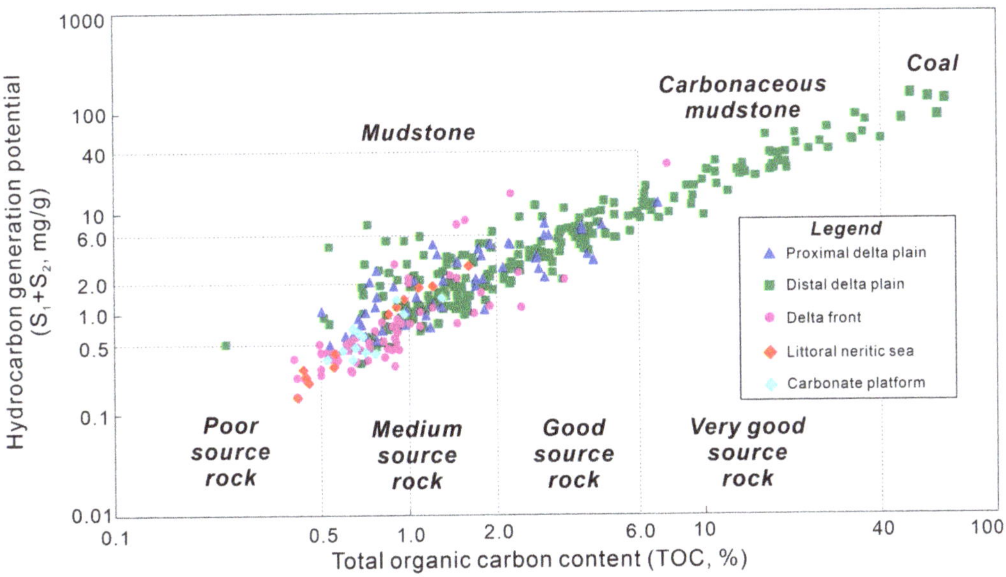

Figure 11. Total organic carbon (TOC) content and hydrocarbon generation potential ($S_1 + S_2$) of the Mungaroo Formation source rocks (mudstone, carbonaceous mudstone, and coal) in different sedimentary facies or subfacies.

Table 1. Statistics of the average content of the organic maceral compositions in different sedimentary facies of the Mungaroo Formation in the NCB.

Sedimentary Subfacies	Average Content of Macerals (%)		
	Inertinite	Vitrinite	Exinite
Proximal delta plain	46%	23%	12%
Distal delta plain	30%	54%	16%
Delta front	38%	21%	41%

Table 2. Total organic carbon (TOC) values of the dark mudstone and coal seam in different delta subfacies of the Mungaroo Formation in the NCB.

Lithology	TOC Value (%)	Proximal Delta Plain	Distal Delta Plain	Delta Front	Prodelta/ Littoral Neritic Sea
Dark mudstone	Average value	1.16%	4.11%	1.05%	0.96%
	Minimum value	0.5%	0.52%	0.43%	0.32%
	Maximum value	4.16%	26.8%	4.26%	1.74%
Coal seam	Average value	33.3%	37.4%	/	/
	Minimum value	28.2%	30.4%	/	/
	Maximum value	35.2%	40.5%	/	/

Generally, the Mungaroo Formation was characterized by coal-bearing source rock with obvious organic matter that derived from terrestrial plants (ferns and seed ferns), showing the organic matter type dominated with type-III kerogen in different facies (Figure 10). The organic matter types of the source rocks in different sedimentary facies zones also had certain differences. Among them, the proximal delta plain and the distal delta plain were characterized by type-II_2 and -III kerogen (Figure 10a,b), while the organic matter types of the source rock in the delta front, prodelta, and carbonate platform were mainly represented by type-III kerogen, and a few samples fell into the type-II_2 area (Figure 10c,d).

The contents of the organic macerals in different sedimentary facies zones also had obvious differences. The organic matter of the coal-bearing delta system was mainly represented by two different types, including the coal and dispersed organic matter. The macerals of coal mainly consisted of vitrinite, and the content of inertinite was relatively low, which mainly developed within the distal delta plain (Table 1). The content of vitrinite in the dispersed organic matter was lower than that in the coal, while the content of the inertinite and exinite was relatively higher, which mainly developed within the proximal delta plain and delta front subfacies (Table 1). From the proximal delta plain to the distal delta plain and then to the delta front and prodelta (littoral neritic sea), the content of vitrinite increased (from 23% to 54%) and then decreased (from 54% to 21%), of which the content of relative vitrinite in the distal delta plain was the highest (54%), while the content of exinite had a significant increasing trend (from 12% to 16% and 41%). Overall, the macerals of organic matter in the proximal and distal delta plains of the Mungaroo Formation in the NCB were mainly vitrinite and inertinite, and the content of exinite was relatively low. The content of vitrinite and inertinite in the macerals of the proximal delta plain and the distal delta plain showed an obvious mutual growth and decline relationship (Table 1). However, the content of exinite in the delta front was relatively high, and some samples even contained exinite alone.

In addition, the terrigenous organic matter content and hydrocarbon generation potential of the dark mudstone and coal seam in different sedimentary facies zones of the Mungaroo Formation were quite different (Figure 11; Table 2). The thin coal seams in the proximal and distal delta plain were well developed with a high average TOC content (33.3% to 37.4%; Table 2). Overall, the distal delta plain was characterized by coal seams and thick carbonaceous mudstones rich in terrigenous organic matter, which can serve as good or very good source rocks. The average total organic carbon (TOC) content of the dark mudstones in the distal delta plain reached 4.11% (Table 2). In the proximal delta plain, the carbonaceous mudstone was not developed, and the TOC content in the dark mudstone was also relatively high, with an average of 1.16%, which could be medium to good hydrocarbon source rock (Figure 11; Table 2). The delta front sedimentary facies belt was relatively narrow with a relatively low TOC content in the dark mudstones (average 1.05%), which might not be a favorable source rock facies belt in this area. The samples of the prodelta/littoral neritic sea and the carbonate platform were also represented by low TOC content features (average 0.96%), which should not be the source rock targets in NCB exploration.

5.3. Coal-Bearing Source Rock Prediction, Controlling Factors, and Implications for Exploration

The Triassic coal-bearing source rocks in the NCB mainly developed during the Mungaroo Formation (SQ2 and SQ3) with typical proximal and distal delta plain deposits (Figures 9 and 12a). Based on the comprehensive analysis of the organic matter types, organic maceral compositions, TOC content, and hydrocarbon generation potential (Figures 10 and 11; Tables 1 and 2) above, the distal delta plain subfacies of the SQ2 and SQ3 should have the most potential coal-bearing source rock that developed in the NCB (Figure 12b). The proximal delta plain could also be a favorable target for source rock evaluation. The narrow delta front, the prodelta, littoral neritic sea, and carbonate platform facies do not have favorable hydrocarbon generation potential.

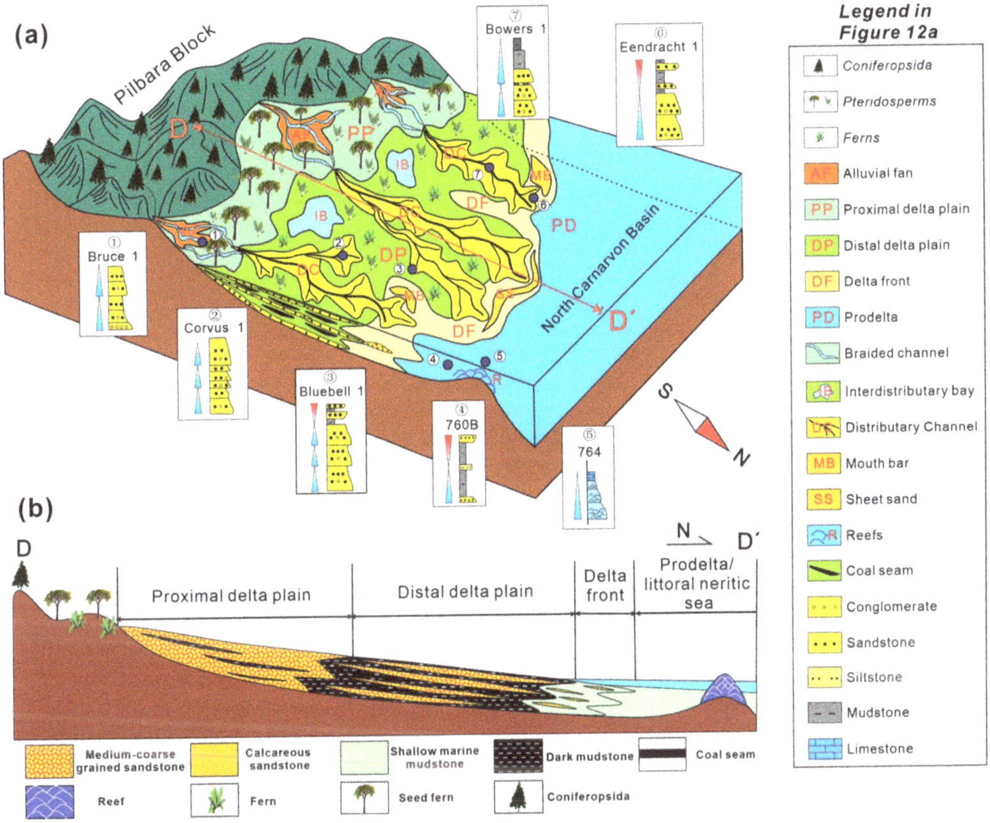

Figure 12. (**a**) The three-dimensional dispersal pattern model of the typical shallow-water coal-bearing delta of the Mungaroo Formation in the NCB. (**b**) Two-dimensional conceptual profile of the distribution pattern of the terrigenous organic matter in the Mungaroo deltaic system, indicating the facies dispersal pattern and lithologies' distribution and combination. See Figure 12a for the position of the profile D–D′.

The depositional evolution and dispersal pattern of the source rocks of the Mungaroo delta were significantly affected by the Triassic megamonsoon, the basin geomorphology, eustatic changes, and provenance from the orogenic belt. Changes in the Triassic climate affected the rainfall, which then influenced the fluvial erosion, sediment supply, and discharge from the source to sink of the drainage system. During the Late Triassic, especially the Carnian stage, the NCB was located in a temperate to warm humid and monsoonal climate, which also benefitted the megamonsoon that developed related to the Pangaea supercontinent. In addition, the gentle topography during the Mungaroo Formation of the NCB was advantageous for the formation of the coal-bearing delta with a broad delta plain. During the Middle–Late Triassic, the major source in the study area was derived from the provenance regions of Antarctica and Southeastern Australia [8], which provided sufficient sediments to form the large-scale delta.

In summary, the broad proximal and distal delta plains of the shallow-water delta system of the Mungaroo Formation (SQ2 and SQ3) with thin coal seams, carbonaceous mudstone, and dark mudstone lithologies' association could be advantageous coal-bearing source rock exploration layers in the NCB of the Northwest Shelf of Australia, which also could be the major targets of reservoir and source rock for the next stage of natural gas field exploration.

6. Conclusions

The sedimentary facies' analysis, distribution and evolution, and coal-bearing source rock features of the Middle–Late Triassic shallow-water delta in the North Carnarvon Basin, Northwest Shelf of Australia, were comprehensively analyzed based on the integrated study of wireline logs, drilled cores, thin sections, seismic facies and attributes, and a series of geological and geochemical data. The major conclusions are presented as follows:

(i) A range of shallow-water delta sedimentary environments was recognized in the Triassic strata of the North Carnarvon Basin, especially during the deposition of the Middle–Late Triassic Mungaroo Formation, including the proximal delta plain channel/interdistributary bay, distal delta plain channel/interdistributary bay, delta front, beach bar, carbonate platform, and prodelta or littoral neritic sea. The coal-bearing shallow-water delta system of the Mungaroo Formation was characterized by a large-scale delta system with relatively broad proximal and distal delta plains and relatively narrow delta front subfacies. The scale of the delta system showed a trend of increasing from SQ1 to SQ2 and then decreasing from SQ2 to SQ4.

(ii) The Triassic coal-bearing source rocks in the NCB mainly developed during the Mungaroo Formation (SQ2 and SQ3) with a relatively high TOC content and hydrocarbon generation potential. The distal delta plain subfacies of the Mungaroo Formation should be the highest potential coal-bearing source rock. The proximal delta plain also could be a favorable target for source rock evaluation. The delta front, prodelta, littoral neritic sea, and carbonate platform facies do not have favorable hydrocarbon generation potential.

(iii) The depositional evolution and dispersal pattern of the source rocks of the Mungaroo delta were affected by the Triassic megamonsoon climate, the gentle basin topographic feature, eustatic changes, and enough sediment supply from the orogenic belt. The proximal and distal delta plains of the shallow-water delta system with thin coal seams, carbonaceous mudstone, and dark mudstone lithologies' association could be advantageous coal-bearing source rock exploration layers, which could serve as a reference for other regions with a similar background.

Author Contributions: Conceptualization, investigation, writing—original draft preparation, review, validation, and editing, Z.Z.; methodology, validation, formal analysis, and editing, W.W.; supervision, methodology, project administration, and data curation, H.Z.; methodology, supervision, project administration, and funding acquisition, X.Y.; software, data curation, and project administration, D.L. All authors have read and agreed to the published version of the manuscript.

Funding: This project was supported by the "CUG Scholar" Scientific Research Funds at China University of Geosciences (Wuhan) (Project No.2022148). This research was also funded by the National Science and Technology Major Projects (No. 2011ZX05030-002-002), the China National Postdoctoral Program for Innovative Talents (Grant No. BX20200310), the China Postdoctoral Science Foundation (Grant No. 2020M682522), and Research funding for postdoctoral innovation in Hubei Province (Grant No. 257236). This research was also funded by the National Natural Science Foundation of China (Grant No. 42172127, Grant No. 41872149).

Data Availability Statement: Not applicable.

Acknowledgments: The Beijing Research Institute of the China National Offshore Oil Corporation is thanked for providing data used in this study and for the permission to publish the results.

Conflicts of Interest: The authors declare no conflict of interest.

References

1. Chen, S.; Steel, R.; Wang, H.; Zhao, R.; Olariu, C. Clinoform growth and sediment flux into late Cenozoic Qiongdongnan shelf margin, South China Sea. *Basin Res.* **2020**, *32*, 302–319. [CrossRef]
2. Zhu, N.; Cao, Y.; Xi, K.; Wu, S.; Zhu, R.; Yan, M.; Ning, S. Multisourced CO_2 injection in fan delta conglomerates and its influence on reservoir quality: Evidence from carbonate cements of the Baikouquan Formation of Mahu Sag, Junggar Basin, Northwestern China. *J. Earth Sci.* **2021**, *32*, 901–918. [CrossRef]

3. Wu, L.; Jiao, Y.; Peng, Y.; Rong, H. Uranium metallogeny in fault-depression transition region: A case study of the Tamusu Uranium deposit in the Bayingobi Basin. *J. Earth Sci.* **2022**, *33*, 409–421. [CrossRef]
4. Martinsen, O.J.; Sømme, T.O.; Thurmond, J.B.; Helland-Hansen, W.; Lunt, I. Source-to-sink systems on passive margins: Theory and practice with an example from the Norwegian continental margin. *Geol. Soc. Lond. Pet. Geol. Conf. Ser.* **2011**, *7*, 913–920. [CrossRef]
5. Cai, G.; Peng, G.; Wu, J.; Bai, H.; Li, Y.; Xu, X.; Gong, W.; Li, K. Sedimentary filling response to detachment structural deformation in shallow-water continental shelf of Pearl River Mouth Basin: A case study of Enping Sag. *Earth Sci.* **2022**, *47*, 2391–2409. (In Chinese)
6. Syvitski, J.; Anthony, E.; Saito, Y.; Zăinescu, F.; Day, J.; Bhattacharya, J.P.; Giosan, L. Large deltas, small deltas: Toward a more rigorous understanding of coastal marine deltas. *Glob. Planet. Chang.* **2022**, *218*, 103958. [CrossRef]
7. Zeng, Z.; Zhu, H.; Yang, X.; Cao, X. Three-dimensional seismic analysis of a polygonal fault system (PFS) in the Northern Carnarvon Basin, Australia: Implications for fluid flow migration and gas hydrate system. *J. Pet. Sci. Eng.* **2022**, *215*, 110602. [CrossRef]
8. Zeng, Z.; Zhu, H.; Yang, X.; Zeng, H.; Hu, X.; Xia, C. The Pangaea Megamonsoon records: Evidence from the Triassic Mungaroo Formation, Northwest Shelf of Australia. *Gondwana Res.* **2019**, *69*, 1–24. [CrossRef]
9. Zhang, Q.; Zhang, X.; Li, H.; Song, P.; Jia, Z.; Zeng, T.; Tao, W. Large submarine fan system controlled by narrow continental shelf-faulted continental slope in Northern South China Sea. *Earth Sci.* **2022**, *47*, 2421–2432. (In Chinese)
10. Johannessen, E.P.; Steel, R. Shelf-margin clinoforms and prediction of deepwater sands. *Basin Res.* **2005**, *17*, 521–550. [CrossRef]
11. Qin, Y.; Zhu, X.; Zhu, S.; Mcelroy, B. Impact of deep-time palaeoclimate on the sedimentary records and morphology of lacustrine shoal-water deltas, Upper Eocene Dongying Depression, Bohai Bay Basin, China. *Basin Res.* **2021**, *68*, 3253–3278. [CrossRef]
12. Olariu, C.; Zhou, C.; Steel, R.; Zhang, Z.; Yuan, X.; Zhang, J.; Chen, S.; Chen, D.; Kim, W. Controls on the stratal architecture of lacustrine delta successions in low-accommodation conditions. *Basin Res.* **2021**, *68*, 1941–1963. [CrossRef]
13. Zeng, Z.; Zhu, H.; Zeng, H.; Yang, X.; Xu, C. Seismic sedimentology analysis of fluvial-deltaic systems in a complex strike-slip fault zone, Bohai Bay Basin, China: Implications for reservoir prediction. *J. Pet. Sci. Eng.* **2022**, *208*, 109290. [CrossRef]
14. Zeng, Z.; Zhu, H.; Mei, L.; Du, J.; Zeng, H.; Xu, X.; Dong, X. Multilevel source-to-sink (S2S) subdivision and application of an ancient uplift system in South China Sea: Implications for further hydrocarbon exploration. *J. Pet. Sci. Eng.* **2019**, *181*, 106220. [CrossRef]
15. Zeng, Z.; Zhu, H.; Yang, X.; Zeng, H.; Zhang, G. Multistage progradational clinoform-set characterisation and evolution analysis of the Early Oligocene in the Baiyun Sag, Pearl River Mouth Basin, South China Sea. *Mar. Pet. Geol.* **2020**, *112*, 104048. [CrossRef]
16. Xiao, M.; Wu, S.; Yuan, X.; Xie, Z. Conglomerate Reservoir Pore Evolution Characteristics and Favorable Area Prediction: A Case Study of the Lower Triassic Baikouquan Formation in the Northwest Margin of the Junggar Basin, China. *J. Earth Sci.* **2021**, *32*, 998–1010. [CrossRef]
17. Fisk, H.N.; Kolb, C.R.; McFarlan, E.; Wilbert, L.J.; Deceased, J.R. Sedimentary framework of the modern Mississippi delta [Louisiana]. *J. Sediment. Petrol.* **1954**, *24*, 76–99. [CrossRef]
18. Guo, W.; Xu, G.; Liu, B.; Xiang, X.; Liu, D.; Zhang, B. Structure-sedimentary response relationship of Wenchang Formation in Baiyun Sag, Pearl River Mouth Basin. *Earth Sci.* **2022**, *47*, 2433–2453. (In Chinese)
19. Liu, B.; Pang, X.; Xie, S.; Mei, L.; Zheng, J.; Sun, H.; Yan, H.; Wu, Y.; Xiang, X.; Feng, X. Control effect of crust-mantle detachment fault activity on deep large delta sedimentary system in Baiyun Sag, Pearl River Mouth Basin. *Earth Sci.* **2022**, *47*, 2354–2373. (In Chinese)
20. Zhang, X.; Xiang, X.; Zhao, M.; Cui, Y.; Zhang, H. Coupling relationship between Pearl River water system evolution and East Asian terrain inversion. *Earth Sci.* **2022**, *47*, 2410–2420. (In Chinese)
21. Ambrosetti, E.; Martini, I.; Sandrelli, F. Shoalwater deltas in high-accommodation settings: Insights from the lacustrine Valimi Formation (Gulf of Corinth, Greece). *Sedimentology* **2017**, *64*, 425–452. [CrossRef]
22. Zhu, X.; Zeng, H.; Li, S.; Dong, Y.; Zhu, S.; Zhao, D.; Huang, W. Sedimentary characteristics and seismic geomorphologic responses of a shallow-water delta in the Qingshankou Formation from the Songliao Basin, China. *Mar. Pet. Geol.* **2017**, *79*, 131–148. [CrossRef]
23. Zhu, X.; Liu, Y.; Fang, Q.; Li, Y.; Liu, Y.; Wang, R.; Song, J.; Liu, S.; Cao, H.; Liu, X. Formation and sedimentary model of shallow delta in large-scale lake: Example from Cretaceous Quantou Formation in Sanzhao Sag, Songliao Basin. *Earth Sci. Front.* **2012**, *19*, 89–99.
24. Winsemann, J.; Lang, J.; Polom, U.; Loewer, M.; Igel, J.; Pollok, L.; Brandes, C. Ice-marginal forced regressive deltas in glacial lake basins: Geomorphology, facies variability and large-scale depositional architecture. *Boreas* **2018**, *47*, 973–1002. [CrossRef]
25. Selim, S. Sedimentological model, architecture, and evolution of a shallow-water Gilbert-type delta from the Lower Miocene, Red Sea Rift, Egypt. *Int. Geol. Rev.* **2019**, *62*, 2053–2076. [CrossRef]
26. Huang, X.; Liu, K.; Zou, C.; Yuan, X.; Gui, L. Forward stratigraphic modelling of the shallow-water delta systemin the Poyang Lake, southern China. *J. Geochem. Explor.* **2014**, *144*, 74–83. [CrossRef]
27. Kroonenberg, S.B.; Rusakov, G.; Svitoch, A. The wandering of the Volga delta: A response to rapid Caspian sea-level change. *Sediment. Geol.* **1997**, *107*, 189–209. [CrossRef]
28. Grenfell, S.; Grenfell, M.; Rowntree, K.; Ellery, W. Fluvial connectivity and climate: A comparison of channel pattern and process in two climatically contrasting fluvial sedimentary systems in South Africa. *Geomorphology* **2014**, *205*, 142–154. [CrossRef]

29. Zhang, L.; Bao, Z.; Dou, L.; Zang, D.; Mao, S.; Song, J.; Zhao, J.; Wang, Z. Sedimentary characteristics and pattern of distributary channels in shallow water deltaic red bed succession: A case from the Late Cretaceous Yaojia formation, southern Songliao Basin, NE China. *J. Pet. Sci. Eng.* **2018**, *171*, 1171–1190. [CrossRef]
30. Went, D.J. Fluvial shoal water deltas: Pre-vegetation sedimentation through the fluvial–marine transition, Lower Cambrian, English Channel region. *Sedimentology* **2020**, *67*, 330–363. [CrossRef]
31. Postma, G. Sea-level-related architectural trends in coarse-grained delta complexes. *Sediment. Geol.* **1995**, *98*, 3–12. [CrossRef]
32. Syvitski, J.P.; Kettner, A.J.; Correggiari, A.; Nelson, B.W. Distributary channels and their impact on sediment dispersal. *Mar. Geol.* **2005**, *222*, 75–94. [CrossRef]
33. Qu, T.; Huang, Z.; Wang, R.; Tan, S.; Li, Z.; Guo, X.; Zhao, J.; Pan, Y. Development characteristics and controlling factors of coal-measure source rocks in the global Tethys region. *Coal Geol. Explor.* **2021**, *49*, 114–131. (In Chinese)
34. Yang, T.; Kang, H.; Liu, D.; Li, X.; Zhao, Z. The sedimentary facies evolution and the development characteristics of source rocks' in North Carnarvon Basin, Australia. *J. Southwest Pet. Univ. (Sci. Technol. Ed.)* **2017**, *39*, 81–91. (In Chinese)
35. Sefein, K.J.; Nguyen, T.X.; Philp, R.P. Organic geochemical and paleoenvironmental characterization of the Brown Shale Formation, Kiliran sub-basin, Central Sumatra Basin, Indonesia. *Org. Geochem.* **2017**, *112*, 137–157. [CrossRef]
36. Osli, L.N.; Shalaby, M.R.; Islam, M.A.; Kalaitzidis, S.; Damoulianou, M.E.; Karim, K.N.P.; Tsikouras, B.; Pasadakis, N. Organic matter characteristics and hydrocarbon generating potential of the Miocene Belait Formation, Brunei-Muara district, Brunei Darussalam. *J. Pet. Sci. Eng.* **2022**, *208*, 109503. [CrossRef]
37. Quan, Y.; Chen, Z.; Jiang, Y.; Diao, H.; Xie, X.; Lu, Y.; Du, X.; Liu, X. Hydrocarbon generation potential, geochemical characteristics, and accumulation contribution of coal-bearing source rocks in the Xihu Sag, East China Sea Shelf Basin. *Mar. Pet. Geol.* **2022**, *136*, 105465. [CrossRef]
38. Huang, B.; Xiao, X.; Li, X. Geochemistry and origins of natural gases in the Yinggehai and Qiongdongnan basins, offshore South China Sea. *Org. Geochem.* **2003**, *34*, 1009–1025. [CrossRef]
39. Xu, C.; Shan, X.; Lin, H.; Hao, G.; Liu, P.; Wang, X.; Shen, M.; Rexiti, Y.; Li, K.; Li, Z.; et al. The formation of early Eocene organic-rich mudstone in the western Pearl River Mouth Basin, South China: Insight from paleoclimate and hydrothermal activity. *Int. J. Coal Geol.* **2022**, *253*, 103957. [CrossRef]
40. Dai, J.; Gong, D.; Ni, Y.; Huang, S.; Wu, W. Stable carbon isotopes of coal-derived gases sourced from the Mesozoic coal measures in China. *Org. Geochem.* **2014**, *74*, 123–142. [CrossRef]
41. Cesar, J.; Grice, K. $\delta^{13}C$ of polycyclic aromatic hydrocarbons to establish the facies variations in a fluvial deltaic Triassic record (Dampier sub-Basin, Western Australia). *Org. Geochem.* **2017**, *107*, 59–68. [CrossRef]
42. Jablonski, D.; Saitta, A.J. Permian to Lower Cretaceous plate tectonics and its impact on the tectonostratigraphic development in the Western Australia margin. *APPEA J.* **2004**, *44*, 287–327. [CrossRef]
43. Mii, H.S.; Shi, G.R.; Wang, C.A. Late Paleozoic middle-latitude Gondwana environment-stable isotope records from Western Australia. *Gondwana Res.* **2013**, *24*, 125–138. [CrossRef]
44. Dal Corso, J.; Mills, B.J.W.; Chu, D.; Newton, R.J.; Song, H. Background Earth system state amplified Carnian (Late Triassic) environmental changes. *Earth Planet. Sci. Lett.* **2022**, *578*, 117321. [CrossRef]
45. Scibiorski, J.; Peyrot, D.; Lang, S.; Payenberg, T.; Charles, A. Depositional settings and palynofacies assemblages of the upper Triassic fluvio-deltaic Mungaroo formation, Northern Carnarvon basin, western Australia. *J. Sediment. Res.* **2020**, *4*, 90. [CrossRef]
46. Ruffell, A.; Simms, M.J.; Wignall, P.B. The Carnian Humid Episode of the late Triassic: A review. *Geol. Mag.* **2016**, *153*, 271–284. [CrossRef]
47. Heldreich, G.; Redfern, J.; Legler, B.; Gerdes, K.; Williams, B.P.J. Challenges in characterizing subsurface paralic reservoir geometries: A detailed case study of the Mungaroo Formation, NorthWest shelf, Australia. *Geol. Soc. Lond. Spec. Publ.* **2017**, *444*, 59–108. [CrossRef]
48. Cathro, D.L.; Karner, G.D. Cretaceous-Tertiary inversion history of the Dampier Subbasin, northwest Australia: Insights from quantitative basin modeling. *Mar. Pet. Geol.* **2006**, *23*, 503–526. [CrossRef]
49. Backhouse, J.; Balme, B.E.; Helby, R.; Marshall, N.G.; Morgan, R. Palynological zonation and correlation of the latest Triassic, Northern Carnarvon Basin. In *Sedimentary Basins of Western Australia 3, Proceedings of the Petroleum Exploration Society of Australia Symposium, Perth, Australia, 20–23 October 2002*; Keep, M., Moss, S.J., Eds.; Petroleum Exploration Society of Australia: Brisbane, Australia, 2002; pp. 13–24.
50. Marshall, N.G.; Lang, S.C. A new sequence stratigraphic framework for the North West Shelf, Australia. In *Sedimentary Basins of Western Australia 4, Proceedings PESA Symposium, Perth, Australia, 18–21 August 2013*; Keep, M., Moss, S.J., Eds.; Petroleum Exploration Society of Australia: Brisbane, Australia, 2013; pp. 1–32.
51. Zeng, W.; Zhang, J.; Ding, W.; Zhao, S.; Zhang, Y.; Liu, Z.; Jiu, K. Fracture development in Paleozoic shale of Chongqing area (South China). Part one: Fracture characteristics and comparative analysis of main controlling factors. *J. Asian Earth Sci.* **2013**, *75*, 251–266. [CrossRef]

MDPI
St. Alban-Anlage 66
4052 Basel
Switzerland
www.mdpi.com

Energies Editorial Office
E-mail: energies@mdpi.com
www.mdpi.com/journal/energies

www.ingramcontent.com/pod-product-compliance
Lightning Source LLC
Chambersburg PA
CBHW061554210926
51850CB00009B/2103
9783036588841